Elastic Stack 8.x Cookbook

Over 80 recipes to perform ingestion, search, visualization, and monitoring for actionable insights

Huage Chen

Yazid Akadiri

Elastic Stack 8.x Cookbook

Group Product Manager: Kaustubh Manglurkar

Publishing Product Manager: Deepesh Patel

Book Project Manager: Aparna Ravikumar Nair

Senior Editor: Tazeen Shaikh

Technical Editor: Seemanjay Ameriya

Copy Editor: Safis Editing

Proofreader: Tazeen Shaikh

Indexer: Rekha Nair

Production Designer: Prashant Ghare

Senior DevRel Marketing Executive: Nivedita Singh

First published: June 2024

Production reference: 1070624

Published by Packt Publishing Ltd.

Grosvenor House

11 St Paul's Square

Birmingham

B3 1RB, UK.

ISBN 978-1-83763-429-3

www.packtpub.com

To my parents, Ying and Shinlang, for their love and unconditional support.

To Noël Jaffré, who has always been an inspiration throughout my career.

– Huage Chen

*To my beloved parents, who have tirelessly shaped a world where I could chase my dreams—
your efforts have been my foundation. To my dear wife and my children, Safaa, Adam, and Zaki,
whose love, patience, and incredible support have been my strength on this incredible journey.
Thank you.*

– Yazid Akadiri

Foreword

As we explore the growing world of data, the skill to understand it and use it to its full strength becomes a key challenge for data practitioners, architects, search specialists, DevOps and SREs, and others.

Since the inception of Elasticsearch and the progressive addition of key components to form what is currently known as the Elastic Stack, we've wanted to help people make sense of their data through the power of search and analytics. The launch of Elastic Stack 8 marks a big milestone in our journey. It is a version enriched with new capabilities, optimized performance, and an ever-stronger foundation for machine learning and AI.

This book serves as a practical resource for anyone who interacts with data and wants to learn how to exploit the power of the Elastic Stack, including Elasticsearch, Kibana, and various integrations, to make data-driven decisions and gain richer insights from their data environments.

As you turn the pages of this cookbook, you will uncover the innovations introduced in version 8.x. Our goal has always been to simplify the complex, and this book aligns perfectly with that ethos— breaking down advanced concepts into easy-to-follow, step-by-step instructions. Whether you are taking your initial steps in Elasticsearch and the Elastic Stack or looking to expand your expertise, the cookbook format provides a unique opportunity to build your skills progressively and systematically.

I am excited about the endless possibilities that Elastic Stack 8.x unlocks, and I look forward to hearing about the innovative ways in which you employ these recipes.

Shay Banon

Creator of Elasticsearch and CTO of Elastic

Contributors

About the authors

Huage Chen is a member of Elastic's customer engineering team and has been with Elastic for over five years, helping users throughout Europe to innovate and implement cloud-based solutions for search, data analysis, observability, and security. Before joining Elastic, he worked for 10 years in web content management, web portals, and digital experience platforms.

Yazid Akadiri has been a solutions architect at Elastic for over four years, helping organizations and users solve their data and most critical business issues by harnessing the power of the Elastic Stack. At Elastic, he works with a broad range of customers, with a particular focus on Elastic observability and security solutions. He previously worked in web services-oriented architecture, focusing on API management and helping organizations build modern applications.

About the reviewers

Evelien Schellekens is a senior solutions architect at Elastic. Evelien enjoys sharing knowledge through public speaking and interacting with the technical community. She's passionate about observability and open source technologies such as Kubernetes.

Giuseppe Santoro is a senior software engineer at Elastic. With deep expertise in Kubernetes, the cloud, and observability, Giuseppe contributes to the tech community through mentoring and technical writing.

Acknowledgments

We would like to express our gratitude to Evelien Schellekens and Giuseppe Santoro for their invaluable contributions and meticulous review of this book. Their expertise and thoughtful feedback have been instrumental in refining our work. We also extend our thanks to our fellow Elasticians for their contributions: Amanda Branch, Bahaaldine Azarmi, Carson Ip, Nicholas Drost, Sean Collin, Yan Savitski, Yannick Fhima, and the entire Elastic South EMEA Solutions Architect and Customer Architect teams.

Table of Contents

2

Ingesting General Content Data 37

3

Building Search Applications 73

4

Timestamped Data Ingestion   107

5

Transform Data   149

6

Visualize and Explore Data   191

7

Alerting and Anomaly Detection   265

8

Advanced Data Analysis and Processing   315

9

Vector Search and Generative AI Integration   363

10

Elastic Observability Solution — 421

11

12

13

Elastic Stack Monitoring   611

Index   647

Other Books You May Enjoy   658

Preface

In this cookbook, you will explore practical recipes and step-by-step instructions for solving real-world data challenges using the latest versions of the Elastic Stack's components, including Elasticsearch, Kibana, Elastic Agent, Logstash, and Beats. This book equips you with the knowledge and skills necessary to unlock the full potential of the Elastic Stack.

The book begins with practical guides on installing the stack through various deployment methods. Subsequently, it delves into the ingestion and search of general content data, illustrating how to develop enhanced search experiences. As you progress, you will explore timestamped data ingestion, data transformation, and enrichment using various components of the Elastic Stack. You will also learn how to visualize, explore, and create dashboards with your data using Kibana. Moving forward, you will refine your skills in anomaly detection and data science, employing advanced techniques in data frame analytics and natural language processing. Equipped with these concepts, you will investigate the latest advancements in search technology, including semantic search and generative AI. Additionally, you will explore Elastic Observability use cases for log, infrastructure, and synthetic monitoring, alongside essential strategies for securing the Elastic Stack. Ultimately, you will gain expertise in Elastic Stack operations, enabling you to monitor and manage your system effectively.

By the end of the book, you will have acquired the necessary knowledge and skills to build scalable, reliable, and efficient data analytics and search solutions with the Elastic Stack.

> **Note**
> The **Elastic Security solution**, a significant component of the Elastic Stack, would have merited considerable attention in this book. However, due to considerations regarding the length of the book and the intended audience, we have opted not to include this section in the current edition.

Who this book is for

This book is intended for Elastic Stack users, developers, observability practitioners, and data professionals of all levels, from beginners to experts, seeking practical experience with the Elastic Stack:

- Developers will find easy-to-follow recipes for utilizing APIs and features to craft powerful applications.
- Observability practitioners will benefit from use cases that cover APM, Kubernetes, and cloud monitoring.

- Data engineers and AI enthusiasts will be provided with dedicated recipes focusing on vector search and machine learning.

No prior knowledge of the Elastic Stack is required.

What this book covers

Chapter 1, Getting Started – Installing the Elastic Stack, explores the installation of the Elastic Stack across environments such as Elastic Cloud and Kubernetes, detailing the setup for Elasticsearch, Kibana, and Fleet along with insights on cluster components and deployment strategies for stack optimization.

Chapter 2, Ingesting General Content Data, dives into the data ingestion process, focusing on indexing, updating, and deleting operations within Elasticsearch, and emphasizes analyzers, index mappings, and templates for effective Elasticsearch index management.

Chapter 3, Building Search Applications, guides you through constructing search experiences using Elasticsearch's Query DSL and new features in Elastic Stack 8, culminating in comprehensive search applications with advanced queries and analytics.

Chapter 4, Timestamped Data Ingestion, delves into data transformation using Elastic Stack tools, instructing on data structuring, enrichment, reorganization, and downsampling, while utilizing ingest pipelines, processors, Transforms, and Logstash.

Chapter 5, Transform Data, delves into data transformation techniques using Elastic Stack tools. You will learn how to structure, enrich, reorganize, and downsample your data to glean actionable insights. This chapter delivers practical know-how on utilizing ingest pipelines, processors, transforms, and Logstash for efficient data manipulation.

Chapter 6, Visualize and Explore Data, shows how to turn transformed data into visualizations, teaching data exploration in Discover, visual creation with Kibana Lens, and the use of dashboards and maps to deeply understand your data.

Chapter 7, Alerting and Anomaly Detection, outlines the setup of alerts and anomaly detection for proactive data management, covering alert creation and monitoring, anomaly investigation, and unsupervised machine learning job implementation.

Chapter 8, Advanced Data Analysis and Processing, delves into machine learning within the Elastic Stack, covering outlier detection, regression, and classification modeling, as well as deploying NLP models for deep data insights.

Chapter 9, Vector Search and Generative AI Integration, explores advanced search technologies and AI integrations, teaching you about vector search, hybrid search, and Generative AI applications for developing sophisticated AI-driven conversational tools.

Chapter 10, *Elastic Observability Solution*, demonstrates how to employ the Elastic Stack for comprehensive system insights, covering application instrumentation, real-user monitoring, Kubernetes observability, synthetic monitors, and incident detection.

Chapter 11, *Managing Access Control*, navigates access control within the Elastic Stack, detailing authentication management, custom role definition, Kibana space security, API key utilization, and single sign-on implementation.

Chapter 12, *Elastic Stack Operation*, provides essential recipes for Elastic Stack management, such as index life cycle, data stream optimization, and snapshot life cycle management, and explores cluster automation with Terraform and cross-cluster search.

Chapter 13, *Elastic Stack Monitoring*, equips you with techniques for Elastic Stack monitoring and troubleshooting, focusing on the stack monitoring setup, custom visualization creation, cluster health assessment, and audit logging strategies.

To get the most out of this book

Before starting this book, you should have a basic understanding of databases, web servers, and data formats such as JSON. No prior Elastic Stack experience is needed, as the book starts with foundational topics. Familiarity with terminal commands and web technologies will be beneficial for following along. Each chapter progresses into more advanced Elastic Stack applications and techniques.

Software/hardware covered in the book	Operating system requirements
Elastic Stack 8.12	
Python 3.11+	
Docker 4.27.0	
Kubernetes 1.24+	
Node.js 19+	
Terraform 1.8.0	Windows, macOS, or Linux
Amazon Web Services (AWS)	
Google Cloud Platform (GCP)	
Okta	
Ollama	
OpenAI/Azure OpenAI	

If you are using the digital version of this book, we advise you to type the code yourself or access the code via the GitHub repository (link available in the next section). Doing so will help you avoid any potential errors related to the copying and pasting of code.

Download the example code files

You can download the example code files for this book from GitHub at `https://github.com/PacktPublishing/Elastic-Stack-8.x-Cookbook`. In case there's an update to the code, it will be updated on the existing GitHub repository.

We also have other code bundles from our rich catalog of books and videos available at `https://github.com/PacktPublishing/`. Check them out!

Conventions used

There are a number of text conventions used throughout this book.

`Code in text`: Indicates code words in text, database table names, folder names, filenames, file extensions, pathnames, dummy URLs, user input, and Twitter handles. Here is an example: "The `or` and `and` operators yield results that are too broad or too strict; you can use the `minimum_should_match` parameter to filter less relevant results."

A block of code is set as follows:

```
GET /movies/_search
{
  "query": {
    "multi_match": {
      "query": "come home",
      "fields": ["title", "plot"]
    }
  }
}
```

When we wish to draw your attention to a particular part of a code block, the relevant lines or items are set in bold:

```
GET movies-dense-vector/_search
{
  "knn": {
    "field": "plot_vector",
    "k": 5,
    "num_candidates": 50,
    "query_vector_builder": {
      "text_embedding": {
```

```
        "model_id": ".multilingual-e5-small_linux-x86_64",
        "model_text": "romantic moment"
      }
    }
  },
  "fields": [ "title", "plot" ]
}
```

Any command-line input or output is written as follows:

```
$ kubectl apply -f elastic-agent-managed-kubernetes.yml
$ sudo metricbeat modules enable tomcat
```

Bold: Indicates a new term, an important word, or words that you see onscreen. For example, words in menus or dialog boxes appear in the text like this. Here is an example: "In Kibana, go to **Observability | APM | Services**, to check whether the different microservices have been correctly instrumented."

> **Tips or important notes**
> Appear like this.

Sections

In this book, you will find several headings that appear frequently (*Getting ready*, *How to do it...*, *How it works...*, *There's more...*, and *See also*).

To give clear instructions on how to complete a recipe, use these sections as follows:

Getting ready

This section tells you what to expect in the recipe and describes how to set up any software or any preliminary settings required for the recipe.

How to do it...

This section contains the steps required to follow the recipe.

How it works...

This section usually consists of a detailed explanation of what happened in the previous section.

There's more...

This section consists of additional information about the recipe in order to make you more knowledgeable about the recipe.

See also

This section provides helpful links to other useful information for the recipe.

Get in touch

Feedback from our readers is always welcome.

General feedback: If you have questions about any aspect of this book, mention the book title in the subject of your message and email us at customercare@packtpub.com.

Errata: Although we have taken every care to ensure the accuracy of our content, mistakes do happen. If you have found a mistake in this book, we would be grateful if you would report this to us. Please visit www.packtpub.com/support/errata, selecting your book, clicking on the Errata Submission Form link, and entering the details.

Piracy: If you come across any illegal copies of our works in any form on the Internet, we would be grateful if you would provide us with the location address or website name. Please contact us at copyright@packt.com with a link to the material.

If you are interested in becoming an author: If there is a topic that you have expertise in and you are interested in either writing or contributing to a book, please visit authors.packtpub.com.

Share Your Thoughts

Once you've read *Elastic Stack 8.x Cookbook*, we'd love to hear your thoughts! Scan the QR code below to go straight to the Amazon review page for this book and share your feedback.

https://packt.link/r/1-837-63429-7

Your review is important to us and the tech community and will help us make sure we're delivering excellent quality content.

Download a free PDF copy of this book

Thanks for purchasing this book!

Do you like to read on the go but are unable to carry your print books everywhere?

Is your eBook purchase not compatible with the device of your choice?

Don't worry, now with every Packt book you get a DRM-free PDF version of that book at no cost.

Read anywhere, any place, on any device. Search, copy, and paste code from your favorite technical books directly into your application.

The perks don't stop there, you can get exclusive access to discounts, newsletters, and great free content in your inbox daily

Follow these simple steps to get the benefits:

1. Scan the QR code or visit the link below

https://packt.link/free-ebook/978-1-83763-429-3

2. Submit your proof of purchase
3. That's it! We'll send your free PDF and other benefits to your email directly

1
Getting Started – Installing the Elastic Stack

The **Elastic Stack** is a suite of components that allows you to ingest, store, search, analyze, and visualize your data from diverse sources. Previously known as the **ELK** Stack, today, it consists of four core components: **Elasticsearch**, **Logstash**, **Elastic Agent**, and **Kibana**.

Elasticsearch is a distributed search and analytics engine that can handle petabytes of unstructured data. Logstash, **Beats**, and Elastic Agent are data ingestion tools that can collect, transform, and load data from various sources into Elasticsearch. Kibana is a web-based interface that allows you to visualize and explore your data, as well as access various solutions built on top of the Elastic Stack. All integrate seamlessly so you can use your data for a variety of use cases such as search, analytics, observability, and security.

The Elastic Stack can be deployed on **Elastic Cloud**, as well as on-premises, and it can be deployed in a hybrid and orchestrated setup. In this chapter, we will guide you through setting up and running Elastic deployments in different environments, including a hosted Elasticsearch service on Elastic Cloud, Kubernetes infrastructure, and self-managed solutions. We will also discuss additional components and nodes within the cluster. By the end of this chapter, you'll have a comprehensive understanding of the various deployment strategies and how to use the Elastic Stack.

Figure 1.1 illustrates the key components of the Elastic Stack and the relationship between different components from a data flow perspective:

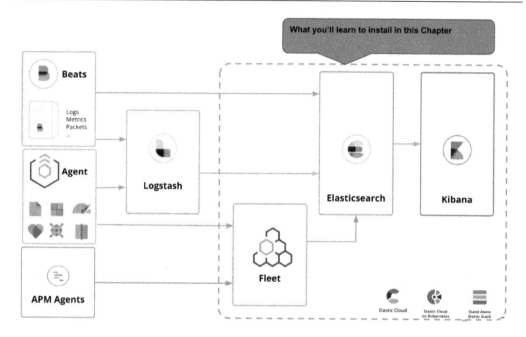

Figure 1.1 – The Elastic Stack components

In this chapter, we are going to learn how to install Elasticsearch, Kibana, and **Fleet** with different deployment options (Elastic Cloud, self-managed, and **Elastic Cloud on Kubernetes** (**ECK**)) highlighted in the right part of the following figure, and then we will proceed to the data ingestion part in the next chapters.

To determine the most suitable deployment option for your needs, *Figure 1.2* provides a comparative summary of the key differences among the various deployment methods:

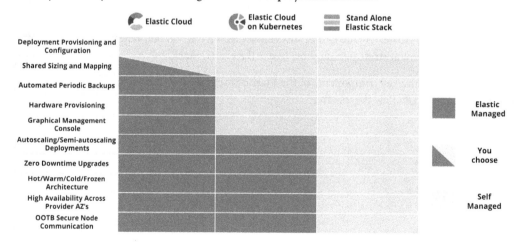

Figure 1.2 – Deployment options comparison

We'll be covering the following recipes:

- Deploying the Elastic Stack on Elastic Cloud
- Installing the Elastic Stack with ECK
- Installing a self-managed Elastic Stack
- Adding data tiering to your deployment
- Setting up additional nodes
- Setting up Fleet Server
- Setting up a snapshot repository

Deploying the Elastic Stack on Elastic Cloud

Elastic Cloud is the most straightforward way to deploy and manage your Elasticsearch, Kibana, **Integrations Server** (a combined component for the **application performance monitoring server** and Fleet Server), and other components of the Elastic Stack. This recipe will guide you through the process of getting started with Elastic Cloud, from signing up for an account to creating your first Elastic deployment.

How to do it...

Before we begin, let's learn how to create a deployment on Elastic Cloud and verify it using this step-by-step guide:

1. We will create an account on Elastic Cloud:

    I.   Visit the Elastic Cloud website at `https://cloud.elastic.co/`.

    II. Click on the **Sign up** button (a 14-day trial without needing a credit card is offered by default).

    III. Fill out the registration form with your details, including your name, email address, and desired password.

    Next, we will create a deployment.

2. On the next screen, you'll be prompted to create your first deployment, you can choose between the following options as shown in *Figure 1.3*:

    - **Cloud provider**: Google Cloud, Azure, or AWS.
    - **Region**: The supported regions for different cloud providers (the list of supported regions can be found here: `https://www.elastic.co/guide/en/cloud/current/ec-reference-regions.html`).

- **Hardware profile**: You can simply start with the **General-purpose** profile. Elastic Cloud allows you to change hardware later.
- **Version**: The latest minor version of Elastic Stack 7 or 8.

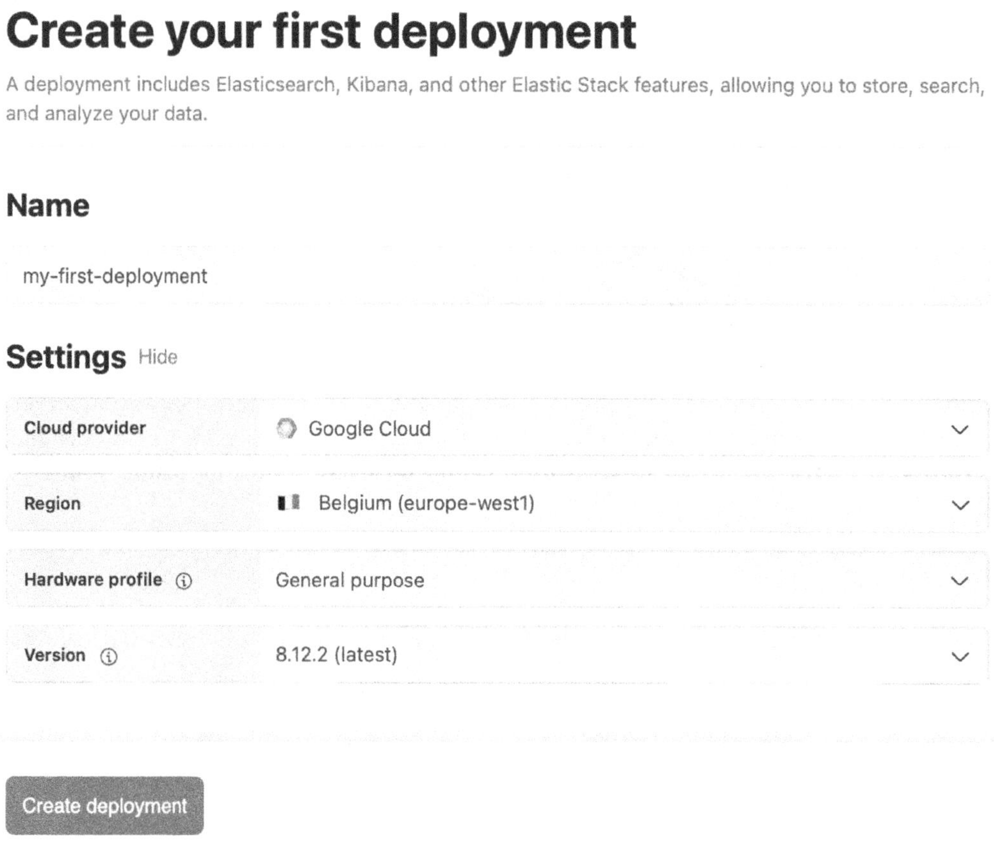

Create your first deployment

A deployment includes Elasticsearch, Kibana, and other Elastic Stack features, allowing you to store, search, and analyze your data.

Name

my-first-deployment

Settings Hide

Cloud provider	Google Cloud	⌄
Region	Belgium (europe-west1)	⌄
Hardware profile ⓘ	General purpose	⌄
Version ⓘ	8.12.2 (latest)	⌄

Create deployment

Figure 1.3 – Creating a cloud deployment

3. On the next screen (shown in *Figure 1.4*), you'll be given a password. Be sure to save it as you'll need it to log in to both Kibana (the application interface) as well as command-line operations:

Creating your deployment (takes about five minutes)   Continue

Save the deployment credentials

These root credentials are shown only once.
They provide super user access to your deployment. Keep them safe.

Username
elastic

Password

9BGKcWV80aloivIGv0Z2uYuz

Download

Skip

Figure 1.4 – Cloud deployment credentials

Finally, let's check the created deployment.

4.   After the deployment creation, you will be redirected to the **Home** page of Kibana, where you can choose one of the data onboarding guides as shown in *Figure 1.5*:

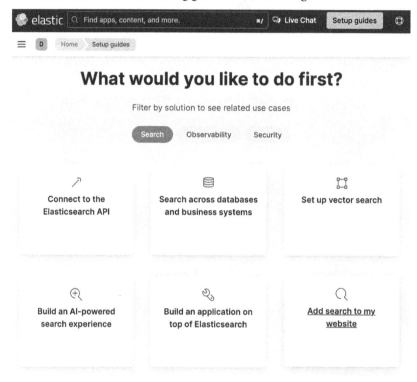

Figure 1.5 – Kibana onboarding screen

You can also check the deployment status from Elastic Cloud's main console (`https://cloud.elastic.co/home`):

Figure 1.6 – Cloud deployment status

5. You can then click on **Manage** to see the details of your deployment and management options as shown in *Figure 1.7*:

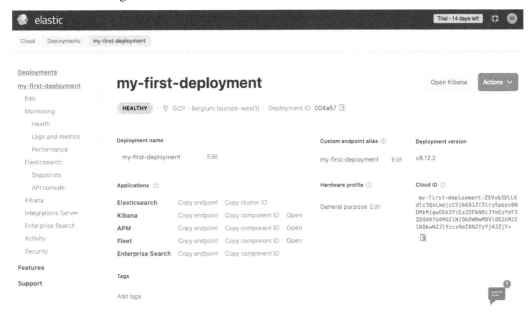

Figure 1.7 – Cloud deployment console

How it works...

At this stage, the following components have been provisioned automatically:

- 2 Elasticsearch hot nodes with 2 GB of RAM

- 1 Elasticsearch master tie-breaker node with 1 GB of RAM

- 1 Kibana node with 1 GB of RAM

- 1 Integrations Server node with 1 GB of RAM

- 1 Enterprise Search node with 2 GB of RAM

You can see the detailed list view of the components that we just mentioned in your deployment as shown in *Figure 1.8*. It gives you valuable information about each component such as **Health**, **Size**, **Role**, **Zone**, **Disk**, and **Memories**:

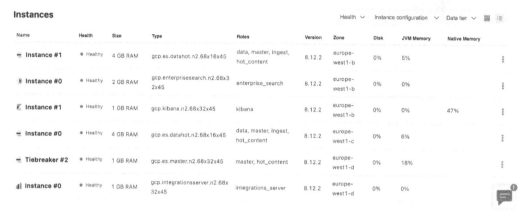

Name	Health	Size	Type	Roles	Version	Zone	Disk	JVM Memory	Native Memory	
Instance #1	● Healthy	4 GB RAM	gcp.es.datahot.n2.68x16x45	data, master, ingest, hot_content	8.12.2	europe-west1-b	0%	5%		
Instance #0	● Healthy	2 GB RAM	gcp.enterprisesearch.n2.68x32x45	enterprise_search	8.12.2	europe-west1-b	0%	0%		
Instance #1	● Healthy	1 GB RAM	gcp.kibana.n2.68x32x45	kibana	8.12.2	europe-west1-b	0%	0%	47%	
Instance #0	● Healthy	4 GB RAM	gcp.es.datahot.n2.68x16x45	data, master, ingest, hot_content	8.12.2	europe-west1-c	0%	6%		
Tiebreaker #2	● Healthy	1 GB RAM	gcp.es.master.n2.68x32x45	master, hot_content	8.12.2	europe-west1-d	0%	18%		
Instance #0	● Healthy	1 GB RAM	gcp.integrationsserver.n2.68x32x45	integrations_server	8.12.2	europe-west1-d	0%	0%		

Figure 1.8 – Cloud deployment components view

We also get different endpoints to access different components of the Elastic Stack:

Figure 1.9 – Cloud deployment endpoints

> **Note**
>
> You will need to save your cloud ID from this screen, as it will be useful and convenient to configure Elasticsearch clients, Beats, Elastic Agent, and so on, with your cloud ID so you can send data to your Elastic deployment.

There's more…

Once you deploy the Elastic Stack on Elastic Cloud, there are different possibilities to manage and configure your deployment. Let us look at a few possibilities.

Here's how to scale and configure your deployment:

1. Scale/autoscale your deployment to meet your growing needs (`https://www.elastic.co/guide/en/cloud/current/ec-autoscaling.html`).

2. Add or remove nodes, change the node type, adjust the node size, and make other configuration changes (More on this in the *Creating and setting up additional Elasticsearch nodes* recipe in this chapter).

3. Configure the data tiering (see the *Creating and setting up data tiering* recipe in this chapter for more information).

4. Monitor, backup your data, and configure your backup repository (More details in the *Setting up snapshot repository* recipe of this chapter).

5. Monitor your deployment health (More details in the *Setting up stack monitoring* recipe in *Chapter 13*).

Here's how you secure and control access to your deployment:

1. Configure authentication methods, such as username/password or **single sign-on** (**SSO**) (See *Chapter 11*).

2. Set up **role-based access control** (**RBAC**) to define user roles and permissions (See *Chapter 11*).

3. Configure a deployment traffic filter (`https://www.elastic.co/guide/en/cloud/current/ec-traffic-filtering-deployment-configuration.html`).

Installing the Elastic Stack with ECK

ECK is the official Kubernetes operator for automating the deployment and management of Elasticsearch and other Elastic components on Kubernetes. ECK enables the use of Kubernetes-native tools and APIs to manage Elasticsearch clusters, offering capabilities for monitoring and securing them. It supports scaling, rolling upgrades, availability zone awareness, and the implementation of hot-warm-cold storage architectures. ECK allows for the exploitation of Elasticsearch's power and flexibility on Kubernetes, both on-premises and in the cloud. In this guide, we will first install the ECK operator in a Kubernetes cluster and then use it to deploy an Elasticsearch cluster and Kibana.

Technical requirements

Ensure you have a Kubernetes cluster ready before deploying ECK and the Elastic Stack. For this recipe, you can use either **minikube** or **Google Kubernetes Engine** (**GKE**). Elastic Cloud on Kubernetes also supports other Kubernetes distributions such as OpenShift, Amazon Elastic Kubernetes Service (Amazon EKS), and Azure Kubernetes Service (Microsoft AKS). To ensure smooth deployment and optimal performance, allocate appropriate resources to your cluster. Your cluster should have at least 16 GB of RAM and 4 CPU cores to provide a seamless experience during the deployment of ECK, Elasticsearch, Kibana, Elastic Agent, and the sample application.

You can find all the related YAML files on the GitHub repository: `https://github.com/PacktPublishing/Elastic-Stack-8.x-Cookbook/tree/main/Chapter1/eck`.

The snippets of this recipe can be found at the following address: `https://github.com/PacktPublishing/Elastic-Stack-8.x-Cookbook/blob/main/Chapter1/snippets.md#installing-elastic-stack-with-elastic-cloud-on-kubernetes`.

Getting ready

Before installing ECK, you need to prepare your Kubernetes environment and ensure that you have the necessary resources and permissions. This recipe presumes that your Kubernetes cluster is already up and running. Your Kubernetes nodes need to have at least 2 GB of free memory. Make sure to check the supported versions of Kubernetes on the official Elastic documentation website: `https://www.elastic.co/support/matrix#matrix_kubernetes`.

How to do it...

Let's start:

1. First, you need to have an ECK operator deployed in your Kubernetes cluster. Let's begin by creating the ECK custom resource definitions:

    ```
    $ kubectl create -f https://download.elastic.co/downloads/
    eck/2.11.0/crds.yaml
    ```

    The following Elastic resources will be created in your Kubernetes cluster:

    ```
    customresourcedefinition.apiextensions.k8s.io/agents.agent.k8s.elastic.co created
    customresourcedefinition.apiextensions.k8s.io/apmservers.apm.k8s.elastic.co created
    customresourcedefinition.apiextensions.k8s.io/beats.beat.k8s.elastic.co created
    customresourcedefinition.apiextensions.k8s.io/elasticmapsservers.maps.k8s.elastic.co created
    customresourcedefinition.apiextensions.k8s.io/elasticsearchautoscalers.autoscaling.k8s.elastic.co created
    customresourcedefinition.apiextensions.k8s.io/elasticsearches.elasticsearch.k8s.elastic.co created
    customresourcedefinition.apiextensions.k8s.io/enterprisesearches.enterprisesearch.k8s.elastic.co created
    customresourcedefinition.apiextensions.k8s.io/kibanas.kibana.k8s.elastic.co created
    customresourcedefinition.apiextensions.k8s.io/logstashes.logstash.k8s.elastic.co created
    customresourcedefinition.apiextensions.k8s.io/stackconfigpolicies.stackconfigpolicy.k8s.elastic.co created
    ```

    Figure 1.10 – Created resources when deploying ECK

2. Now that the custom resources definitions have been created, proceed with the installation of the ECK operator:

    ```
    $ kubectl apply -f https://download.elastic.co/downloads/
    eck/2.11.0/operator.yaml
    ```

Executing the previous command will give you the following output:

```
) kubectl apply -f https://download.elastic.co/downloads/eck/2.11.0/operator.yaml
namespace/elastic-system created
serviceaccount/elastic-operator created
secret/elastic-webhook-server-cert created
configmap/elastic-operator created
clusterrole.rbac.authorization.k8s.io/elastic-operator created
clusterrole.rbac.authorization.k8s.io/elastic-operator-view created
clusterrole.rbac.authorization.k8s.io/elastic-operator-edit created
clusterrolebinding.rbac.authorization.k8s.io/elastic-operator created
service/elastic-webhook-server created
statefulset.apps/elastic-operator created
validatingwebhookconfiguration.admissionregistration.k8s.io/elastic-webhook.k8s.elastic.co created
```

Figure 1.11 – Installing the ECK operator in a Kubernetes cluster

> **Important note**
> The best practice is to use a dedicated Kubernetes namespace for all workloads related to ECK, which offers enhanced isolation for various applications and robust security with RBAC permissions by default. The provided manifest uses the `elastic-system` namespace by default.

3. We can then monitor the operator logs:

    ```
    $ kubectl -n elastic-system logs -f statefulset.apps/elastic-
    operator
    ```

4. Now, let's deploy a three-node Elasticsearch cluster by applying the YAML file provided in the GitHub repository:

    ```
    $ kubectl apply -f elasticsearch.yaml
    ```

5. To check the status of Elasticsearch, you can get an overview of the clusters with the following `kubectl` command:

    ```
    $ kubectl get elasticsearch
    ```

> **Note**
> This might take a couple of minutes if you need to pull the images.

Figure 1.12 shows the results of this command when the cluster has been successfully deployed:

```
NAME                    HEALTH    NODES    VERSION    PHASE    AGE
elasticsearch-sample    green     3        8.12.2     Ready    2m58s
```

Figure 1.12 – Checking the cluster status

6. Now, deploy the Kibana instance by applying the following `kibana.yaml` file in your cluster:

    ```
    $ kubectl apply -f kibana.yaml
    ```

7. Similar to Elasticsearch, you can find details about Kibana instances with the following command:

    ```
    $ kubectl get kibana
    ```

```
) kubectl get kibana
NAME             HEALTH    NODES    VERSION    AGE
kibana-sample    green     1        8.12.2     5m45s
```

Figure 1.13 – Checking Kibana status

Finally, let's connect to Kibana. This is quite straightforward, as ECK automatically creates a `ClusterIP` service for Kibana. Follow the next steps to log in to your Kibana instance.

8. Get the `ClusterIP` service created for Kibana:

    ```
    $ kubectl get service kibana-sample-kb-http
    ```

    You should expect to see an output like *Figure 1.14*:

```
) kubectl get service kibana-sample-kb-http
NAME                    TYPE         CLUSTER-IP     EXTERNAL-IP    PORT(S)     AGE
kibana-sample-kb-http   ClusterIP    10.24.7.205    <none>         5601/TCP    70m
```

Figure 1.14 – Printing the Kibana ClusterIP

9. Now, use `kubectl port-forward` to access Kibana from your host:

    ```
    $ kubectl port-forward service/kibana-sample-kb-http 5601
    ```

10. Before visiting the Kibana login page, we'll need to retrieve the password of the elastic user provisioned by the operator with the following command:

```
$ kubectl get secret elasticsearch-sample-es-elastic-user
-o=jsonpath='{.data.elastic}' | base64 --decode; echo
```

Copy the output of the command.

Now that you've forwarded the port, open it in your web browser and use the credentials obtained in the previous steps to log in to Kibana as shown in the following figure:

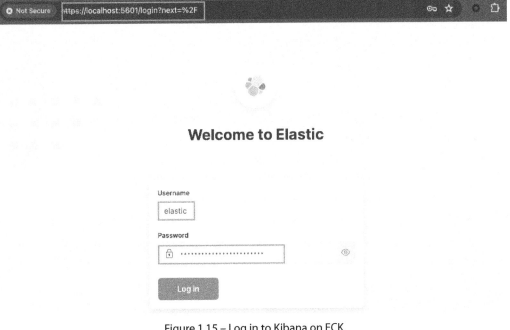

Figure 1.15 – Log in to Kibana on ECK

Important note

When accessing Kibana, you might see a security warning due to self-signed certificates not being trusted by the browser. You can safely bypass this warning and proceed to Kibana's URL. For production environments, it's recommended to use certificates from your own **certificate authority** (**CA**) to ensure security.

How it works...

As you have seen, ECK greatly simplifies the setup of Elasticsearch and Kibana, getting you up and running in a few minutes. It accomplishes this by managing a variety of tasks on our behalf. Let's review what ECK has done for us in the cluster:

- **Security**: Security features are enabled in ECK, ensuring robust protection for all deployed Elastic Stack resources. By default, all resources deployed through ECK are secured. The system provisions a built-in basic authentication user named `elastic`. **Transport Layer Security (TLS)** is configured to secure network traffic within and to your Elasticsearch cluster.

- **Certificates**: A self-signed, internally generated CA certificate is used by default for each cluster, providing secure communication within the Elasticsearch cluster. For advanced configurations, you have the option to use externally signed certificates or other custom certificate setups.

- **Default service exposure**: Your cluster is automatically set up with a `ClusterIP` service, which offers internal network connectivity. You also have the option to configure these services to be of the `LoadBalancer` type, making them accessible from external networks.

- **Elasticsearch connection**: You may have noticed by looking at the provided `kibana.yaml` file that there are no explicit Elasticsearch connection details. The information is provided to Kibana with the `ElasticsearchRef` specification defined in the ECK operator.

There's more...

As an alternative installation method, ECK can also be installed using a **Helm chart** from the Elastic Helm repository:

```
$ helm repo add elastic https://helm.elastic.co
$ helm repo update
```

Starting with ECK version 2.8, Logstash can be managed as a custom resource using the operator.

See also

- Here is an excellent blog article on ECK in production: `https://www.elastic.co/blog/eck-in-production-environment`

- If you're interested in deploying ECK with Terraform, check out the following: `https://www.elastic.co/blog/installing-eck-with-terraform-on-gcp`

- Also, there is this practical article on running ECK with Helm: `https://www.elastic.co/blog/using-eck-with-helm`

Installing a self-managed Elastic Stack

In this recipe, you will learn how to install and manage the Elastic Stack on your local machine, focusing primarily on the essential components: Elasticsearch and Kibana.

Getting ready

Before proceeding with the installation, make sure your system meets the minimum requirements for running Elasticsearch, Kibana, and Fleet Server. Check the official documentation for the specific version you want to install to ensure compatibility with your operating system (`https://www.elastic.co/support/matrix`).

How to do it...

Let's first look at how to download Elasticsearch:

1. Visit the Elasticsearch download page (`https://www.elastic.co/downloads/elasticsearch`).

2. By default, the official Elasticsearch download page provides you with the download links for the latest release. Choose the right package for your operating system.

3. Once the download is complete, extract the contents of the package to a working directory of your choice.

Next, let's configure Elasticsearch:

1. Open the Elasticsearch configuration file located in the extracted directory. For example, in Linux, it's found at `config/elasticsearch.yml`.

2. Adjust the settings as needed, such as the cluster name, network settings, and heap size.

3. Save the configuration file.

Now, let's see how you start Elasticsearch:

1. Open a terminal or command prompt and navigate to the Elasticsearch directory.

2. Run the Elasticsearch executable or script that is appropriate for your operating system:

   - For Linux/Mac, it is the following:

   ```
   $ ./bin/elasticsearch
   ```

   - For Windows, it is the following:

   ```
   $ bin\elasticsearch.bat
   ```

On the first launch, Elasticsearch will perform an initial security configuration, which includes generating a password for the built-in `elastic` user, an enrollment token for Kibana (valid for 30 minutes), and certificates and keys for transport and HTTP layers:

```
The Elasticsearch node is up and running and reachable at HTTPS
port 9200< You can check the Elasticsearch node with curl
command line curl --cacert <PATH_TO_CERTIFICATE> -u elastic
https://localhost:9200
```

Next, we will download and install Kibana:

1. Go to the official Kibana download page and go to the **Downloads** section.

2. By default, the official Kibana download page provides you with the download links for the latest release of Kibana. Download the appropriate package for your operating system (`tar.gz`/`zip`, `deb`, or `rpm`).

3. Extract the downloaded Kibana package to a directory of your choice.

4. Open a terminal or command prompt and navigate to the Kibana directory.

5. Run the Kibana executable file (e.g., `bin/kibana` for Unix-like systems or `bin\kibana.bat` for Windows) to start Kibana.

6. In your browser, access Kibana with the `https://localhost:5601` default URL, use the enrollment token from the earlier step when Kibana starts, and click the button to confirm the connection with Elasticsearch.

7. Use the `elastic` superuser and the previously generated password to log in to Kibana.

How it works...

Starting with Elastic 8.0, security features such as TLS for both inter-node communication and HTTP layer security are enabled by default in self-managed clusters. As a result, certificates and keys are automatically generated during the Elasticsearch installation process. This allows stack-level security, activating, by default, both node-to-node TLS and Elasticsearch API TLS, which we have seen during the installation of Kibana.

There's more...

You can also use Docker as a self-managed deployment option – please refer to the official documentation: `https://www.elastic.co/guide/en/elasticsearch/reference/current/docker.html`.

Creating and setting up data tiering

A data tier consists of several Elasticsearch nodes that have the same data role and usually run on similar hardware. Often, different hardware is configured for each tier; for example, the hot tier might use the most powerful and expensive hardware, while the cold or frozen tiers could utilize less expensive, storage-oriented hardware. Using data tiers is an efficient strategy for reducing hardware requirements in an Elasticsearch cluster while maintaining access to data and the ability to search through it. To illustrate, a single frozen node can keep up to 100 TB of data compared to 2 TB of data for a hot node.

However, there is a caveat: as data moves to colder tiers, query performance can decrease. This is expected since the data is less frequently queried.

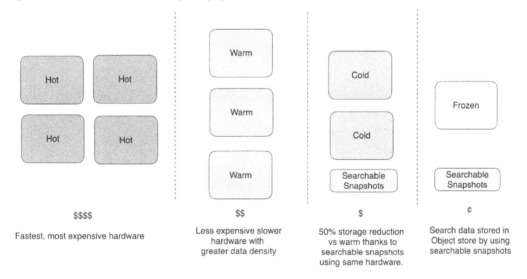

Figure 1.16 – Elasticsearch data tiering

As we can see in *Figure 1.16*, there are four data tiers provided by Elasticsearch:

- **Hot tier**: This tier handles mostly indexing and query for timestamped data (most recent and frequently accessed data). This tier can also be referenced as the **content tier** for non-timestamped data.

- **Warm tier**: This tier is used for less recent timestamped data (more than seven days) that does not need to be updated. Extends storage capacity up to five times compared to the hot tier.

- **Cold tier**: This tier is used for timestamped data that is not so frequently accessed and not updated anymore. This tier is built on **searchable snapshots technology** and can store twice as much data compared to the warm tier.

- **Frozen tier**: This tier is used for timestamped data that is never updated and queried rarely but needs to be kept for regulation, compliance, or security use cases such as forensics. The frozen tier stores most of the data on searchable snapshots and only the necessary data based on query is pulled and cached on a local disk inside the node.

In this recipe, you'll learn how to set up data tiering in a self-managed Elasticsearch cluster. We will also discuss implementation on Elastic Cloud.

Getting ready

Make sure your self-managed cluster for the earlier recipe is up and running. For the sake of simplicity, we will create two additional nodes on the same local machine. We'll add two data tiers to our cluster:

- A node for the cold tier
- A node for the frozen tier

The code snippets for this recipe can be found at the following link: `https://github.com/ PacktPublishing/Elastic-Stack-8.x-Cookbook/blob/main/Chapter1/ snippets.md#creating-and-setting-up-data-tiering`.

How to do it on your local machine...

On your local machine, execute the following steps:

1. Open the `elasticsearch.yaml` file of the cluster you've previously set up and uncomment the `transport.host` setting at the end.

2. Create two new directories for the new nodes, and let's call those directories the following:

   - `node-cold`
   - `node-frozen`

3. Download and extract the content of Elasticsearch package in each directory. Make sure to use the same version and operating system as previously used in the *Installing a self-managed Elastic Stack* recipe.

4. In a separate terminal from where your cluster from the previous recipe is running, navigate to the directory where Elasticsearch is installed and run the following command:

   ```
   $ ./bin/elasticsearch --enrollment-token -s node
   ```

   This command generates an enrollment token that you'll copy and use to enroll new nodes with your Elasticsearch cluster.

5. Go to the cold node directory and open the `elasticseach.yaml` file and add the following settings:

    ```
    node.name: node-cold
    node.roles: ["data_cold"]
    ```

6. From the installation directory of the cold node, start Elasticsearch and pass the enrollment token with `--enrollment-token`:

    ```
    $ ./bin/elasticsearch --enrollment-token <enrollment-token>
    ```

    Check that your node has successfully started.

7. Now, let's do the same for the frozen node. Open the `elasticseach.yaml` file and add the following settings:

    ```
    node.name: node-frozen
    node.roles: ["data_frozen"]
    ```

8. From the installation directory of the frozen node, start Elasticsearch and pass the enrollment token with `--enrollment-token`:

    ```
    $ bin/elasticsearch --enrollment-token <enrollment-token>
    ```

Check that the new frozen node has successfully started.

How it works (on self-managed)...

Elasticsearch now provides specific roles that match the different data tiers (hot, warm, cold, frozen). It means we can add one of the `data_hot`, `data_warm`, `data_cold`, or `data_frozen` node roles to the roles setting in the configuration file. Once the appropriate roles are defined in the configuration file, new nodes are introduced into the cluster using an enrollment token. The `-s node` argument specifies that we're creating a token to enroll an Elasticsearch node into a cluster.

How to do it on Elastic Cloud...

Adding data tiers on an Elastic Cloud deployment is a more straightforward and streamlined process. There is no configuration file to edit and no infrastructure to provision; just head to your deployment and follow these steps:

1. On the Elastic Cloud deployment page, click **Manage**.
2. Click on **Edit** on the left navigation pane.
3. Click on **Add capacity** for any data tiers you wish to add.

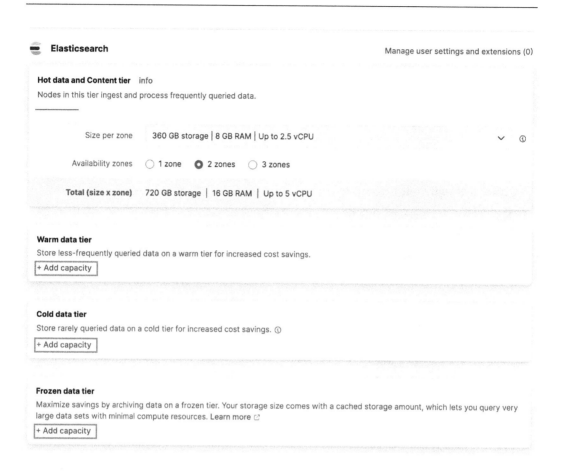

Figure 1.17 – Cloud deployment data tiering

How to do it on ECK...

In ECK, you define your cluster's topology using a concept called nodeSets. Within the nodeSets attribute, each entry represents a group of Elasticsearch nodes sharing the same Kubernetes and Elasticsearch configurations. For instance, you might have one nodeSets attribute for master nodes, another for your hot tier nodes, and so forth. You can find an example configuration in the GitHub repository: https://github.com/PacktPublishing/Elastic-Stack-8.x-Cookbook/blob/main/Chapter1/eck/elasticsearch-data-tiers.yaml.

When examining the provided configuration, it's clear that there are three `nodeSets` attributes named `hot`, `cold`, and `frozen`, as illustrated in the following code block. Please note that for readability, the code has been abbreviated; the complete code is accessible at the specified GitHub repository location:

```
spec:
  version: 8.12.2
  nodeSets:
    - name: hot
      config:
        node.store.allow_mmap: false
      podTemplate:
        ...
      count: 3
    - name: cold
      config:
        node.roles: ["data_cold"]
        node.store.allow_mmap: false
      podTemplate:
        ...
      count: 1
    - name: frozen
      config:
        node.roles: [ "data_frozen" ]
        node.store.allow_mmap: false
        podTemplate:
          ...
      count: 1
```

In a real production scenario, additional configuration such as **Kubernetes node affinity** is necessary. Kubernetes node affinity uses `NodeSelector` to ensure Elasticsearch workloads are confined to selected Kubernetes nodes. Under the hood, Elasticsearch shard allocation awareness is used to allocate shards to the specified Kubernetes nodes.

There's more...

In the production scenario, adding new data tiers to the cluster on a self-managed Elastic Stack is a bit more complex. For high availability and resilience, you'll need to deploy nodes on separate machines and thus it requires additional configuration steps that were not covered in this recipe, such as binding to an address other than localhost.

Data tiers are the first steps of your data management strategy with Elasticsearch. The next step is to define an **index life cycle management** (**ILM**) policy that'll automate the migration of your data between the different tiers. This will be covered in the *Setting up index life cycle policy* recipe in *Chapter 12*.

Data tiering is primarily intended for timestamped data. To fully leverage data tiers, matching infrastructure resources must be allocated for each tier. For instance, warm and cold tiers can use spinning disks rather than SSDs and have a larger RAM-to-disk ratio, enabling them to store more data. These tiers are ideal for frequent read access to your data. Meanwhile, the frozen tier depends entirely on searchable snapshots, making it most suitable for long-term retention and infrequent searches.

See also

- For more information about data tiering, see `https://www.elastic.co/guide/en/elasticsearch/reference/current/data-tiers.html`

- You can also read the blog, *Data lifecycle management with data tiers*, by Lee Hinman: `https://www.elastic.co/blog/elasticsearch-data-lifecycle-management-with-data-tiers`

Creating and setting up additional Elasticsearch nodes

An Elasticsearch cluster can have a variety of node roles, besides data tiers, to function efficiently. *Figure 1.18* outlines the several types of nodes available in a cluster:

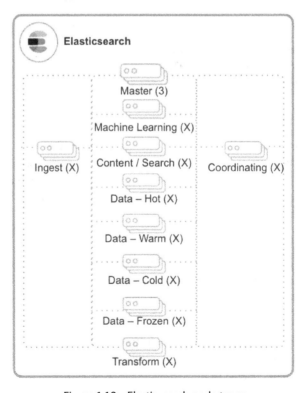

Figure 1.18 – Elasticsearch node types

Roles such as *Master*, *Machine Learning*, or *Ingest* can be dedicated to specific Elasticsearch instances, and this is often a best practice in a production environment.

In this recipe, we will learn how to configure dedicated nodes for both self-managed deployments and Elastic Cloud.

Getting ready

Ensure that your self-managed cluster from the previous recipe is operational. For simplicity, we will create additional nodes on the same local machine. The nodes will undertake the following roles:

- A dedicated master eligible node
- A machine learning node

The snippets for this recipe are available at `https://github.com/PacktPublishing/Elastic-Stack-8.x-Cookbook/blob/main/Chapter1/snippets.md#creating-and-setting-up-additional-elasticsearch-nodes`.

How to do it...

On your local machine, proceed with the following steps:

1. Create two new directories for the new nodes, which we will name the following:

   - `node-master`
   - `node-ml`

2. Repeat *Steps 1* and *2* of the *Installing a self-managed Elastic Stack* recipe in each directory.

3. In a separate terminal from where your cluster from the previous recipe is running, navigate to the directory where Elasticsearch is installed and run the following command:

   ```
   $ bin\elasticsearch-create-enrollment-token -s node
   ```

4. Copy the enrollment token. You will use it to enroll the new nodes with your Elasticsearch cluster.

5. Navigate to the `node-master` directory and open the `elasticsearch.yaml` file and add the following settings:

   ```
   node.name: node-master
   node.roles: ["master"]
   ```

6. From the installation directory of the cold node, start Elasticsearch and pass the enrollment token with -enrollment-token:

```
$ bin/elasticsearch --enrollment-token <enrollment-token>
```

Verify that the node has started successfully.

Now, let's follow the same steps to add a dedicated **machine learning node**.

7. Open the elasticsearch.yaml file in the node-ml directory and add the following settings:

```
node.name: node-ml
node.roles: ["ml"]
```

8. Start the node with the following command:

```
$ bin/elasticsearch --enrollment-token <enrollment-token>
```

Check the new machine learning node has successfully started and joined the cluster.

How it works...

As explained in the previous recipe, we're basically using the node.roles attributes to specify the roles.

How to do it on Elastic Cloud...

On Elastic Cloud, dedicated master nodes are provisioned based on the number of Elasticsearch nodes available in your deployment. If your deployment has more than six Elasticsearch nodes, dedicated master nodes are automatically created. If your deployment has less than six Elasticsearch nodes, a tie-breaker node is set up behind the scenes to ensure high availability.

For machine learning and the other node types, follow the steps outlined here:

1. On the Elastic Cloud deployment page, click **Manage**.
2. Click on **Edit** on the left navigation pane.
3. Click on **Add capacity** for the node type you wish to add (coordinating and ingest, machine learning).

How to do it on ECK...

In ECK, expanding your cluster with additional node types requires you to update your YAML specification. As discussed, when setting up data tiering, you introduce the concept of nodeSets. By simply adding a nodeSets attribute with the necessary role (e.g., ml, master, ingest, etc.), you instruct the operator to allocate those resources within your cluster. A sample YAML file is available at the following link: https://github.com/PacktPublishing/Elastic-Stack-8.x-Cookbook/blob/main/Chapter1/eck/elasticsearch-dedicated-master-ml.yaml.

There's more...

In a production scenario, it's always best to have dedicated hardware and hosts for specific node roles. You can also configure voting-only nodes that participate in the election for the master node but don't serve as the master. A configuration with at least two dedicated master nodes and one voting-only node can be a suitable alternative to three full master nodes.

See also

- For more information on the machine learning setup and requirements, see the following link: `https://www.elastic.co/guide/en/machine-learning/current/setup.html`

- For an in-depth exploration of the various node roles and their specific functions, see the following link: `https://www.elastic.co/guide/en/elasticsearch/reference/8.13/modules-node.html#node-roles`

Creating and setting up Fleet Server

Fleet Server is a key component of the new ingest architecture in the Elastic Stack, which revolves around the Elastic Agent. Before delving into this recipe, let's review some important concepts about Fleet and the Agent.

Fleet serves as the central management component, providing a UI within Kibana that manages Agents and their configurations at scale. The Elastic Agent is a single, unified binary responsible for data collection tasks – gathering logs, metrics, security events, and more, running on your hosts.

Fleet Server connects the Elastic Agent to Fleet and acts as a control plane for Elastic Agents. It is an essential piece if you intend to use Fleet for centralized management. The schema in *Figure 1.19* illustrates the various components and their interactions:

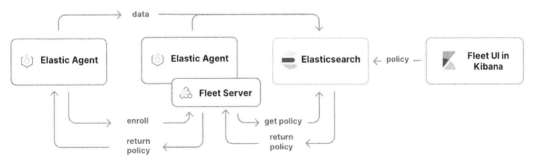

Figure 1.19 – Architecture including Elastic Agent and Fleet Server

In this recipe, we'll cover the setup of Fleet Server for self-managed deployments and Elastic Cloud.

Getting ready

Make sure you have an Elasticsearch cluster up and running with Kibana connected to the cluster.

For self-managed setups, this recipe assumes that you will be installing Fleet Server on the same local machine as your cluster.

> **Note**
>
> This configuration is not recommended for production environments.

How to do it on a self-managed Elastic Stack...

We will use the quick-start wizard in Kibana for our setup:

1. In Kibana, on the left menu pane, go to **Management | Fleet**.

2. Click on **Add Fleet Servers**. This will present instructions for adding a Fleet Server with two options: **Quick Start** and **Advanced**. We'll use the **Quick Start** option:

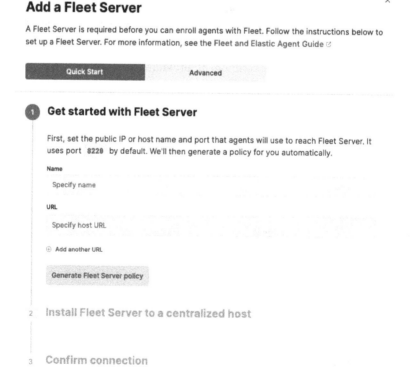

Figure 1.20 – Fleet Server configuration

3. Fill in the name and the URL.

4. Click on **Generate Fleet Server policy**.

5. Copy the generated command and paste it into your terminal.

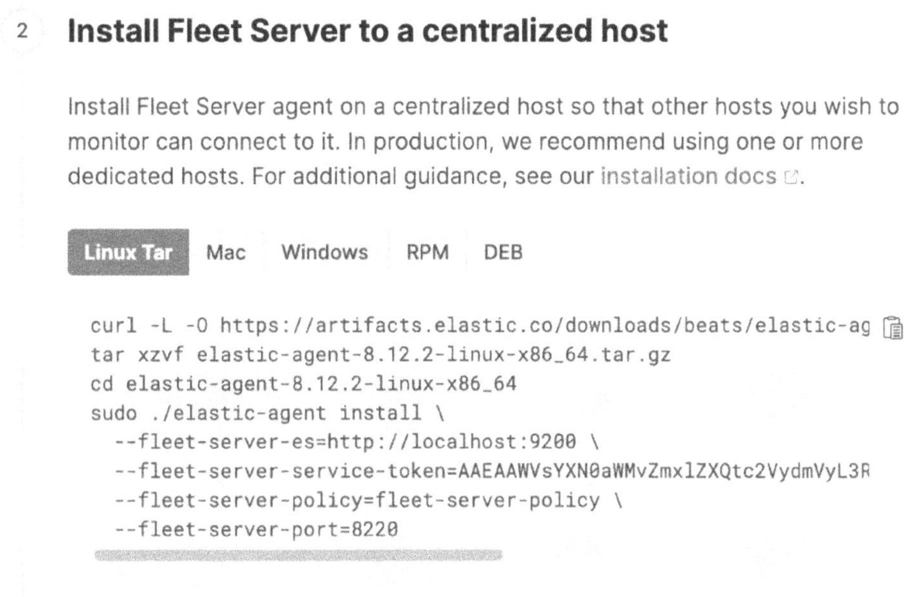

2   **Install Fleet Server to a centralized host**

Install Fleet Server agent on a centralized host so that other hosts you wish to monitor can connect to it. In production, we recommend using one or more dedicated hosts. For additional guidance, see our installation docs ⬀.

| Linux Tar | Mac | Windows | RPM | DEB |

```
curl -L -O https://artifacts.elastic.co/downloads/beats/elastic-ag
tar xzvf elastic-agent-8.12.2-linux-x86_64.tar.gz
cd elastic-agent-8.12.2-linux-x86_64
sudo ./elastic-agent install \
  --fleet-server-es=http://localhost:9200 \
  --fleet-server-service-token=AAEAAWVsYXN0aWMvZmxlZXQtc2VydmVyL3R
  --fleet-server-policy=fleet-server-policy \
  --fleet-server-port=8220
```

Figure 1.21 – Fleet Server installation

If the installation is successful, you will see a confirmation showing that Fleet Server is operational and connected.

How it works...

By using the **Quick Start** option, Fleet automatically creates a Fleet Server instance and an enrollment token object in the background. Note that this option relies on self-signed certificates and is not suitable for production environments. For more details on how to set up Fleet using the **Advanced** mode, refer to the *See also* section of this recipe.

Setting up on Elastic Cloud

Elastic Cloud offers a hosted Integrations Server that includes Fleet Server, simplifying the setup process considerably.

To verify the availability of Fleet Server in your cloud deployment, do the following:

1. In Kibana, on the left menu pane, go to **Management | Fleet**.

2. Look for **Elastic Cloud agent policy** on the **Agents** tab:

Figure 1.22 – Centralized management for Elastic Agents

3. Check that the agent status is healthy.

See also

For configuration samples to set up Fleet Server on ECK, see the following examples:

- To gain deeper insights into Fleet and Agent you also look at this recorded webinar: `https://www.elastic.co/webinars/introducing-elastic-agent-and-fleet`

- To set up Fleet Server for production, check the official documentation: `https://www.elastic.co/guide/en/fleet/current/add-fleet-server-on-prem.html`

Setting up snapshot repository

After you've set up a functional Elastic cluster, we recommend setting up a snapshot repository according to your deployment method. This allows you to back up your valuable data. Elasticsearch features a native capability for data backup and restoration.

When you create a deployment on Elastic Cloud, it comes also with a default repository called `found repository`. In this recipe, you'll learn how to register and manage a snapshot repository for an Amazon S3 bucket with Elastic Cloud, a popular option. The setup concepts can also apply to other cloud repositories, such as Google Cloud Storage or Azure Blob Storage, and self-managed repositories.

Later in the book, we will provide a guide on how to configure and execute snapshot and restore operations.

Getting ready

Make sure that your Elastic Cloud deployment is up and running and that you have sufficient permissions to create and configure S3 buckets on AWS.

How to do it...

In the first step, we will create a S3 bucket:

1. First, let us go to **AWS Console | S3 | Create Bucket**. Provide a name for the bucket, for instance: `elasticsearch-s3-bucket-repository`. Make sure to choose **Block all public access** before proceeding to create the bucket:

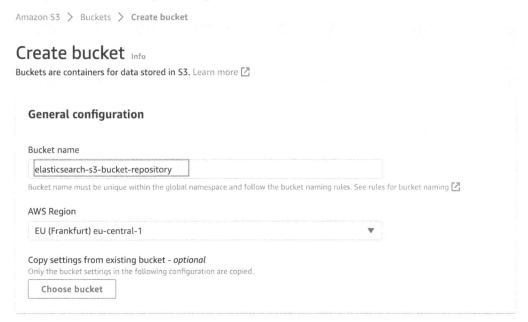

Figure 1.23 – Creating an S3 bucket

Create an AWS policy to allow the Identity and Access Management user (IAM user) to access the S3 Bucket.

2. Navigate to **AWS Management Console**, then go to **IAM | Policies**.

3. Click on **Create Policy**.

4. Switch to the JSON editor and set up the policy with the following snippet (the snippet can be found at this address: `https://github.com/PacktPublishing/Elastic-Stack-8.x-Cookbook/blob/main/Chapter1/snippets.md#sample-aws-s3-policy`):

```
{
  "Version": "2012-10-17",
  "Statement": [
  {
      "Sid": "VisualEditor0",
      "Effect": "Allow",
      "Action": "s3:*",
      "Resource": [
      "arn:aws:s3:::elasticsearch-s3-bucket-repository",
      "arn:aws:s3:::elasticsearch-s3-bucket-repository/*"
      ]
  }
  ]
}
```

5. On the next screen, give the policy the name `elasticsearch-s3-bucket-policy` and click on **Create Policy**.

   Create an IAM user and attach the policy we created.

6. Navigate to the **AWS Management Console** and then go to **IAM | Access Management | Users**.

7. Click **Create User**, and provide `elastic-s3-default-user` as the username:

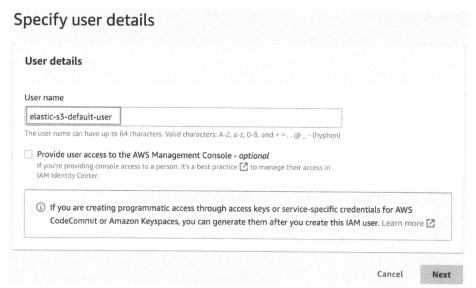

Figure 1.24 – Creating Elastic S3 default user

8. On the next screen (*Figure 1.25*), choose **Attach policies directly** and attach the policy that you previously generated:

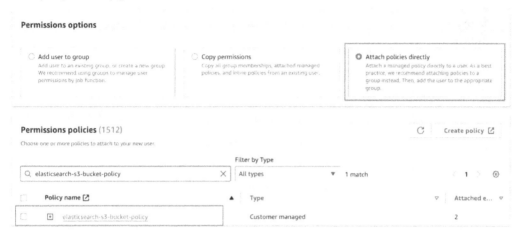

Figure 1.25 – Attaching permission policy to the user

9. On the next screen (*Figure 1.26*), click on **Create user** to complete the user creation:

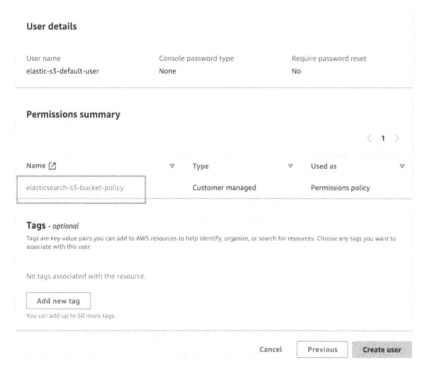

Figure 1.26 – Finalizing the user creation

Now, we will generate an access key and secret access.

10. Open the **Security credentials** tab, and then choose **Create access key**.

11. On the next screen (*Figure 1.27*), you will need to choose **Third-party service** for the use case, confirm the recommendation, and click on **Next**:

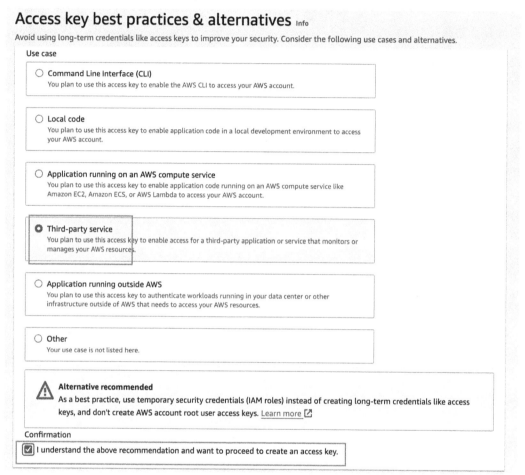

Figure 1.27 – Access key configuration

12. On the next screen, click on **Create access key**.

13. Download the key pair on the last screen of the wizard and choose **Download .csv file**. Store the `.csv` file with keys in a secure location that we will use in the next step.

14. Store the access secrets in the Elastic Cloud deployment keystore (if you are configuring for an on-premises Elasticsearch cluster, you will have to use the `elasticsearch-keystore` command-line tool: `https://www.elastic.co/guide/en/elasticsearch/reference/current/elasticsearch-keystore.html`).

15. Go to the Elastic Cloud console and navigate to the management console of your deployment then go to the security page. Add settings for the Elasticsearch keystore with **Type** set to **Single string** and add the keys and values with the access key and secret access from the previous step:

```
s3.client.secondary.access_key
s3.client.secondary.secret_key
```

16. Make sure you get the following security keys on the security page of your deployment and restart the deployment to apply the changes:

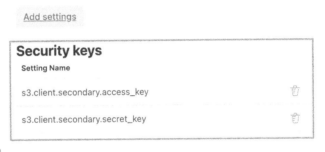

Figure 1.28 – Elastic Cloud keystore setting

17. For a self-managed deployment, you can set up the same keys with the following commands:

```
$ bin/elasticsearch-keystore add s3.client.secondary.access_key
$ bin/elasticsearch-keystore add s3.client. secondary.secret_key
```

18. We can now register the repository with Kibana. Let's go to **Kibana | Management | Stack Management | Snapshot & Restore | Repositories | Register a repository**, name it `my-custom-s3-repo`, and choose **AWS S3** as the **Repository type** option, as shown in *Figure 1.29*:

Register repository

Repository name

A unique name for the repository.

Name

`my-custom-s3-repo`

Repository type

Storage location for your snapshots. Learn more about repository types. ⌁

Azure	Google Cloud Storage	AWS S3
① Learn more	① Learn more	① Learn more
Select	Select	✓ Selected

Figure 1.29 – Snapshot repository creation

19. Set **Client** to **secondary**; this is part of your s3.client.secondary.access_key keystore secrets. Make sure to use the exact same bucket name that you created on AWS, elasticsearch-s3-bucket-repository, as shown in *Figure 1.30*:

Register repository

'my-custom-s3-repo' settings

⊙ AWS S3 repository docs

Client

The name of the AWS S3 client.

Client

`secondary`

Bucket

The name of the AWS S3 bucket to use for snapshots.

Bucket (required)

`elasticsearch-s3-bucket-repository`

Figure 1.30 – Snapshot repository client configuration

20. Click to verify the repository and make sure that your S3 bucket is successfully connected as a snapshot repository:

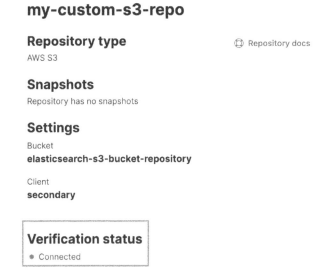

Figure 1.31 – Snapshot repository status

How it works...

To create and register a snapshot repository, the most important step is choosing the repository type and configuring the corresponding settings in Elasticsearch. We've detailed the process of creating an S3 bucket and setting up the S3 access policy, generating the access key, and using this information to register the snapshot repository via Kibana. Like most cluster-level operations, snapshot repository registration can also be performed via the API (`https://www.elastic.co/guide/en/elasticsearch/reference/current/put-snapshot-repo-api.html`).

There's more...

Although we used AWS S3 and Elastic Cloud deployments for this recipe, you can learn about other options including Azure, Google Cloud Storage, shared file systems, and read-only/source-only repositories in the Elastic documentation: `https://www.elastic.co/guide/en/elasticsearch/reference/current/snapshots-register-repository.html`.

If you're choosing ECK for your deployment, please consult the documentation specific to ECK for registering a snapshot repository: `https://www.elastic.co/guide/en/cloud-on-k8s/current/k8s-snapshots.html`.

Later in the book, in the *Managing the snapshot life cycle* recipe in *Chapter 12*, you'll find guidance on how to configure and execute the snapshot/restore process.

2
Ingesting General Content Data

This chapter, along with *Chapter 4*, will focus on data ingestion. Generally, we can categorize data into two groups – **general content** (data from APIs, HTML pages, catalogs, data from **Relational Database Management System** (**RDBMS**), PDFs, spreadsheets, etc.), and **time series** (data indexed in chronological order, such as logs, metrics, traces, and security events). In this chapter, we will ingest general content to illustrate the basic concepts of data ingestion, including fundamental data operations (index, delete, and update), analyzers, static and dynamic index mappings, and index templates.

Figure 2.1 illustrates the connections between various components, and in this chapter, we will explore recipes dedicated to the **Client APP**, **Analyzer**, **Mapping**, and **Index template** components (you can view the color image when you download the free PDF version of this book):

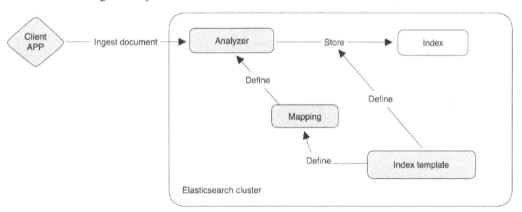

Figure 2.1 – Elasticsearch index management components

In this chapter, we are going to cover the following main topics:

- Adding data from the Elasticsearch client
- Updating data in Elasticsearch
- Deleting data in Elasticsearch

- Using an analyzer

- Defining index mapping

- Using dynamic templates in document mapping

- Creating an index template

- Indexing multiple documents using Bulk API

Introducing the Wikipedia Movie Plots dataset

For the general content sample data that we will use in this chapter, we will use the `Wikipedia Movie Plots` dataset from `kaggle.com`, authored by JustinR. (`https://www.kaggle.com/datasets/jrobischon/wikipedia-movie-plots`). The dataset contains interesting metadata of more than 34,000 movies scraped from Wikipedia.

> **Dataset citation**
>
> Wikipedia Movie Plots. (2018, October 15). Kaggle: `https://www.kaggle.com/datasets/jrobischon/wikipedia-movie-plots`.
>
> Note that this dataset is under the CC BY-SA 4.0 license (`https://creativecommons.org/licenses/by-sa/4.0/`).

Technical requirements

To follow the different recipes in this chapter, you will need an Elastic Stack deployment that includes the following:

- Elasticsearch to search and store data

- Kibana for data visualization and Dev Tools access

In addition to the Elastic Stack deployment, you'll also need to have Python 3+ installed on your local machine.

Adding data from the Elasticsearch client

To ingest general content such as catalogs, HTML pages, and files from your application, Elastic provides a wide range of Elastic language clients to easily ingest data via Elasticsearch `REST` APIs. In this recipe, we will learn how to add sample data to Elasticsearch hosted on Elastic Cloud using a Python client.

To use Elasticsearch's `REST` APIs through various programming languages, a client application chooses a suitable client library. The client initializes and sends HTTP requests, directing them to the Elasticsearch cluster for data operations. Elasticsearch processes the requests and returns HTTP responses containing results or errors. The client application parses these responses and acts on the data accordingly. *Figure 2.2* shows the summarized data flow:

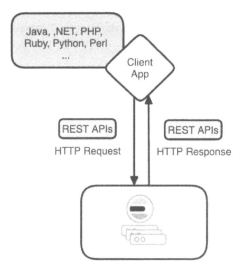

Figure 2.2 – Elasticsearch's client request and response flow

Getting ready

To simplify the package management, we recommend you install `pip` (`https://pypi.org/project/pip/`).

The snippets of this recipe are available here: `https://github.com/PacktPublishing/Elastic-Stack-8.x-Cookbook/blob/main/Chapter2/snippets.md#adding-data-from-the-elasticsearch-client`.

How to do it...

First, we will install the Elasticsearch Python client:

1. Add `elasticsearch`, `elasticsearch-async`, and `load_dotenv` to the `requirements.txt` file of your Python project (the sample `requirements.txt` file can be found at this address: `https://github.com/PacktPublishing/Elastic-Stack-8.x-Cookbook/blob/main/Chapter2/python-client-sample/requirements.txt`).

2. Run the following command to install the Elasticsearch Python client library:

    ```
    $ pip install -r requirements.txt
    ```

    Now, let's set up a connection to Elasticsearch.

3. Prepare a `.env` file to store the access information, `Cloud ID("ES_CID")`, user name `("ES_USER")`, and password `("ES_PWD")`, for the basic authentication. You can find the sample `.env` file at this address: `https://github.com/PacktPublishing/Elastic-Stack-8.x-Cookbook/blob/main/Chapter2/python-client-sample/.env`.

Remember that we saved the password for our default user, `elastic`, in the *Deploying Elastic Stack on Elastic Cloud* recipe in *Chapter 1*, and the instructions to find the cloud ID can be found in the same recipe.

4. Import the libraries in a Python file (`sampledata_index.py`), which we will use for this recipe:

```
import os
from elasticsearch import Elasticsearch
from dotenv import load_dotenv
```

5. Load the environment variables and initiate an Elasticsearch connection:

```
load_dotenv()

ES_CID = os.getenv('ES_CID')
ES_USER = os.getenv('ES_USER')
ES_PWD = os.getenv('ES_PWD')

es = Elasticsearch(
    cloud_id=ES_CID,
    basic_auth=(ES_USER, ES_PWD)
)

print(es.info())
```

6. Now, you can run the script to check whether the connection is successful. Run the following command:

```
$ python sampledata_index.py
```

You should see an output that looks like the following screenshot:

```
) python sampledata_index.py
{'name': 'instance-0000000001', 'cluster_name': '280897b13d1d477ba3bf7d485580fe64', 'cluster_uuid': '
ueN9oYBzQZmTGLIXr__Rmw', 'version': {'number': '8.12.2', 'build_flavor': 'default', 'build_type': 'do
cker', 'build_hash': '48a287ab9497e852de30327444b0809e55d46466', 'build_date': '2024-02-19T10:04:32.7
742731902', 'build_snapshot': False, 'lucene_version': '9.9.2', 'minimum_wire_compatibility_version':
 '7.17.0', 'minimum_index_compatibility_version': '7.0.0'}, 'tagline': 'You Know, for Search'}
```

Figure 2.3 – Connected Elasticsearch information

7. We can now extend the script to ingest a document. Prepare a sample JSON document from the movie dataset:

```
mymovie = {
    'release_year': '1908',
    'title': 'It is not this day.',
    'origin': 'American',
    'director': 'D.W. Griffith',
    'cast': 'Harry Solter, Linda Arvidson',
    'genre': 'comedy',
    'wiki_page':'https://en.wikipedia.org/wiki/A_Calamitous_
Elopement',
    'plot': 'A young couple decides to elope after being caught
in the midst of a romantic moment by the woman.'
}
```

8. Index the sample data in Elasticsearch. Here, we will choose the index name `'movies'` and print the index results. Finally, we will store the `document ID` in a tmp file that we will reuse for the following recipes:

```
response = es.index(index='movies', document=mymovie)
print(response)
# Write the '_id' to a file named tmp.txt
with open('tmp.txt', 'w') as file:
    file.write(response['_id'])


# Print the contents of the file to confirm it's written
correctly
with open('tmp.txt', 'r') as file:
    print(f"document id saved to tmp.txt: {file.read()}")
time.sleep(2)
```

9. Verify the data in Elasticsearch to ensure that it has been successfully indexed; wait two seconds after the indexing, query Elasticsearch using the _search API, and then print the results:

```
response = es.search(index='movies', query={"match_all": {}})
print("Sample movie data in Elasticsearch:")
for hit in response['hits']['hits']:
print(hit['_source'])
```

10. Execute the script again with the following script:

```
$ python sampledata_index.py
```

You should have the following result in the console output:

```
{'_index': 'movies', '_id': 'ArZrWI4B6R5W3CvT1Dxa', '_version': 1, 'result': 'created', '
_shards': {'total': 2, 'successful': 1, 'failed': 0}, '_seq_no': 0, '_primary_term': 1}
document id saved to tmp.txt: ArZrWI4B6R5W3CvT1Dxa
Sample movie data in Elasticsearch:
{'release_year': '1908', 'title': 'It is not this day.', 'origin': 'American', 'director'
: 'D.W. Griffith', 'cast': 'Harry Solter, Linda Arvidson', 'genre': 'comedy', 'wiki_page'
: 'https://en.wikipedia.org/wiki/A_Calamitous_Elopement', 'plot': 'A young couple decides
 to elope after being caught in the midst of a romantic moment by the woman .'}
```

Figure 2.4 – The output of the sampledata_index.py script

The full code sample can be found at `https://github.com/PacktPublishing/Elastic-Stack-8.x-Cookbook/blob/main/Chapter2/python-client-sample/sampledata_index.py`.

How it works...

In this recipe, we learned how to use the Elastic Python client to securely connect to a hosted deployment on Elastic Cloud.

Elasticsearch created the `movies` index by default during the first ingestion, and the fields were created with default mapping.

Later in this chapter, we will learn how to define static and dynamic mapping to customize field types with the help of concrete recipes.

It's also important to note that as we did not provide a document ID, Elasticsearch automatically generated an ID during the indexing phase as well.

The following diagram (*Figure 2.5*) shows the index processing flow:

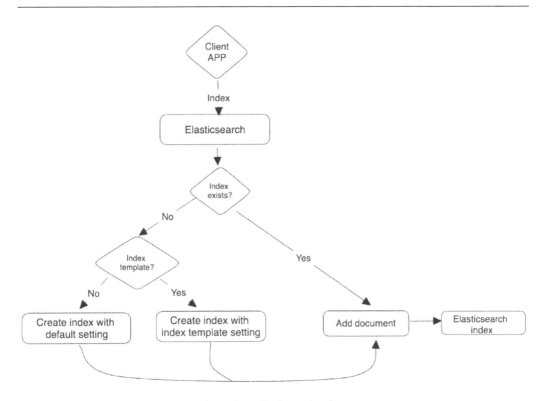

Figure 2.5 – The ingestion flow

There's more...

In this recipe, we used the HTTP basic authentication method. The Elastic Python client provides authentication methods such as **HTTP Bearer authentication** and **API key authentication**. Detailed documentation can be found at the following link: https://www.elastic.co/guide/en/elasticsearch/client/python-api/current/connecting.html#auth-bearer.

We chose to illustrate the simplicity of general content data ingestion by using the Python client. Detailed documentation for other client libraries can be found at the following link: https://www.elastic.co/guide/en/elasticsearch/client/index.html

During the development and testing phase, it's also very useful to use the Elastic REST API and test either with an HTTP client, such as CURL/Postman, or with the Kibana Dev Tools console (https://www.elastic.co/guide/en/kibana/current/console-kibana.html).

Updating data in Elasticsearch

In this recipe, we will explore how to update data in Elasticsearch using the Python client.

Getting ready

Ensure that you have installed the Elasticsearch Python client and have successfully set up a connection to your Elasticsearch cluster (refer to the *Adding data from the Elasticsearch client* recipe). You will also need to have completed the previous recipe, which involves ingesting a document into the movies index.

> **Note**
>
> The following three recipes will use the same set of requirements.

How to do it...

In this recipe, we're going to update the director field of a particular document within the movies index. The director field will be changed from its current value, D.W. Griffith, to a new value, Clint Eastwood. The following are the steps you'll need to follow in your Python script to perform this update and confirm that it has been successfully applied. Let's inspect the Python script that we will use to update the ingested document (https://github.com/PacktPublishing/Elastic-Stack-8.x-Cookbook/blob/main/Chapter2/python-client-sample/sampledata_update.py):

1. First, we need to retrieve the document ID of the previously ingested document from the tmp.txt file, which we intend to update. The field to update here is director; we are going to update the value from D.W. Griffith to Clint Eastwood:

```
index_name = 'movies'
document_id = ''
# Read the document_id the ingested document of the previous
recipe
with open('tmp.txt', 'r') as file:
    document_id = file.read()

document = {
    'director': 'Clint Eastwood'
}
```

2. We can now check `document_id`, verify that the document exists in the index, and then perform the `update` operation:

```
# Update the document in Elasticsearch if document_id is valid
if document_id != '':
    if es.exists(index=index_name, id=document_id):
        response = es.update(index=index_name, id=document_id,
                             doc=document)
        print(f"Update status: {response['result']}")
```

3. Once the document is updated, to verify that the update is successful, you can retrieve the updated document from Elasticsearch and print the modified fields:

```
updated_document = es.get(index=index_name, id=document_id)
print("Updated document:")
print(updated_document)
```

4. After inspecting the script, let's run it with the following command:

```
$ python sampledata_update.py
```

```
> python sampledata_update.py
Update status: updated
Updated document:
{'_index': 'movies', '_id': 'ArZrWI4B6R5W3CvT1Dxa', '_version': 2, '_seq_no': 1, '_primar
y_term': 1, 'found': True, '_source': {'release_year': '1908', 'title': 'It is not this d
ay.', 'origin': 'American', 'director': 'Clint Eastwood', 'cast': 'Harry Solter, Linda Ar
vidson', 'genre': 'comedy', 'wiki_page': 'https://en.wikipedia.org/wiki/A_Calamitous_Elop
ement', 'plot': 'A young couple decides to elope after being caught in the midst of a rom
antic moment by the woman .'}}
```

Figure 2.6 – The output of the sampledata_update.py script

You should see that the `_version` and `director` fields are updated.

How it works...

Each document includes a `_version` field in Elasticsearch. Elasticsearch documents cannot be modified directly, as they are immutable. When you update an existing document, a new document is generated with an incremented version, while the previous document is flagged for deletion.

There's more...

We have just seen how to update a single document in Elasticsearch; in general, this is not optimal from a performance point of view. To update multiple documents that match a specific query, you can use the Update By Query API. This allows you to define a query to select the documents you want to update and specify the changes to be made; here is an example of how to do it via Elasticsearch's REST API:

```
q = {
    "script": {
        "source": "ctx._source.genre = 'comedies'",
        "lang": "painless"
    },
    "query": {
        "bool": {
            "must": [
                {
                    "term": {
                        "genre": "comedy"
                    }
                }
            ]
        }
    }
}
es.update_by_query(body=q, index=index_name)
```

The full Python script is available here: https://github.com/PacktPublishing/Elastic-Stack-8.x-Cookbook/blob/main/Chapter2/python-client-sample/sampledata_update_by_query.py.

> **Note**
> The script used here is based on a painless script; we will see more examples in *Chapter 6*.

The other way to update multiple documents in a single request is via **Elasticsearch's Bulk API**. The Bulk API can be used to insert, update, and delete multiple documents efficiently. We will learn how to use the Bulk API to ingest multiple documents at the end of this chapter. For more detailed information, refer to the following documentation: https://www.elastic.co/guide/en/elasticsearch/reference/current/docs-bulk.html.

To retrieve the ID of the document we want to update, we rely on a `tmp.txt` file where the ID of a previously created document was saved. Alternatively, you can retrieve the document's ID by using the Dev Tools in Kibana, perform a search on the movies index, go to **Kibana | Dev Tools**, and execute the following command:

```
GET movies/_search
```

This query should return a list of hits that display all documents in the index, along with their respective IDs, as shown in *Figure 2.7*. Using these results, locate and record the ID of the document you would like to update:

```
{
  "took": 1,
  "timed_out": false,
  "_shards": {
    "total": 1,
    "successful": 1,
    "skipped": 0,
    "failed": 0
  },
  "hits": {
    "total": {
      "value": 1,
      "relation": "eq"
    },
    "max_score": 1,
    "hits": [
      {
        "_index": "movies",
        "_id": "PL_BJ40BEiCEihtzgPbf",
        "_score": 1,
        "_source": {
          "release_year": "1908",
          "title": "It is not this day.",
          "origin": "American",
          "director": "Clint Eastwood",
          "cast": "Harry Solter, Linda Arvidson",
          "genre": "comedy",
          "wiki_page": "https://en.wikipedia.org/wiki/A_Calamitous_Elopement",
          "plot": "A young couple decides to elope after being caught in the midst of a romantic
            moment by the woman ."
        }
      }
    ]
  }
}
```

Figure 2.7 – Checking the document ID

Deleting data in Elasticsearch

In this recipe, we will explore how to delete a document from an Elasticsearch index.

Getting ready

Refer to the requirements for the *Updating data in Elasticsearch* recipe.

Make sure to download the following Python script from the GitHub repository: `https://github.com/PacktPublishing/Elastic-Stack-8.x-Cookbook/blob/main/Chapter2/python-client-sample/sampledata_delete.py`.

The snippets of the recipe are available at `https://github.com/PacktPublishing/Elastic-Stack-8.x-Cookbook/blob/main/Chapter2/snippets.md#deleting-data-in-elasticsearch`.

How to do it...

1. First, let us inspect the `sampledata_delete.py` Python script. Like the process in the previous recipe, we need to retrieve `document_id` from the `tmp.txt` file:

    ```
    with open('tmp.txt', 'r') as file:
            document_id = file.read()
    ```

2. We can now check `document_id`, verify that the document exists in the index, and then perform the `delete` operation by using the previously obtained `document_id`:

    ```
    if document_id != '':
        if es.exists(index=index_name, id=document_id):
            # delete the document in Elasticsearch
            response = es.delete(index=index_name, id=document_id)
            print(f"delete status: {response['result']}")
    ```

3. After reviewing the `delete` script, execute it with the following command:

    ```
    $ python sampledata_delete.py
    ```

    You should see the following output:

    ```
    ) python sampledata_delete.py
    delete status: deleted
    ```

    Figure 2.8 – The output of the sampledata_delete.py script

4. For further verification, return to the Dev Tools in Kibana and execute the search request again on the `movies` index:

```
GET movies/_search
```

This time, the result should reflect the deletion:

```
{
  "took": 2,
  "timed_out": false,
  "_shards": {
    "total": 1,
    "successful": 1,
    "skipped": 0,
    "failed": 0
  },
  "hits": {
    "total": {
      "value": 0,
      "relation": "eq"
    },
    "max_score": null,
    "hits": []
  }
}
```

Figure 2.9 – The search results in the movies index after deletion

The total hits will now be 0, confirming that the document has been successfully deleted.

How it works...

When a document is deleted in Elasticsearch, it is not immediately removed from the index. Instead, Elasticsearch marks the document as deleted. These documents remain in the index until a *merging* process occurs during routine optimization tasks, when Elasticsearch physically expunges the deleted documents from the index.

This mechanism allows Elasticsearch to handle deletions efficiently. By marking documents as deleted rather than expunging them outright, Elasticsearch avoids costly segment reorganizations within the index. The removal occurs during optimized, controlled background tasks.

There's more...

While we have discussed deleting documents by `document_id`, this might not be the most efficient approach for deleting multiple documents. For such scenarios, the `Delete By Query` API is more suitable, such as the following:

> **Note**
>
> Before executing the upcoming query, it is necessary to re-index the document, since it was deleted earlier in the recipe. Ensure that you have re-added the document to the `movies` index by executing the `sampledata_index.py` Python script.

```
POST /movies/_delete_by_query
{
  "query": {
    "match": {
      "genre": "comedy"
    }
  }
}
```

The preceding query will delete all movies matching the `comedy` genre in our index.

Also, when deleting many documents, the best practice is to use the `Delete By Query` with the `slices` parameter to improve performance. The `Delete by Query` feature with the `slices` parameter in Elasticsearch offers considerable advantages, especially when dealing with the deletion of numerous documents. This best practice enhances performance by splitting a large deletion task into smaller, parallel operations. This method not only boosts the efficiency and reliability of the deletion process but also lessens the burden on the cluster. By dividing the task, you ensure a more balanced and effective approach to managing large-scale deletions in Elasticsearch.

See also

For more details on the `Delete By Query` feature, refer to the official documentation: `https://www.elastic.co/guide/en/elasticsearch/reference/current/docs-delete-by-query.html`.

Using an analyzer

In this recipe, we are going to learn how to set up and use a specific analyzer for text analysis. Indexing data in Elasticsearch, especially for search use cases, requires that you define how text should be processed before indexation; this is what analyzers accomplish.

Analyzers in Elasticsearch handle tokenization and normalization functions. Elasticsearch offers a variety of ready-made analyzers for common scenarios, as well as language-specific analyzers for English, German, Spanish, French, Hindi, and so on.

In this recipe, we will see how to configure the standard analyzer with the English stopwords filter.

Getting ready

Make sure that you completed the *Adding data from the Elasticsearch client* recipe. Also, make sure to download the following sample Python script from the GitHub repository: `https://github.com/PacktPublishing/Elastic-Stack-8.x-Cookbook/blob/main/Chapter2/python-client-sample/sampledata_analyzer.py`.

The command snippets of this recipe are available at `https://github.com/PacktPublishing/Elastic-Stack-8.x-Cookbook/blob/main/Chapter2/snippets.md#using-analyzer`.

How to do it...

In this recipe, you will learn how to configure your Python code to interface with an Elasticsearch cluster, define a custom English text analyzer, create a new index with the analyzer, and verify that the index uses the specified settings.

Let's look at the provided Python script:

1. At the beginning of the script, we create an instance of the Elasticsearch client:

   ```
   es = Elasticsearch(
       cloud_id=ES_CID,
       basic_auth=(ES_USER, ES_PWD)
   )
   ```

2. To ensure that we do not use an existing `movies` index, the script includes code that deletes any such index:

   ```
   if es.indices.exists(index="movies"):
       print("Deleting existing movies index...")
       es.options(ignore_status=[404, 400]).indices.
   delete(index="movies")
   ```

3. Next, we define the analyzer configuration:

   ```
   index_settings = {
       "analysis": {
           "analyzer": {
   ```

```
                           "standard_with_english_stopwords": {
                               "type": "standard",
                               "stopwords": "_english_"
                   }
               }
           }
       }
```

4. We then create the index with settings that define the analyzer:

    ```
    es.indices.create(index='movies', settings=index_settings)
    ```

5. Finally, to verify the successful addition of the analyzer, we retrieve the settings:

    ```
    settings = es.indices.get_settings(index='movies')
    analyzer_settings = settings['movies']['settings']['index']
    ['analysis']
    print(f"Analyzer used for the index: {analyzer_settings}")
    ```

6. After reviewing the script, execute it with the following command, and you should see the
    output shown in *Figure 2.10*:

    ```
    $ python sampledata_analyzer.py
    ```

```
) python sampledata_analyzer.py
Deleting existing movies index...
{'_index': 'movies', '_id': 't7Zyao4B6R5W3CvTP1th', '_version': 1, 'result': 'created'
, '_shards': {'total': 2, 'successful': 1, 'failed': 0}, '_seq_no': 0, '_primary_term'
: 1}
Analyzer used for the index: {'analyzer': {'standard_with_stopwords': {'type': 'standa
rd', 'stopwords': '_english_'}}}
```

Figure 2.10 – The output of the sampledata_analyzer.py script

Alternatively, you can go to **Kibana | Dev Tools** and issue the following request:

```
GET /movies/_settings
```

In the response, you should see the settings currently applied to the `movies` index with the
configured analyzer, as shown in *Figure 2.11*:

```json
{
  "movies": {
    "settings": {
      "index": {
        "routing": {
          "allocation": {
            "include": {
              "_tier_preference": "data_content"
            }
          }
        },
        "number_of_shards": "1",
        "provided_name": "movies",
        "creation_date": "1710931861363",
        "analysis": {
          "analyzer": {
            "standard_with_stopwords": {
              "type": "standard",
              "stopwords": "_english_"
            }
          }
        },
        "number_of_replicas": "1",
        "uuid": "7pebWAQXTV2kt3LXa7RiUQ",
        "version": {
          "created": "8500010"
        }
      }
    }
  }
}
```

Figure 2.11 – The analyzer configuration in the index settings

How it works...

The settings block of the index configuration is where the analyzer is set. As we are modifying the built-in standard analyzer in our recipe, we will give it a unique name (standard_with_english_stopwords) and set the type to standard. Text indexed from this point will undergo analysis by the modified analyzer. To test this, we can use the _analyze endpoint on the index:

```
POST movies/_analyze
{
   "text": "A young couple decides to elope.",
   "analyzer": "standard_with_stopwords"
}
```

It should yield the results shown in *Figure 2.12*:

```
{
  "tokens": [
    {
      "token": "young",
      "start_offset": 2,
      "end_offset": 7,
      "type": "<ALPHANUM>",
      "position": 1
    },
    {
      "token": "couple",
      "start_offset": 8,
      "end_offset": 14,
      "type": "<ALPHANUM>",
      "position": 2
    },
    {
      "token": "decides",
      "start_offset": 15,
      "end_offset": 22,
      "type": "<ALPHANUM>",
      "position": 3
    },
    {
      "token": "elope",
      "start_offset": 26,
      "end_offset": 31,
      "type": "<ALPHANUM>",
      "position": 5
    }
  ]
}
```

Figure 2.12 – The index result of a text with the stopword analyzer

There's more...

While Elasticsearch offers many built-in analyzers for different languages and text types, you can also define custom analyzers. These allow you to specify how text is broken down and modified for indexing or searching, using components such as tokenizers, token filters, and character filters – either those provided by Elasticsearch or custom ones you create. For example, you can design an analyzer that converts text to lowercase, removes common words, substitutes synonyms, and strips accents.

Reasons for needing a custom analyzer may include the following:

- Handling various languages and scripts that require special processing, such as Chinese, Japanese, and Arabic

- Enhancing the relevance and comprehensiveness of search results using synonyms, stemming, lemmatization, and so on

- Unifying text by removing punctuation, whitespace, and accents and making it case-insensitive

Defining index mapping

In Elasticsearch, **mapping** refers to the process of defining the schema or structure of an index. It defines how documents and their fields are stored and indexed within Elasticsearch. Mapping allows you to specify the data type of each field, such as text, a keyword, a numeric character, and a date, and configure various properties for each field, including indexing options and analyzers. By defining a mapping, you provide Elasticsearch with crucial information about the data you intend to index, enabling it to efficiently store, search, and analyze the documents.

Mapping plays a critical role in delivering precise search results, efficient data storage, and effective handling of different data types within Elasticsearch.

When no mapping is predefined, Elasticsearch attempts to dynamically infer data types and create the mapping; this is what has occurred with our movie dataset thus far.

In this recipe, we will apply an explicit mapping to the `movies` index.

Getting ready

Make sure that you have completed the *Updating data in Elasticsearch* recipe.

All the command snippets for the Dev Tools in this recipe are available at `https://github.com/PacktPublishing/Elastic-Stack-8.x-Cookbook/blob/main/Chapter2/snippets.md#defining-index-mapping`.

How to do it...

You can define mappings during index creation or update them in an existing index.

> **An important note on mappings**
> When updating the mapping of an existing index that already contains documents, the mapping of those existing documents will not change. The new mapping will only apply to documents indexed afterward.

In this recipe, you are going to create a new index with explicit mapping, and then re-index the data from the `movie` index, assuming that you have already created that index beforehand:

1. Head to **Kibana | Dev Tools**.

2. Next, let's check the mapping of the previously created index with the following command:

```
GET /movies/_mapping
```

You will get the results shown in the following figure. Note that, for readability, some fields were collapsed.

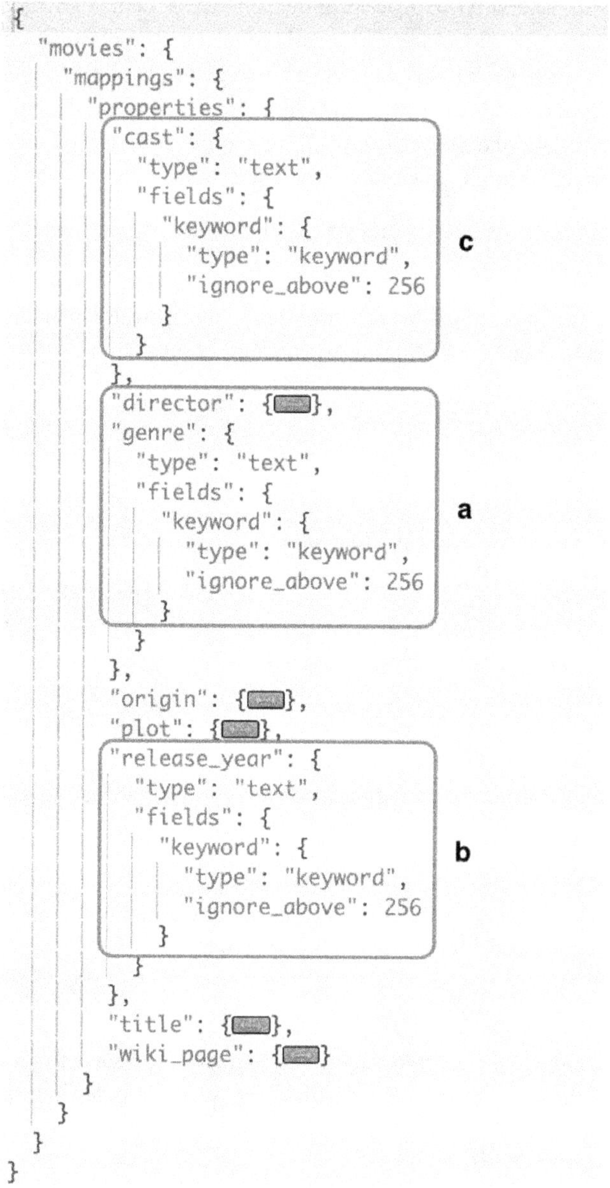

```
{
  "movies": {
    "mappings": {
      "properties": {
        "cast": {
          "type": "text",
          "fields": {
            "keyword": {
              "type": "keyword",
              "ignore_above": 256
            }
          }
        },                                          c
        "director": {▭},
        "genre": {
          "type": "text",
          "fields": {
            "keyword": {
              "type": "keyword",
              "ignore_above": 256
            }
          }
        },                                          a
        "origin": {▭},
        "plot": {▭},
        "release_year": {
          "type": "text",
          "fields": {
            "keyword": {
              "type": "keyword",
              "ignore_above": 256
            }
          }
        },                                          b
        "title": {▭},
        "wiki_page": {▭}
      }
    }
  }
}
```

Figure 2.13 – The default mapping on the movies index

Let's review what's going on in the figure:

a.   Examining the current mapping of the genre field reveals a multi-field mapping technique. This approach allows a single field to be indexed in several ways to serve different purposes. For example, the genre field is indexed both as a text field for full-text search and as a keyword field for sorting and aggregation. This dual approach to mapping the genre field is actually beneficial and well-suited for its intended use cases.

b.   Examining the release_year field reveals that indexing it as a text field is not optimal, since it represents numerical data, which could be beneficial for range queries, as well as other numeric-specific operations. Retaining the keyword mapping for this field is advantageous for sorting and aggregation purposes. To address this, applying an explicit mapping to treat release_year appropriately as a numerical field is the next step.

c.   There are two other fields that will require mapping adjustments – plot and cast. Given their nature, these fields should be indexed solely as text, considering it is unlikely there will be a need to sort or aggregate on these fields. However, this indexing strategy still allows for effective searching against them.

3.   Now, let's create a new index with the correct explicit mapping for the cast, plot, and release_year fields:

```
PUT movies-with-explicit-mapping
{
  "mappings": {
    "properties": {
      "release_year": {
        "type": "short",
        "fields": {
          "keyword": {
            "type": "keyword",
            "ignore_above": 256
          }
        }
      },
      "cast": {
        "type": "text"
      },
      "plot": {
        "type": "text"
      }
    }
  }
}
```

4. Next, reindex the original data in the new index so that the new mapping is applied:

```
POST /_reindex
{
  "source": {
    "index": "movies"
  },
  "dest": {
    "index": "movies-with-explicit-mapping"
  }
}
```

5. Check whether the new mapping has been applied to the new index:

```
GET movies-with-explicit-mapping/_mapping
```

Figure 2.14 shows the explicit mapping applied to the index:

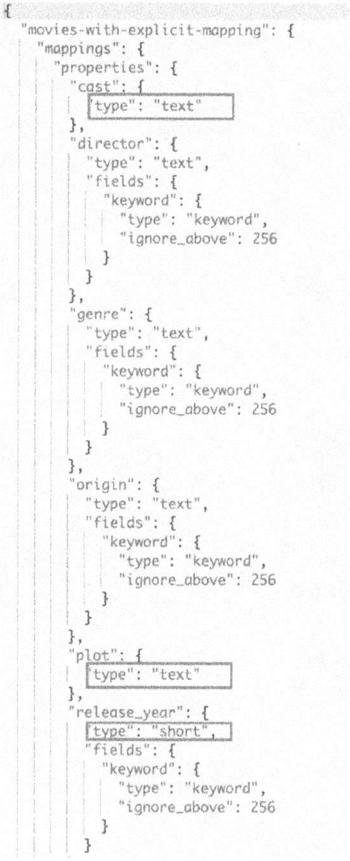

Figure 2.14 – Explicit mapping

How it works...

Explicit mapping in Elasticsearch allows you to define the schema or mapping for your index explicitly. Instead of relying on dynamic mapping, which automatically detects and creates the mapping based on the first indexed document, explicit mapping gives you full control over the data types, field properties, and analysis settings for each field in your index, as shown in *Figure 2.15*:

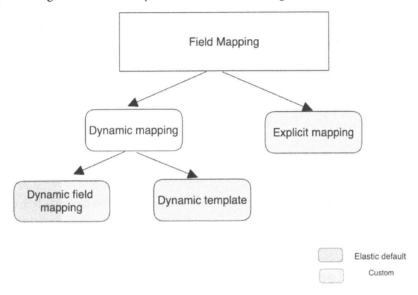

Figure 2.15 – The field mapping options

There's more...

Mapping is a key aspect of data modeling in Elasticsearch. Avoid relying on dynamic mapping and try, when possible, to explicitly define your mappings to have better control over the field types, properties, and analysis settings. This helps maintain consistency and avoids unexpected field mappings.

You should consider using multi-field mapping to index the same field in different ways, depending on the use cases. For instance, for a full-text search of a string field, text mapping is necessary. If the same string field is mostly used for aggregations, filtering, or sorting, then mapping it to a keyword field is more efficient. Also, consider using mapping limit settings to prevent a mapping explosion (`https://www.elastic.co/guide/en/elasticsearch/reference/current/mapping-settings-limit.html`). A situation where every new ingested document introduces new fields, as with dynamic mapping, can result in defining too many fields in an index. This can cause a mapping explosion. When each new field is continually added to the index mapping, it can grow excessively and lead to memory shortages and recovery challenges.

When it comes to mapping limit settings, there are several best practices to keep in mind. First, limit the number of field mappings to prevent documents from causing a mapping explosion. Second, limit the maximum depth of a field. Third, restrict the number of different nested mappings an index can have. Fourth, set a maximum for the count of nested JSON objects allowed in a single document, across all nested types. Finally, limit the maximum length of a field name. Keep in mind that setting higher limits can affect performance and cause memory problems.

For many years now, Elastic has been developing a specification called **Elastic Common Schema** (**ECS**) that provides a consistent and customizable way to structure data in Elasticsearch. Adopting this mapping has a lot of benefits (data correlation, reuse, and future-proofing, to name a few), and as a best practice, always refer to the ECS convention when you consider naming your fields. We will see more examples using ECS in the next chapters.

See also

- For a comprehensive introduction to Elasticsearch mapping, see the following: `https://www.elastic.co/blog/found-elasticsearch-mapping-introduction`

- For insights on preventing mapping explosion, read this informative blog article: `https://www.elastic.co/blog/3-ways-to-prevent-mapping-explosion-in-elasticsearch`

Using dynamic templates in document mapping

In this recipe, we will explore how to leverage dynamic templates in Elasticsearch to automatically apply mapping rules to fields, based on their data types. Elasticsearch allows you to define dynamic templates that simplify the mapping process by dynamically applying mappings to new fields as they are indexed.

Getting ready

Make sure that you have completed the previous recipes:

- *Using an analyzer*

- *Defining index mapping*

The snippets of the recipe are available at this address: `https://github.com/PacktPublishing/Elastic-Stack-8.x-Cookbook/blob/main/Chapter2/snippets.md#using-dynamic-templates-in-document-mapping`.

How to do it...

1. In our example, the default mapping of the `year` field is set to the `long` field type, which is suboptimal for storage. We also want to prepare the document mapping so that if additional `year` fields such as `review_year` and `award_year` are introduced, they will have a dynamically applied mapping. Let's go to **Kibana | Dev Tools**, where we can extend the previous mapping as follows:

```
PUT movies/_mapping
{
  "dynamic_templates": [{
    "years_as_short": {
      "match_mapping_type": "long",
        "match": "*year",
          "mapping": {
            "type": "short"
          }
      }
  }]
}
```

2. Next, we ingest a new document with a `review_year` field using the following command:

```
POST movies/_doc/
{
  "review_year": 1993,
  "release_year": 1992,
  "title": "Reservoir Dogs",
  "origin": "American",
  "director": "Quentin Tarantino",
  "cast": "Harvey Keitel, Tim Roth, Steve Buscemi, Chris Penn,
Michael Madsen, Lawrence Tierney",
  "genre": "crime drama",
  "wiki_page": "https://en.wikipedia.org/wiki/Reservoir_Dogs",
  "plot": "a group of criminals whose planned diamond robbery
goes disastrously wrong, leading to intense suspicion and
betrayal within their ranks."
}
```

3. We can now check the mapping with the following command, and we can see that the `movies` mapping now contains the dynamic template, and the `review_year` field correctly maps to `short`, as shown in *Figure 2.16*.

```
GET /movies/_mapping
```

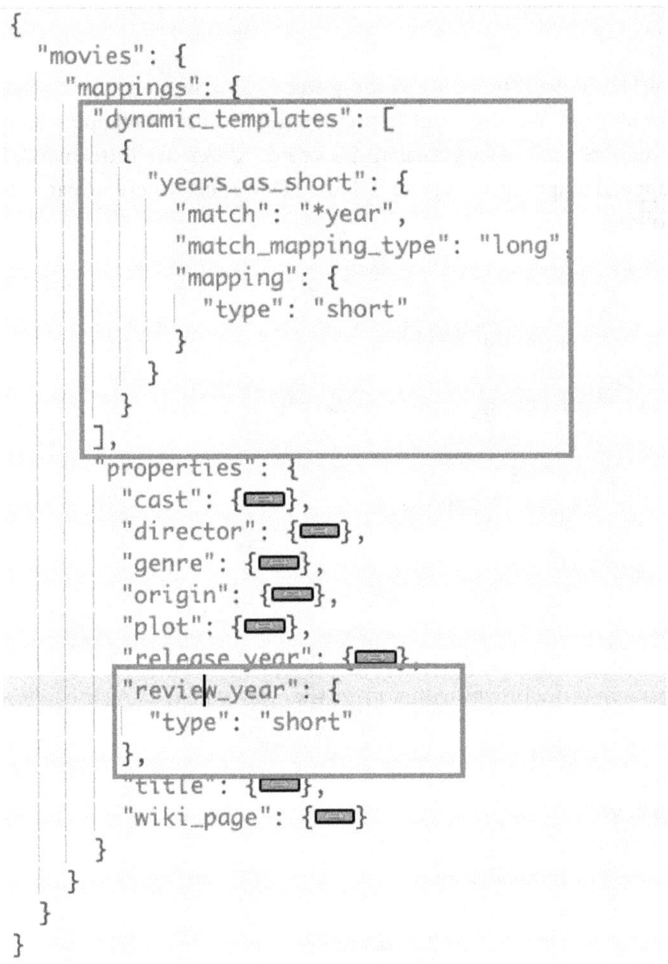

```
{
  "movies": {
    "mappings": {
      "dynamic_templates": [
        {
          "years_as_short": {
            "match": "*year",
            "match_mapping_type": "long",
            "mapping": {
              "type": "short"
            }
          }
        }
      ],
      "properties": {
        "cast": {▭},
        "director": {▭},
        "genre": {▭},
        "origin": {▭},
        "plot": {▭},
        "release_year": {▭},
        "review_year": {
          "type": "short"
        },
        "title": {▭},
        "wiki_page": {▭}
      }
    }
  }
}
```

Figure 2.16 – Updated mapping for the movies index with a dynamic template

How it works...

In our example for the years_as_short dynamic template, we configured custom mapping as follows:

- The match_mapping_type parameter is used to define the data type to be detected. In our example, we try to define the data type for long values.

- The match parameter is used to define the wildcard for the filename ending with year. It uses a pattern to match the field name. (It is also possible to use the unmatch parameter, which uses one or more patterns to exclude fields matched by match.)

- `mapping` is used to define the mapping the match field should use. In our example, we map the target field type to `short`.

There's more...

Apart from the example that we have seen in this recipe, dynamic templates can also be used in the following scenarios:

- Only with a `match_mapping_type` parameter that applies to all the fields of a single type, without needing to match the field name
- With `patch_match` or `patch_unmatch` for a full dotted patch to the field such as `"path_match"`: `"myfield_prefix.*"` or `"path_unmatch"`: `"*.year"`.

For timestamped data, it is common to have many numeric fields such as metrics. In such cases, filtering on those fields is rarely required and only aggregation is useful. Therefore, it is recommended to disable indexing on those fields to save disk space. You can find a concrete example in the following documentation: `https://www.elastic.co/guide/en/elasticsearch/reference/current/dynamic-templates.html#_time_series`.

The default dynamic field mapping in Elasticsearch is convenient to get started, but it is beneficial to consider defining field mappings more strategically to optimize storage, memory, and indexing/search speed. The workflow to design new index mappings can be as follows:

1. Index a sample document containing the desired fields in a dummy index.
2. Retrieve the dynamic mapping created by Elasticsearch.
3. Modify and optimize the mapping definition.
4. Create your index with the custom mapping, either explicit or dynamic.

See also

There are some more resources in Elastic's official documentation, such as the following:

- **Mapping**: `https://www.elastic.co/guide/en/elasticsearch/reference/current/mapping.html`
- **Dynamic templates**: Templates can also be created by calling REST API with an HTTP client such as CURL/Postman; here is the documentation: `https://www.elastic.co/guide/en/elasticsearch/reference/current/indices-put-mapping.html`

Creating an index template

In this recipe, we will explore how to use index templates in Elasticsearch to define mappings, settings, and other configurations for new indices. Index templates automate the index creation process and ensure consistency across your Elasticsearch cluster.

Getting ready

Before we begin, familiarize yourself with creating component and index templates by using Kibana Dev Tools as explained in this documentation:

Make sure that you have completed the previous recipes:

- *Using an analyzer*
- *Defining index mapping*

All the commands for the Dev Tools in this recipe are available at this address: `https://github.com/PacktPublishing/Elastic-Stack-8.x-Cookbook/blob/main/Chapter2/snippets.md#creating-an-index-template`.

How to do it...

In this recipe, we will create two component templates – one for the `genre` field and another for `*year` fields with dynamic mapping – and then combine them in an index template:

1. Create the first component template for the `genre` field:

```
PUT _component_template/movie-static-mapping
{
  "template": {
    "mappings": {
      "properties": {
        "genre": {
          "type": "keyword"
        }
      }
    }
  }
}
```

2. Create the second component template for the dynamic `*year` field:

```
PUT _component_template/movie-dynamic-mapping
{
   "template": {
     "mappings": {
       "dynamic_templates": [{
         "years_as_short": {
           "match_mapping_type": "long",
           "match": "*year",
           "mapping": {
             "type": "short"
           }
         }
       }]
     }
   }
}
```

3. Create the index template, which consists of the component templates that we just created; additionally, we define an explicit mapping `director` field directly in the index template:

```
PUT _index_template/movie-template
{
   "index_patterns": ["movie*"],
   "template": {
     "settings": {
       "number_of_shards": 1
     },
     "mappings": {
       "_source": {
         "enabled": true
       },
       "properties": {
         "director": {
         "type": "keyword"
         }
       }
     },
     "aliases": {
       "mydata": { }
     }
   },
   "priority": 500,
```

```
  "composed_of": ["movie-static-mapping", "movie-dynamic-
mapping"],
  "version": 1,
  "_meta": {
    "description": "movie template"
  }
}
```

4. Now, we can index another new movie with a field called `award_year`, as follows:

```
POST movies/_doc/
{
  "award_year": 1998,
  "release_year": 1997,
  "title": "Titanic",
  "origin": "American",
  "director": "James Cameron",
  "cast": "Leonardo DiCaprio, Kate Winslet, Billy Zane, Frances
Fisher, Victor Garber, Kathy Bates, Bill Paxton, Gloria Stuart,
David Warner, Suzy Amis",
  "genre": "historical epic",
  "wiki_page": "https://en.wikipedia.org/wiki/Titanic_(1997_
film)",
  "plot": "The ill-fated maiden voyage of the RMS Titanic,
centering on a love story between a wealthy young woman and a
poor artist aboard the luxurious, ill-fated R.M.S. Titanic"
}
```

5. Let's check the mapping after the document ingestion with the following command:

```
GET /movies/_mapping
```

6. Note the updated mapping, as illustrated in *Figure 2.17*, with `award_year` dynamically mapped to `short`. Additionally, both the `genre` and `director` fields are mapped to `keyword`, thanks to our field definitions in the `movie-static-mapping` component template and the `movie-template` index template.

```
{
  "movies": {
    "mappings": {
      "dynamic_templates": [
        {
          "years_as_short": {
            "match": "*year",
            "match_mapping_type": "long",
            "mapping": {
              "type": "short"
            }
          }
        }
      ],
      "properties": {
        "award_year": {
          "type": "short"
        },
        "cast": {
          "type": "text",
          "fields": {
            "keyword": {
              "type": "keyword",
              "ignore_above": 256
            }
          }
        },
        "director": {
          "type": "keyword"
        },
        "genre": {
          "type": "keyword"
        },
        "origin": {
          "type": "text",
          "fields": {
            "keyword": {
              "type": "keyword",
              "ignore_above": 256
            }
          }
        },
```

Figure 2.17 – The updated mapping for the movies index

How it works...

Index templates include various configuration settings, such as shard and replica initialization parameters, mapping configurations, and aliases. They also allow you to assign priorities to templates, with a default priority of 100.

Component templates act as building blocks for index templates, which can comprise settings, aliases, or mappings and can be combined in an index template, using the composed_of parameter.

Legacy index templates were deprecated upon the release of Elasticsearch 7.8.

Figure 2.18 gives you an overview of the relationship between index templates, component templates, and legacy templates:

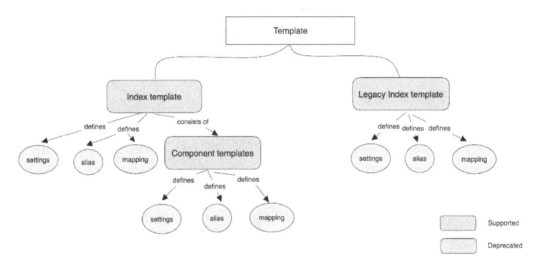

Figure 2.18 – Index templates versus legacy index templates

There's more...

Elasticsearch provides predefined index templates that are associated with index and data stream patterns (you can find more details in *Chapter 4*), such as logs-*-*, metrics-*-*, and synthetics-*-*, with a default priority of 100. If you wish to create custom index templates that override the predefined ones but still use the same patterns, you can assign a priority value higher than 100. If you want to disable the built-in index and component templates altogether, you can set the stack.templates.enabled configuration parameter to false; the detailed documentation can be found here: https://www.elastic.co/guide/en/elasticsearch/reference/current/index-templates.html.

Indexing multiple documents using Bulk API

In this recipe, we will explore how to use the Elasticsearch client to ingest an entire movie dataset using the bulk API. We will also integrate various concepts we have covered in previous recipes, specifically related to mappings, to ensure that the correct mappings are applied to our index.

Getting ready

For this recipe, we will work with the sample *Wikipedia Movie Plots* dataset introduced at the beginning of the chapter. The file is accessible in the GitHub repository via this URL: `https://github.com/PacktPublishing/Elastic-Stack-8.x-Cookbook/blob/main/Chapter2/dataset/wiki_movie_plots_deduped.csv`.

Make sure that you have completed the previous recipes:

- *Using an analyzer*
- *Creating index template*

How to do it...

Head to the GitHub repository to download the Python script at `https://github.com/PacktPublishing/Elastic-Stack-8.x-Cookbook/blob/main/Chapter2/python-client-sample/sampledata_bulk.py` and then follow these steps:

1. Update the `.env` file with the `MOVIE_DATASET` variable, which specifies the path to the downloaded movie dataset CSV file:

   ```
   MOVIE_DATASET=<the-path-of-the-csv-file>
   ```

2. Once the `.env` file is correctly configured, run the `sampledata_bulk.py` Python script. During execution, you should see output similar to the following (note that, for readability, the output image has been truncated):

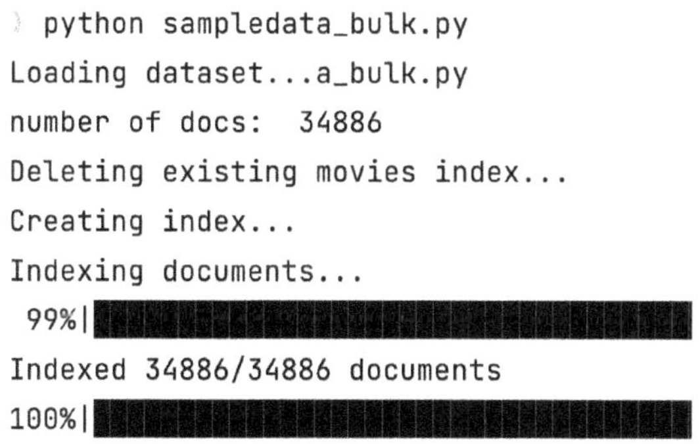

   ```
   ) python sampledata_bulk.py
   Loading dataset...a_bulk.py
   number of docs:  34886
   Deleting existing movies index...
   Creating index...
   Indexing documents...
    99%|████████████████████████████████
   Indexed 34886/34886 documents
   100%|████████████████████████████████
   ```

   Figure 2.19 – The output of the sampledata_bulk.py script

3. To verify that the new `movies` index has the appropriate mappings, head to **Kibana | Dev Tools** and execute the following command:

```
GET /movies/_mapping
```

```
{
  "movies": {
    "mappings": {
      "dynamic_templates": [
        {
          "years_as_short": {
            "match": "*year",
            "match_mapping_type": "long",
            "mapping": {
              "type": "short"
            }
          }
        }
      ],
      "properties": {
        "cast": {
          "type": "text"
        },
        "director": {
          "type": "text",
          "fields": {
            "keyword": {
              "type": "keyword",
              "ignore_above": 256
            }
          }
        },
        "genre": {
          "type": "keyword"
        },
        "origin": {
          "type": "keyword"
        },
        "plot": {
          "type": "text"
        },
        "release_year": {
          "type": "short",
          "fields": {
            "keyword": {
              "type": "keyword",
              "ignore_above": 256
            }
          }
        },
        "title": {
          "type": "text"
        },
        "wiki_page": {
          "type": "keyword"
        }
      }
    }
  }
}
```

Figure 2.20 – The movies index with a new mapping

4. As illustrated in *Figure 2.20*, dynamic mapping on the `release_year` field was applied to the newly created `movies` index, despite a mapping being explicitly specified in the script. This occurred because an index template was defined in the *Using dynamic templates in document mapping* recipe, with the index pattern set to `movies*`. As a result, any index that matches this pattern will automatically inherit the settings from the template, including its dynamic mapping configuration.

5. Next, to verify that the entire dataset has been indexed, execute the following command:

```
GET /movies/_count
```

The command should produce the output illustrated in *Figure 2.21*. According to this output, your `movies` index should contain 34,886 documents:

```
{
  "count": 34886,
  "_shards": {
    "total": 1,
    "successful": 1,
    "skipped": 0,
    "failed": 0
  }
}
```

Figure 2.21 – A count of the documents in bulk-indexed movies

We have just set up an index with the right explicit mapping and loaded an entire dataset by using the Elasticsearch Python client.

How it works...

The script we've provided contains several sections. First, we delete any existing `movies` indexes to make sure we start from a clean slate. This is the reason you did not see the `award_year` and `review_year` fields in the new mapping shown in *Figure 2.20*. We then use the `create_index` method to create the `movies` index and specify the settings and the mappings we wish to apply to the documents that will be stored in this index.

Then, there is the `generate_actions` function that yields a document for each row in our CSV dataset. This function is then used by the `streaming_bulk` helper method.

The `streaming_bulk` helper function in the Elasticsearch Python client is used to perform bulk indexing of documents in Elasticsearch. It is like the `bulk` helper function, but it is designed to handle large datasets.

The `streaming_bulk` function accepts an iterable of documents and sends them to Elasticsearch in small batches. This strategy allows you to efficiently process substantial datasets without exhausting system memory.

There's more...

The Elasticsearch Python Client provides several helper functions for the bulk API, which can be challenging to use directly because of its specific formatting requirements and other considerations. These helpers accept an instance of the es class and an iterable action, which can be any iterable or generator.

The most common format for the iterable action is the same as that returned by the search() method. The bulk() API accepts the index, create, delete, and update actions. The _op_type field is used to specify an action, with _op_type defaulting to index. There are several bulk helpers available, including bulk(), parallel_bulk(), streaming_bulk(), and bulk_index(). The following table outlines these helpers and their preferred use cases:

Bulk helper functions	Use cases
bulk()	This helper is used to perform bulk operations on a single thread. It is ideal for small- to medium-sized datasets and is the simplest of the bulk helpers.
parallel_bulk()	This helper is used to perform bulk operations on multiple threads. It is ideal for large datasets and can significantly improve indexing performance.
streaming_bulk()	This helper is used to perform bulk operations on a large dataset that cannot fit into memory. It is ideal for large datasets and can be used to stream data from a file or other source.
bulk_index()	This helper is used to perform bulk indexing operations on a large dataset that cannot fit into memory. It is ideal for large datasets and can be used to stream data from a file or other source.

Table 2.1 – Bulk helper functions and their associated use cases

See also

If you are interested in more examples of bulk ingest using the Python client, check out the official Python client repository: https://github.com/elastic/elasticsearch-py/tree/main/examples/bulk-ingest.

3

Building Search Applications

Now that you are familiar with getting your data into Elasticsearch, using the flexible tools provided by Elastic Stack components, we will dive into how you can build a great search experience with that data.

Traditionally, users create Search Applications based on Elasticsearch indices to leverage the full power, flexibility, and ecosystem of Elasticsearch and its Query DSL (a powerful and flexible JSON-style domain-specific language with which you can define and execute queries in Elasticsearch). In the later years, Elastic also introduced App Search and the Search Application client for users who prefer ease of use and rapid implementation over more control and flexibility. *Figure 3.1* shows you different ways to ingest content and build Search Applications with Elastic:

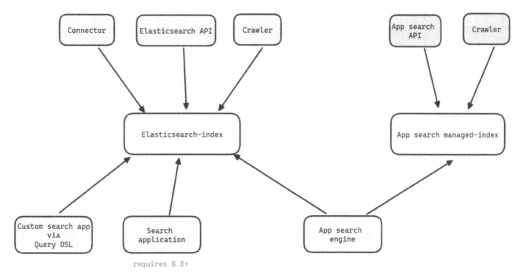

Figure 3.1 – Ingestion and search portfolio

This chapter focuses on Elasticsearch indices. We will begin with recipes for Query DSL and then explore the new capabilities presented in Elastic 8 for Search Applications, designed to streamline the process of building search experiences. By the end of the chapter, you will have a functional Search Application equipped with essential features such as full-text search, faceted search, pagination, and search behavior analytics.

In this chapter, we will cover the following recipes:

- Searching with Query DSL

- Building advanced search queries with Query DSL

- Using search templates to pre-render search requests

- Getting started with Search Applications for your Elasticsearch index

- Building a search experience with the Search Application client

- Measuring the performance of your Search Applications with Behavioral analytics

Technical requirements

To follow different recipes in this chapter, you will need an Elastic Stack deployment that includes the following components:

- Elasticsearch for searching and storing the data

- Kibana for data visualization and management of the stack (a Kibana user with **'All'** privileges for Fleet and Integrations is required)

- Integrations Server (included by default in the Elastic Cloud deployment)

- An Enterprise Search node to provide a refined search experience on top of Elasticsearch

To quickly set up a deployment on Elastic Cloud, refer to the *Installing Elastic Stack on Elastic Cloud* recipe in *Chapter 1*.

Searching with Query DSL

Using **Query DSL** in Elasticsearch opens the door to a powerful and flexible method for retrieving precise information from your indexed data. The **Query Domain-Specific Language** (**Query DSL**) provides an advanced approach for building queries with a JSON-like syntax. It enables the construction of search queries that include conditions, filters, aggregations, and more. In the first recipe, we will explore some basic Query DSL examples for our movie dataset.

Getting ready

To follow along with this recipe, you need a functional Elastic deployment that includes Elasticsearch and Kibana. To set up a cluster on Elastic Cloud, refer to the *Installing Elastic Stack on Elastic Cloud* recipe in *Chapter 1*.

Additionally, you'll need the movie dataset we've previously indexed using the *Indexing multiple documents using Bulk API* recipe in *Chapter 2*.

You can find the code snippets for this recipe at `https://github.com/PacktPublishing/Elastic-Stack-8.x-Cookbook/blob/main/Chapter3/snippets.md#searching-with-the-query-dsl`

How to do it...

1. In Kibana, navigate to **Dev Tools | Console**. We can begin with a match query for the title field. Paste the following code snippet (the snippets can be found at `https://github.com/PacktPublishing/Elastic-Stack-8.x-Cookbook/blob/main/Chapter3/snippets.md#sample-query-with-or-operator`):

    ```
    GET movies/_search
    {
      "query": {
        "match": {
          "title": "Come home"
        }
      }
    }
    ```

    The query should return `139` hits. By default, the two words `"Come  home"` are searched for with the `"OR"` operator.

2. We can then refine the query by introducing the `"and"` operator, indicating that the search should only return documents where the title contains both the words `"Come"` and `"home"` (the snippets can be found at `https://github.com/PacktPublishing/Elastic-Stack-8.x-Cookbook/blob/main/Chapter3/snippets.md#sample-query-with-and-operator`):

    ```
    GET movies/_search
    {
      "query": {
        "match": {
          "title": {
            "query": "Come home",
            "operator": "and"
          }
    ```

```
        }
      }
    }
```

The query should return 3 hits.

3. The "or" or "and" operators yield results that are too broad or too strict. You can use the minimum_should_match parameter to filter less relevant results (the snippets can be found at https://github.com/PacktPublishing/Elastic-Stack-8.x-Cookbook/blob/main/Chapter3/snippets.md#minumum-sould-match-query):

```
GET movies/_search
{
  "query": {
    "match": {
      "title": {
        "query": "Come sweet home",
        "minimum_should_match": 2
      }
    }
  }
}
```

The query should return 10 hits.

4. The match query does not consider the order of words, so you can use the match_phrase query if you need to find terms that are close to each other (the snippets can be found at https://github.com/PacktPublishing/Elastic-Stack-8.x-Cookbook/blob/main/Chapter3/snippets.md#match-phrase-query):

```
GET movies/_search
{
  "query": {
    "match_phrase": {
      "title": "sweet home"
    }
  }
}
```

The query should return 5 hits.

How it works...

In general, Query DSL enables you to perform operations such as querying, aggregating, sorting, paginating, and manipulating responses.

In this recipe, we focused on basic query operations:

```
1 ▾ {
2     "took": 17,
3     "timed_out": false,
4 ▾   "_shards": {
5         "total": 1,
6         "successful": 1,
7         "skipped": 0,
8         "failed": 0
9 ▾   },
10 ▾  "hits": {
11 ▾      "total": {
12           "value": 139,
13           "relation": "eq"
14 ▾      },
15       "max_score": 11.93121,
16 ▾      "hits": [
17 ▾          {
18               "_index": "movies",
19               "_id": "9NY0KY0BEiCEihtz4ShI",
20               "_score": 11.93121,
21 ▾            "_source": {
22                   "release_year": "1943",
23                   "title": "Lassie Come Home",
24                   "origin": "American",
25                   "director": "Fred M. Wilcox",
26                   "cast": "Roddy McDowall, Donald Crisp, May Whitty",
27                   "genre": "family",
28                   "wiki_page": "https://en.wikipedia.org/wiki/Lassie_Come_Home",
29                   "plot": """Set in Depression-era Yorkshire, England, Mr. and Mrs. Carraclough (Donald Crisp and Elsa
                         Lanchester) are hit by hard times and forced to sell their collie, Lassie (Pal), to the rich Duke of
                         Rudling (Nigel Bruce), who has always admired her. Young Joe Carraclough (Roddy McDowall) grows
```

Figure 3.2 – Query results

As shown in *Figure 3.2*, the result of each query provides us with the number of total *hits*, which represent the matching documents. Elasticsearch calculates a relevance score for each document that is a hit, indicating its importance in relation to the specific query. The default BM25 algorithm determines this relevance score. The following elements affect the document score:

- **Term frequency (TF)**: The importance of a term within a field increases with its frequency in the document

- **Inverse document frequency (IDF)**: The significance of a term decreases as the number of documents containing it increases

- **Field length**: Fields with fewer words tend to be more relevant than those with more words

By default, the query response returns the top 10 matching documents sorted by _score in descending order, and the documents include every field, including the original document, in the "_source" field.

If you want to change the default behavior:

- You can set from and size to paginate through the search results.

- By using sort, you can order the results based on one or more fields instead of score.

- The "fields" clause allows you to retrieve only specific fields. Setting "_source" to false excludes the full source document from the response.

Here is an example:

```
GET movies/_search
{
  "_source": false,
  "fields": [
    "genre",
    "title"
  ],
  "from": 10,
  "size": 50,
  "sort": [
    {
      " release_year ": {
        "order": "desc"
      }
    },
    "_score"
  ],
  "query": {
    ...
  }
}
```

There's more...

Besides the examples discussed in this recipe, Query DSL offers various types of queries as outlined in *Table 3.1*:

Full-text queries	Term-level queries	Advanced queries
• `match` • `match_phrase` • `multi_match` • `query_string` • ...	• `term` • `range` • `exists` • `fuzzy` • `regexp` • `wildcard` • ...	• `script` • `percolate` • `span_ queries` • `geo_ queries` • `nested` • ...

Table 3.1 – Different query types of Query DSL

We will continue to explore additional Query DSL capabilities in the next recipe.

Building advanced search queries with Query DSL

In this recipe, we'll explore more advanced Query DSL techniques to enhance your search and analysis tasks.

Getting ready

Ensure that you have completed the previous *Searching with Query DSL* recipe.

You can find the snippets for this recipe at `https://github.com/PacktPublishing/Elastic-Stack-8.x-Cookbook/blob/main/Chapter3/snippets.md#building-advanced-search-query-with-query-dsl`.

How to do it...

In the previous recipe, we mostly saw queries for full-text search, ideal for searching words within bodies of text. But how do we query for numbers, dates, and IPs? This is where the range query becomes useful. Let's explore how to find movies released between 1925 and 1927:

1. Start with a range query for the `"release_year"` field: Go to **Dev Tools | Console** in Kibana and try the query with the following code snippet (you can find the snippet at `https://github.com/PacktPublishing/Elastic-Stack-8.x-Cookbook/blob/main/Chapter3/snippets.md#range-query`):

```
GET movies/_search
{
  "query": {
    "range": {
      "release_year ": {
        "gte": "1925",
        "lte": "1927"
      }
    }
  }
}
```

2. Next, how would you find movies that mention `"come  home"` in both the `"title"` and `"plot"` fields? We can use a multi-match query. Try the following code (the snippet can be found at `https://github.com/PacktPublishing/Elastic-Stack-8.x-Cookbook/blob/main/Chapter3/snippets.md#multi-match-query`):

```
GET /movies/_search
{
  "query": {
    "multi_match": {
      "query": "come home",
      "fields": ["title", "plot"]
    }
  }
}
```

3. Multi-match queries can be varied by adding a type to the query, such as `"most_fields"`, allowing the score to be the sum of the scores of the individual fields. By using the `"most_fields"` type, the more fields that contain the words `"come home"`, the higher the score. Try the following query (the snippet can be found at `https://github.com/PacktPublishing/Elastic-Stack-8.x-Cookbook/blob/main/Chapter3/snippets.md#multi-match-most-queries`):

```
GET /movies/_search
{
  "query": {
    "multi_match": {
      "type": "most_fields",
      "query": "come home",
      "fields": ["title", "plot"]
    }
  }
}
```

4. If you want to find movies with titles or plot descriptions that contain a specific phrase or sequence of words, such as `"come home"`, you can use the match phrase query. Try the query with the following snippet (the snippet can be found at `https://github.com/PacktPublishing/Elastic-Stack-8.x-Cookbook/blob/main/Chapter3/snippets.md#multi-match-phrase`):

```
GET /movies/_search
{
  "query": {
    "multi_match": {
      "type": "phrase",
      "query": "come home",
      "fields": ["title", "plot"]
    }
  }
}
```

5. Now, suppose you want to find comedy movies with titles that include `"home"`, you can combine queries using a bool query. Let's start by combining two match queries. Try the query with the following code snippet (the snippet can be found at `https://github.com/PacktPublishing/Elastic-Stack-8.x-Cookbook/blob/main/Chapter3/snippets.md#boolean-query`):

    ```
    GET /movies/_search
    {
      "query": {
        "bool": {
          "must": [
              { "match": { "title": "home" } },
              { "match": { "genre": "comedy" } }
          ]
          }
        }
    }
    ```

6. To ask the same question, we could have used the `filter` clause within the `bool` query to specifically filter for the `"genre"` as comedy. Try the following snippet. (The other possible clauses for the bool query are `"must not"` and `"should"`.) The snippet can be found at `https://github.com/PacktPublishing/Elastic-Stack-8.x-Cookbook/blob/main/Chapter3/snippets.md#boolean-query-with-filter`:

    ```
    GET /movies/_search
    {
      "query": {
        "bool": {
          "must": [ {"match": {"title": "home"}}],
          "filter": [ {"match": {"genre": "comedy"}}]
          }
        }
    }
    ```

7. Alternatively, you can ask a similar question using the `"should"` clause. Try the following snippet (the snippet can be found at `https://github.com/PacktPublishing/Elastic-Stack-8.x-Cookbook/blob/main/Chapter3/snippets.md#boolean-query-with-should`):

    ```
    GET /movies/_search
    {
      "query": {
        "bool": {
          "must": [ {"match": {"title": "home"}}],
          "should": [ {"match": {"genre":"comedy"}} ]
    ```

```
                }
            }
        }
```

You should see differences in total hits and result sorting because the `"should"` clause works differently from `"must"` or `"filter"`, as we will explain in the next section.

How it works...

In this recipe, you've learned some advanced query techniques, including how to combine queries using the bool query. The bool query allows you to combine multiple clauses such as `"must"`, `"filter"`, `"must_not"`, and `"should"`.

You may have noticed that the hits returned by the `"filter"` clause and the match clause when filtering by genre are the same. However, a key difference is that queries within a `"filter"` clause do not contribute to the document score.

From a performance perspective, the `"filter"` clause is preferred because skipping the score calculation not only makes the query faster, but frequently used filters can be cached as well.

Meanwhile, changing from a `"filter"` clause to a `"should"` clause alters the query results. While documents that do not match the queries in the `"should"` clause are still returned as hits, those that do match receive a score boost.

There's more...

We've just scratched the surface of the power of Query DSL. We'll continue to explore more examples of aggregations and filters in the upcoming recipes of this chapter.

> **Note**
>
> Regarding the range query, you may have noticed that our `release_year` field has been mapped as a numeric field instead of a date field. We intentionally left it as a numeric field to simplify the import steps during CSV data ingestion. Additionally, we have provided a sample snippet for a Boolean query with a range filter at `https://github.com/PacktPublishing/Elastic-Stack-8.x-Cookbook/blob/main/Chapter3/snippets.md#boolean-query-with-range-filter`.

See also

The full documentation for the range query can be found at the following link: `https://www.elastic.co/guide/en/elasticsearch/reference/current/query-dsl-range-query.html`

For a deeper understanding of Query DSL, refer to the full documentation available here: `https://www.elastic.co/guide/en/elasticsearch/reference/current/query-dsl.html`

When building alerting rules in Kibana, Query DSL is a very useful rule type, we will see some examples in *Chapter 7* in the *Creating alerts in Kibana* recipe.

Using search templates to pre-render search requests

Search templates are a way to define templates for search queries. These templates act as placeholders for variables defined inside the search queries. You can store search templates as Mustache scripts in the cluster state or provide them inline in the request body.

In this recipe, we will show you how to create and use search templates in Elasticsearch and explain how they work behind the scenes. We will also explore some of the benefits and use cases of search templates.

Getting ready

To follow along in this recipe, you'll need an Elastic deployment up and running with Elasticsearch and Kibana. To spin up a cluster on Elastic Cloud refer to the *Installing Elastic Stack with Elastic Cloud on Kubernetes* recipe in *Chapter 1*.

Also, you'll need the sample movie dataset we've previously used in the *Indexing multiple documents using Bulk API* recipe in *Chapter 2*.

The code used in this recipe is available at `https://github.com/PacktPublishing/Elastic-Stack-8.x-Cookbook/blob/main/Chapter3/snippets.md#using-search-template-to-pre-render-search-requests`.

How to do it...

Consider the goal of designing a search template intended to execute a query on a selected group of fields within our movies index, culminating in the aggregation of results focused on the most prominent directors and genres.

Make sure that you have the movies dataset to hand to build the search template.

1. In the GitHub repo, copy the snippet available at `https://github.com/PacktPublishing/Elastic-Stack-8.x-Cookbook/blob/main/Chapter3/snippets.md#add-search-template`.

2. Go to **Dev Tools** | **Console** and paste the snippet you got from the previous step:

```
PUT _scripts/movies-search-template
{
  "script": {
    "lang": "mustache",
    "source": {
      "query": {
        "bool": {
          "must": [
            {
              "multi_match": {
                "query": "{{query}}",
                "fields": [
                  "title^4",
                  "plot",
                  "cast",
                  "director"
                ]
              }
            },
            {
              "multi_match": {
                "query": "{{query}}",
                "type": "phrase_prefix",
                "fields": [
                  "title^4",
                  "plot",
                  "director"
                ]
              }
            }
          ]
        }
      },
      "aggs": {
        "genre_facet": {
          "terms": {
            "field": "genre",
            "size": "{{agg_size}}"
          }
        },
        "director_facet": {
          "terms": {
```

```
            "field": "director.keyword",
            "size": "{{agg_size}}"
          }
        }
      },
      "sort": [
        {
          "release_year": "desc"
        }
      ],
      "fields": [
        "title",
        "release_year",
        "director",
        "origin"
      ],
      "_source": "false"
    }
  }
}
```

We can use the "_render/template" API to test our template and see what the actual query looks like (the snippet is available at https://github.com/PacktPublishing/ Elastic-Stack-8.x-Cookbook/blob/main/Chapter3/snippets.md#render- search-template-with-sample-parameters):

```
GET _render/template
{
  "id": "movies-search-template",
  "params": {
    "query": "space",
    "agg_size": 3
  }
}
```

As you see, we provide the values for the variable in the **params** object. In our example, we're searching for space movies. This results in the values being replaced in the actual query (for readability, the following figure only displays the "query" statement of the output):

```
{
  "template_output": {
    "query": {
      "bool": {
        "must": [
          {
            "multi_match": {
              "query": "space",
              "fields": [
                "title^4",
                "plot",
                "cast",
                "director"
              ]
            }
          },
          {
            "multi_match": {
              "query": "space",
              "type": "phrase_prefix",
              "fields": [
                "title^4",
                "plot",
                "director"
              ]
            }
          }
        ]
      }
    },
    "aggs": {...},
    "sort": [...],
    "fields": [...],
    "_source": "false"
  }
}
```

Figure 3.3 – Rendering search template

3. Now, let's try out the template with the following request (the snippet is available at https://
    github.com/PacktPublishing/Elastic-Stack-8.x-Cookbook/blob/main/
    Chapter3/snippets.md#query-with-search-template):

    ```
    GET movies/_search/template
    {
      "id": "movies-search-template",
      "params": {
        "query": "space",
        "agg_size": 3
      }
    }
    ```

Running the query will return documents matching our query criteria, as shown in *Figure 3.4*:

```
1 ▾ {                                                  },
2      "took": 6,                                      "aggregations": {
3      "timed_out": false,                               "director_facet": {
4 ▾    "_shards": {                                        "doc_count_error_upper_bound": 0,
5        "total": 1,                                       "sum_other_doc_count": 632,
6        "successful": 1,                                  "buckets": [
7        "skipped": 0,                                       {
8        "failed": 0                                           "key": "Unknown",
9 ▾    },                                                      "doc_count": 37
10 ▾   "hits": {                                             },
11 ▾     "total": {                                          {
12         "value": 681,                                       "key": "Honda, IshirōIshirō Honda",
13         "relation": "eq"                                    "doc_count": 7
14 ▾     },                                                  },
15       "max_score": null,                                  {
16 ▾     "hits": [                                             "key": "Michael Bay",
17 ▾       {                                                   "doc_count": 5
18           "_index": "movies",                             }
19           "_id": "jKU1Io0BybzHeQkWMq34",               ]
20           "_score": null,                            },
21           "_source": {},                             "genre_facet": {
22 ▾         "fields": {                                  "doc_count_error_upper_bound": 0,
23 ▾           "director": [                              "sum_other_doc_count": 421,
24               "Peter Chelsom"                          "buckets": [
25 ▾           ],                                           {
26 ▾           "origin": [                                    "key": "science fiction",
27               "American"                                   "doc_count": 109
28 ▾           ],                                           },
29 ▾           "release_year": [                            {
30               2017                                        "key": "unknown",
31 ▾           ],                                            "doc_count": 83
32 ▾           "title": [                                   },
33               "The Space Between Us"                     {
34 ▾           ]                                              "key": "comedy",
35 ▾         },                                              "doc_count": 68
36 ▾         "sort": [                                     }
37             2017                                     ]
38 ▾         ]                                         }
39 ▾       },                                        }
40 ▾       {
41           "_index": "movies",
42           "_id": "qaU1Io0BybzHeQkWMq34",
43           "_score": null,
44           "_source": {},
45 ▾         "fields": {
46 ▾           "director": [
47               "Dean Israelite"
```

Figure 3.4 – Querying with search template

How it works...

Search templates are a way to define and store search queries using the Mustache scripting language. You can run a search template with different variables, such as user input or parameters, and get the results without exposing the Elasticsearch query syntax.

Let's break down the components of our search template:

- **Script language** (**Mustache**): The query uses Mustache, a template scripting language, allowing dynamic insertion of variables (`{{query}}`, `{{agg_size}}`) at query time.

- **Query** (bool with must and multi-match):

  - The `bool` query with the `must` clause ensures that both conditions within it must be met.

- The first `multi_match` query searches the input (`{{query}}`) across multiple fields (`title`, `plot`, `cast`, and `director`). The `title` field is given higher importance (`^4` signifies four times the normal relevance).

- The second `multi_match` query, also using `{{query}}`, is a `phrase_prefix` type. This means it looks for phrases that start with the query string, again across the `title`, `plot`, and `director` fields, with a higher emphasis on `title`.

- **Aggregations (Aggs)**:

  - `genre_facet`: Aggregates the data by genre, showing the most common genres in the search results. The number of genres returned is controlled by `{{agg_size}}`.

  - `director_facet`: Similar to `genre_facet`, but aggregates by `director` keyword, focusing on exact matches of director names.

- **Sorting**: The results are sorted by `release_year` in descending order, showing newer movies first.

- **Fields and source**: Specifies that the `title`, `release_year`, `director`, and `origin` fields should be included in the response.

  - Sets `_source` to false, indicating that the full source document should not be returned in the response.

Search templates work behind the scenes by replacing the variables in the template with the actual values and then executing the query against Elasticsearch. You can store search templates in a special index called `.scripts` or in a file on every data node in the cluster.

The benefits of using search templates are as follows:

- They make your queries more readable and maintainable by separating the structure from the parameters

- They allow you to reuse common queries across different applications or indices

- They provide security and validation by preventing users from injecting malicious queries or accessing restricted fields

Some use cases for search templates are the following:

- Building a search engine for your website or app, where you can pass user input from a search bar as parameters for a search template

- Creating dynamic dashboards or reports, where you can use search templates to generate different visualizations based on user selections or filters

- Performing complex or frequent analysis, where you can use search templates to simplify your queries and avoid repetition

There's more...

We have only scratched the surface here in terms of what you can accomplish with search templates. There are many interesting features, among which are default values and conditionals.

You can define the default values for parameters if there are no values set at query time. For example, we can tweak our template by updating our range query like this:

```
"range": {
  "release_year": {
    "gte": "{{start_date}}",
    "lt": "{{end_date}}{{^end_date}}now{{/end_date}}"
  }
}
```

The preceding will be executed by replacing the `"lt"` value with now if it's not specified.

We can also add conditional values to search templates by using the following syntax:

```
{{#condition}}value{{/condition}}
```

Let's say we want a search template that displays the movies for the last 10 years; when the `last_10y` parameter is set to true, we can use the following template (the snippet is available at `https://github.com/PacktPublishing/Elastic-Stack-8.x-Cookbook/blob/main/Chapter3/snippets.md#search-template-with-conditions`):

```
GET _render/template
{
  "source": "{ \"query\": { \"bool\": { \"filter\": [ {{#last_10y}}
{ \"range\": { \"release_year\": { \"gte\": \"now-10y/d\",
\"lt\": \"now/d\" } } }, {{/last_10y}} { \"term\": { \"origin\":
\"{{origin}}\" }}]}}}",
  "params": {
    "last_10y": true,
    "origin": "American"
  }
}
```

See also

For further information on search templates and the full range of options they offer, consult the official documentation: `https://www.elastic.co/guide/en/elasticsearch/reference/current/search-template.html`

Getting started with Search Applications for your Elasticsearch index

Elastic Search Applications are a new concept introduced in Elastic version 8.8.0 to quickly build search-powered applications with Elasticsearch and the full breadth of its query capabilities. It combines many of the core concepts including Elasticsearch indices, Search templates that we just covered in the previous recipe, and an easy-to-use UI within Kibana for previewing search results. Additionally, there is a Search Application API available for seamless integration within your applications.

In this recipe, we will set up a Search Application on top of our movie dataset and take advantage of the search template we have built in the previous recipe.

The snippets used in this recipe are available at `https://github.com/PacktPublishing/Elastic-Stack-8.x-Cookbook/blob/main/Chapter3/snippets.md#getting-started-with-search-applications-for-your-elasticsearch-index`.

Getting ready

To follow along in this recipe, you'll need an Elastic deployment up and running with Elasticsearch, Kibana, and Fleet Server.

Make sure you have completed the previous *Using search templates to pre-render search requests* recipe.

How to do it...

To create a Search Application, apply the following steps:

1. Go to **Kibana** | **Search** | **Search Applications**, then, click on the **Create** button to open a flyout for configuration (be aware that this feature was in beta testing at the time of writing this book. Its functionality may change in future updates).

2. In the flyout, locate the **Select Indices** field and choose the movie dataset index.

3. Enter a name for your Search Application, such as `movies-search-application`.

4. Finally, click the **Create** button located at the bottom right of the panel to create the Search Application:

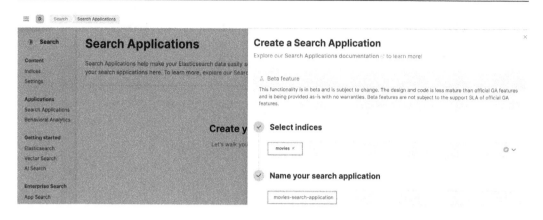

Figure 3.5 – Creating a Search Application

5. Your Seach Application will be created and listed as shown in *Figure 3.6*:

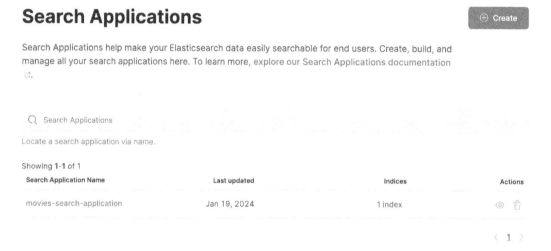

Figure 3.6 – Search Application list

6. By clicking on the name of your application, you will be directed to a dedicated search page where you can query the indices connected to your Search Application.

    From there, you can run a search query and view the results. For example, try searching for **Comedy** as shown in *Figure 3.7*:

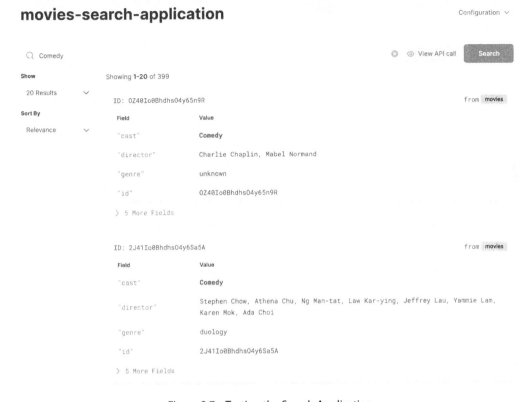

Figure 3.7 – Testing the Search Application

7. Now that our Search Application is in place, let's modify its default search template using the one we created in the previous *Using search templates to pre-render search queries* recipe. For that, we will head to **Dev Tools** and execute the command that can be found at https://github.com/PacktPublishing/Elastic-Stack-8.x-Cookbook/blob/main/Chapter3/snippets.md#create-search-application-with-search-template.

8. Once the Search Application is updated, let's try the following search query using the template we just applied to our Search Application (the snippet is available at https://github.com/PacktPublishing/Elastic-Stack-8.x-Cookbook/blob/main/Chapter3/snippets.md#test-the-search-application-in-dev-tools):

```
GET _application/search_application/movies-search-application/_
search
{
  "params": {
    "query": "space",
    "agg_size": "5"
```

```
        }
    }
```

You should see the results as displayed in *Figure 3.8*. As expected, the number of aggregation buckets has increased to 5:

```
{
    "took": 11,
    "timed_out": false,
    "_shards": {
        "total": 1,
        "successful": 1,
        "skipped": 0,
        "failed": 0
    },
    "hits": {
        "total": {
            "value": 681,
            "relation": "eq"
        },
        "max_score": null,
        "hits": [▭]
    },
    "aggregations": {
        "director_facet": {
            "doc_count_error_upper_bound": 0,
            "sum_other_doc_count": 624,
            "buckets": [
                {
                    "key": "Unknown",
                    "doc_count": 37
                },
                {
                    "key": "Honda, IshirōIshirō Honda",
                    "doc_count": 7
                },
                {
                    "key": "Michael Bay",
                    "doc_count": 5
                },
                {
                    "key": "Alex Proyas",
                    "doc_count": 4
                },
                {
                    "key": "George Lucas",
                    "doc_count": 4
                }
            ]
        },
```

Figure 3.8 – Search Application template search results

How it works...

Search Applications provide a simple way to integrate a search experience within your application, eliminating the need for deep expertise in Query DSL and Elasticsearch's complexities. It functions as a server-side component and as a layer for data persistence. Integrating the Search Application with your frontend or client application is straightforward—simply use the dedicated endpoint automatically generated for each Search Application. This feature also emphasizes security by utilizing API keys. These parameters can be configured by accessing the **Connect** menu in the Search Application, as shown in *Figure 3.9*:

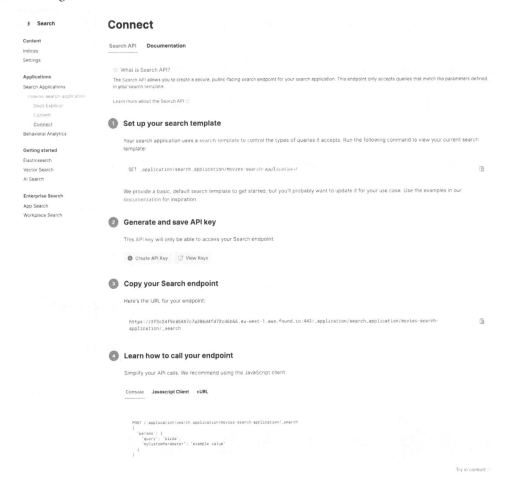

Figure 3.9 – Connecting to the Search Application

In the next recipes, we'll learn how to build an effective search UI on top of your Search Application. Additionally, we'll discuss analyzing user behavior and improving the search experience.

Building a search experience with the Search Application client

In this recipe, we'll see how to leverage the Elastic-provided Search Application client to build a very simple and effective frontend for our movie index. We'll build upon what we did in the previous recipe with the Search Application, acting as the backend of our client.

Getting ready

To follow along in this recipe, you'll need an Elastic deployment up and running with Elasticsearch, Kibana, and Fleet Server. Make sure you have followed the previous recipe, *Getting started with Search Application for your Elasticsearch index*, as we'll be reusing the Search Application.

Make sure to have an Enterprise Search node in your cluster. If you are running an Elastic Cloud deployment you should have one by default. For self-managed clusters, follow the instructions outlined here: `https://www.elastic.co/downloads/enterprise-search`.

You will also need to have Node.js installed on your machine to install the Search Application. We recommend using Node.js version 20 or later.

The snippets used in this recipe are available at `https://github.com/PacktPublishing/Elastic-Stack-8.x-Cookbook/blob/main/Chapter3/snippets.md#building-search-experience-with-search-application-client`.

How to do it...

For simplicity, we have already prepared a sample application available in the GitHub repository: `https://github.com/PacktPublishing/Elastic-Stack-8.x-Cookbook/tree/main/Chapter3/sample-search-app`.

In order to use the application, work through the following steps:

1. In your terminal, go to the `sample-search-app` directory and install the dependencies with the following command:

    ```
    $ yarn install
    ```

2. Head to **Kibana | Search | Search Applications**. Then, click on **movies-search-application** and select the **Connect** tab from the left menu. To collect the parameters you will need to configure the application client: at this point, you should have the `movies-search-application` Search Application configured if you followed the *Getting started with Search Applications for your Elasticsearch index* recipe. Then, do the following:

    A. Generate an API key and save it locally.

    B. Copy the Search endpoint (just the `https://<host>:<port>` part).

3. To make a search API call from the application, the browser needs to make requests to the Elasticsearch API directly. Elasticsearch supports **Cross-Origin Resource Sharing** (**CORS**), but this feature is disabled by default. So, we need to activate it by adding the following settings to `elasticsearch.yml`:

    I.   Go to the Elastic Cloud console, and once you've successfully logged in, locate your cloud deployment and select **Manage** on the right.

    II.   Go to the **Edit** page from the deployment menu and select **Manage user settings and extension** next to **Elasticsearch**.

    III. Update the user settings with the following:

    ```
    http.cors.allow-origin: "*"
    http.cors.enabled: true
    http.cors.allow-credentials: true
    http.cors.allow-methods: OPTIONS, HEAD, GET, POST, PUT, DELETE
    http.cors.allow-headers: X-Requested-With, X-Auth-Token,
    Content-Type, Content-Length, Authorization, Access-Control-
    Allow-Headers, Accept
    ```

    Click on the **Back** button located at the bottom, then scroll to find and click the **Save** button. This will restart the deployment.

4. Update the following parameters for your Search Application client in the `App.tsx` file with the information collected in the previous step:

    *   `elasticsearch_endpoint`

    *   `ApiKey`

    Update the file accordingly:

    ```
    const request = SearchApplicationClient(
        'movies-search-application'
        /*your_elasticsearch_endpoint*/,
        /*your_api_key*/
    )
    ```

5. You will need to modify the existing search template so that it aligns with the requirements of our Search Application frontend. This involves adapting the template we used previously in the *Using search templates to pre-render search requests* recipe to fit the sample Search Application's `QueryBuilder` client. We have already set this up for you. Simply copy the command at `https://github.com/PacktPublishing/Elastic-Stack-8.x-Cookbook/blob/main/Chapter3/snippets.md#update-search-application-with-search-template`, then paste and execute it in **Kibana | Dev Tools**.

6.   We can now start the Search Application client with the following command from the terminal:

```
$ yarn start
```

Open your preferred browser and enter this URL: `http://localhost:3000`. You should see a functional React Search Application as shown in *Figure 3.10*:

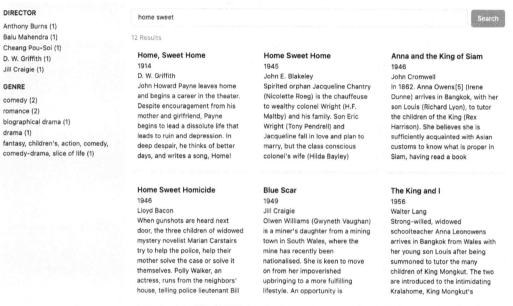

Figure 3.10 – Search Application UI

How it works...

The Search Application client is a JavaScript library that provides the capability to interact with a Search Application. It can serve as a starting point for quickly interacting with your data through a UI. *Figure 3.11* provides an overview of this architecture:

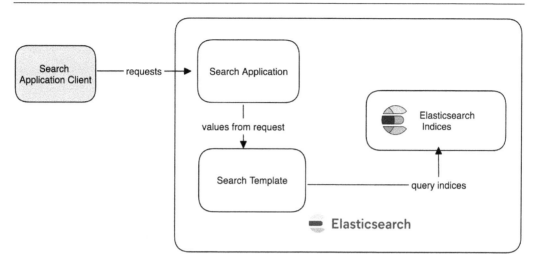

Figure 3.11 – Search Application result flow

Let's examine some essential parts of the code in the `App.tsx` file.

First, we start by creating an instance of the client and initialize the facets filtering on the `director` and `genre` fields with the following code:

```
const request = SearchApplicationClient(
  'applicationName',
  'your_elasticsearch_endpoint',
  'your_api_key',
  {
    facets: {
      director: {
        type: 'terms',
        field: 'director.keyword',
        size: 10,
        disjunctive: true,
      },
      genre: {
        type: 'terms',
        size: 10,
        field: 'genre,
      },
    },
  }
)
```

To perform a search using the `SearchApplicationClient` class, you can use the `search` method of the `request` object:

```
const r = request()
.setSort(['_score'])
        .query(query)
        .setPageSize(8)
        .setFrom(12 * (page - 1))
const results = await r.search()
```

It takes in a search request object as an argument and returns a promise that resolves to a response object. The search request object contains information about the search query such as the query string, filters, and aggregations. The search response object contains information about the search results such as the total number of hits, the documents that matched the query, and aggregations.

In this recipe, as we've seen, the client for Search Applications also depends on a search template, but with an important distinction: the values for the parameters in these templates are supplied by your client application's request. Elastic simplifies and streamlines the use of these templates by offering a boilerplate template as a useful starting point.

There's more...

Search Applications offer a robust way to combine a search template with a safe and secure search endpoint, but there is another way if you want to build a search experience on Elasticsearch directly: Search UI.

Search UI offers the building blocks to create the overall UI on top of an Elasticsearch index. Unlike the Search Application client, which allows you to combine many indices behind a search template, Search UI is built upon the `elasticsearch-connector` paradigm. It acts as a gateway between your data stored in an Elasticsearch index and your frontend in Search UI as shown in *Figure 3.12*:

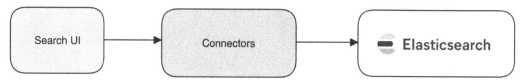

Figure 3.12 – Search UI and Connectors integration to Elasticsearch

In case you're wondering how to decide between Search Application clients and Search UI, here is a decision tree to help you pick the right framework according to your requirements:

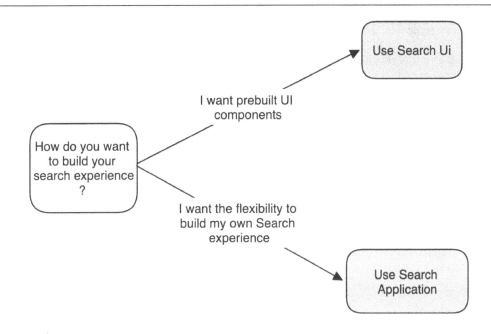

Figure 3.13 – Search experience decision tree

See also

You can find the official code of the Search Application client in this GitHub repository: `https://github.com/elastic/search-application-client#usage`

For more information about the Search UI, see: `https://docs.elastic.co/search-ui/tutorials/elasticsearch`

You can also check the official documentation on the Connector framework: `https://www.elastic.co/guide/en/enterprise-search/current/connectors.html`

Measuring the performance of your Search Applications with Behavioral Analytics

In the previous recipe, we set up our Search Application. The next step is to collect statistics to understand how the Search Application is being used: identifying popular search terms, clicked results, and searches that yield no results. Collecting user behavior statistics is crucial to measure the effectiveness of a Search Application and to gain insights that can improve the overall search experience. To achieve this, we will use the **Behavioral Analytics** platform (introduced in version 8.7.0) to gather statistics on our application.

Getting ready

Make sure that your Search Application from the previous *Building search experience with the Search Application client* recipe is up and running.

The snippets of this recipe are available at `https://github.com/PacktPublishing/Elastic-Stack-8.x-Cookbook/blob/main/Chapter3/snippets.md#measuring-the-performance-of-your-search-applications-with-behaviour-analytics`.

How to do it...

To set up Behavioral Analytics, follow these steps:

1. Go to **Kibana | Search | Search Applications**. In the side navigation bar, click on **Behavioral Analytics,** then on **Create collection**, and name your collection `movie-stats` as shown in *Figure 3.14*:

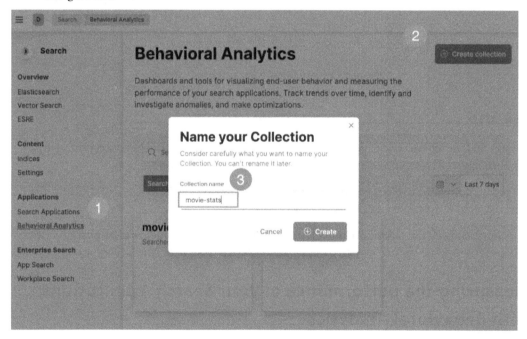

Figure 3.14 – Creating a Behavioral Analytics collection

2. Click on **Integration** on the sidebar, then click **Create API Key**, and finally click on **Generate key**. Store the generated key as we will need it later.

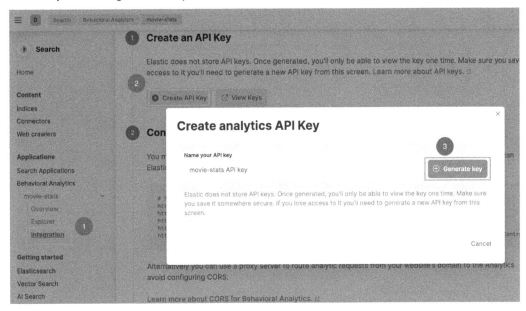

Figure 3.15 – Creating an analytics API key

3. To collect search behavior, the browser needs to make requests to the Elasticsearch API directly. This step is done (in *step 3*) in the previous recipe (you don't have to repeat this step).

4. In the `src` folder of our `sample-search-app` that we built in the previous recipe, open `App.tsx` file, and search for `/* Behavior analytics */` in the file. By default, these four code blocks are commented out. We need to uncomment them and make the adjustment with connection parameters as follows:

    I. Locate and uncomment the `createTracker` function, replace the endpoint with your Elasticsearch endpoint URL, and replace the value of `apiKey` (`"your_analytics_collection_api_key"`) with the one you just created for the `movie-stats` collection in *step 2*:

```
createTracker({
    endpoint: "your_elasticsearch_endpoint",
    collectionName: "movie-stats",
    apiKey: "your_analytics_collection_api_key",
});
```

II.   Locate and uncomment the `trackPageView` function:

```
trackPageView()
```

III.   Locate and uncomment the `trackSearch` function:

```
trackSearch({
  search: {
  query: query,
  results: {
    ...
  },
});
```

IV.   Locate and uncomment the `trackClick` function:

```
trackSearchClick({
  document: { ... },
  search: {
    ...
  },
  page: {
    url: url,
    ...
  },
});
```

5. After uncommenting all four functions and adjusting the connection parameters within the `App.tsx` file, from your terminal, start the Search Application with the `yarn start` command.

6. Once the application is started, visit the Search Application in your browser at `http://localhost:3000`. Then, make some queries in the Search Application and click on the results to generate some traffic.

7. Now we can check the behavioral analytics data by going to **Kibana | Search | Behavioral Analytics** and clicking on the `movie-stats` collection. We should see our first analytics data related to sessions, search numbers, search terms, and no-result search count as shown in *Figure 3.16*:

Figure 3.16 – Behavioral Analytics UI

How it works...

We used the Behavioral Analytics feature, introduced in Elastic version 8.7.0, to collect statistics and thereby gain insights to refine the relevance of our search results and identify content gaps.

In our sample React Search Application, we integrated the tracker using the JavaScript client to collect statistics on sessions, searches, and clicks.

Data and analytics are stored in Elasticsearch indices for advanced analysis and visualization. Importantly, no personal data is collected to ensure GDPR compliance.

There's more...

The data and analytics are collected and stored in Elasticsearch indices. We can navigate to **Kibana |
Discover**, where we can find a data view called `behavioral_analytics.events-movie-stats` that allows us to visualize the raw data of the index corresponding to our `movie-stats` collection. See *Figure 3.17*:

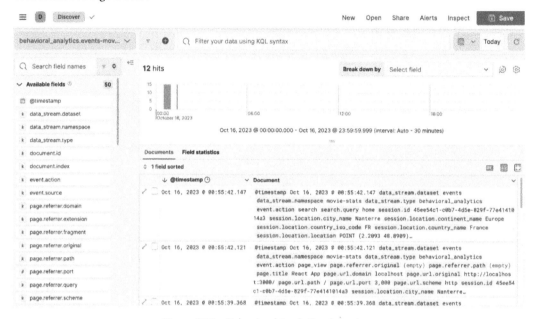

Figure 3.17 – Behavioral Analytics data view

From here, we can build any custom visualizations and dashboards that we need to help us better understand our behavioral analytics data. We will cover a lot of concrete recipes for data visualization in *Chapter 6*.

See also

Behavioral analytics are also integrated with Searchkit (`https://www.searchkit.co/`), which is an open source library to build search UIs with Elasticsearch.

Timestamped Data Ingestion

The Elastic Stack provides many ways for ingesting timestamped data. In this chapter, we're going to focus on two of them: **Elastic Agent** and **Beats**. Elastic Agent acts as an integrated solution for monitoring a variety of data types on each host, including logs and metrics. It additionally safeguards hosts against security risks and enables querying of operating system data. Elastic Agent can also be centrally managed by leveraging Fleet. Meanwhile, Beats have been around for a long time and are lightweight data shippers that transmit operational data to Elasticsearch. Depending on what data you want to collect, you may need to install multiple Beats shippers on a single host.

Figure 4.1 presents a decision tree to help you select the optimal ingestion strategy:

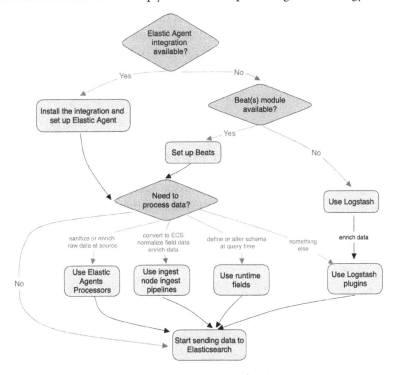

Figure 4.1 – Ingestion strategy decision tree

In this chapter, we're going to cover the following recipes:

- Deploying Elastic Agent with Fleet
- Monitoring Apache HTTP logs and metrics using the Apache integration
- Deploying standalone Elastic Agent
- Adding data using Beats
- Setting up a data stream manually
- Setting up a time series data stream manually

Technical requirements

To follow the different recipes in this chapter, you'll need an Elastic Stack deployment that includes the following:

- Elasticsearch for data storage and search capabilities.
- Kibana for data visualization and stack management (a Kibana user with full privileges for Fleet and integrations is required).
- Integrations Server (included by default in Elastic Cloud deployments).
- As we will be deploying Elastic Agent throughout this chapter, setting up a **virtual machine** (**VM**) is necessary. For instance, to set up a Linux VM on Google Cloud, you can follow the guidelines provided in this tutorial: `https://www.cloudskillsboost.google/ focuses/56596?parent=catalog`.

To quickly create a deployment on Elastic Cloud, you can refer to the *Installing Elastic Stack on Elastic Cloud* recipe in *Chapter 1*.

Deploying Elastic Agent with Fleet

In this recipe, we're going to deploy a Fleet-managed Elastic Agent on a VM. Fleet is a UI application in Kibana for centrally managing Elastic Agent (for example, enrolling, monitoring, and upgrading). To achieve this, we'll create a set of configurations and rules that govern how the agent behaves and operates – called a **policy** – and assign the Elastic Agent to it. By the end of the recipe, you will have a running Elastic Agent sending logs and metrics data from the system.

Getting ready

On the Elastic Stack side, make sure your Elastic deployment with Elasticsearch, Kibana, and Fleet Server is up and running and that you have access to Kibana with the *elastic* user.

> **Important note**
>
> If you want to try out this recipe with a self-managed deployment, please ensure you have completed the steps described in the *Installing self-managed Elastic Stack* and *Creating and setting up Fleet server* recipes in *Chapter 1*. Remember to grant all privileges on Fleet and integrations to your Kibana user.

How to do it...

Let's get started:

1. In Kibana, go to **Management | Fleet | Agent Policies**. This will open the Fleet web UI where we will perform many of the configuration steps; we will start by creating a new policy. This is necessary as the policy will be applied to the agent we're going to deploy.

2. Click on **Create agent policy** to bring up the **Create agent policy** flyout.

3. Name the policy and check the **Collect system logs and metrics** option, as shown in *Figure 4.2*. Then, click on **Create agent policy** to save the policy:

Create agent policy

Agent policies are used to manage settings across a group of agents. You can add integrations to your agent policy to specify what data your agents collect. When you edit an agent policy, you can use Fleet to deploy updates to a specified group of agents.

Name

linux

☑ Collect system logs and metrics ⓘ

Figure 4.2 – Creating the agent policy

> **Important note**
>
> The **Collect system logs and metrics** option allows the agent to automatically ship its telemetry data, making it very useful for debugging and troubleshooting.

4. With our new policy in place, we can choose it and move forward with setting up Elastic Agent. You will see that the system integration is already included in the policy we just created. This integration is standard for all policies and agents. We will utilize this to confirm that our agent is transmitting data once the setup is finished. On the policy page, click on the **Actions** option dropdown seen in the top-right corner of the UI, and then click on the **Add agent** option:

Figure 4.3 – Adding an agent from agent policy

This will open the flyout menu for Elastic Agent setup and you will notice that three steps are displayed:

I.   The first step involves the authentication settings. The enrollment token associated with our policy is already filled in, so there is nothing to do here.

II. *Step 2* addresses the enrollment strategy of our agent. For this recipe, we aim to enroll in Fleet, which is the default option. If you're interested in a standalone option, we've got you covered in the *Deploying standalone Elastic Agent* recipe of this chapter.

III. *Step 3* highlights the agent installation procedure:

3 Install Elastic Agent on your host

Select the appropriate platform and run commands to install, enroll, and start Elastic Agent. Reuse commands to set up agents on more than one host. For aarch64, see our downloads page ☐. For additional guidance, see our installation docs ☐.

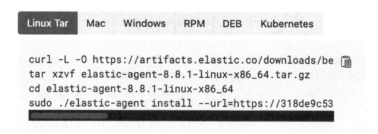

Figure 4.4 – Installing Elastic Agent

5. Copy the instructions found under the **Linux Tar** tab. Keep the Kibana setup page open and, at the same time, connect to your host using SSH or your preferred client. Then, copy and paste the instructions into the terminal to install Elastic Agent. After completing the installation on your host, head back to Kibana. If everything went well, you should see that the agent enrollment is confirmed, as illustrated in *Figure 4.5*:

Figure 4.5 – Elastic Agent enrollment confirmation

Et voila! You've successfully installed Elastic Agent with a policy that you can entirely manage from Fleet in Kibana.

How it works...

Elastic Agent is a single agent for logs, metrics, and security data collection. In this recipe, we've deployed the agent enrolled in Fleet. In this mode, Elastic Agent policies and lifecycles are centrally managed by the Fleet application in Kibana.

Fleet Server, a key element of the Elastic Stack, is utilized for the central management of Elastic Agents. It operates within an Elastic Agent on a designated server host, and a single Fleet Server process can handle numerous connections from various Elastic Agents. Fleet Server also acts on the security side of things by validating enrollment keys and generating API keys used for authentication and data ingestion in Elasticsearch. It's important to note that, in this recipe, we have leveraged the Fleet Server component provided and managed by Elastic Cloud. For on-premises deployment, you need to set up your own Fleet Server.

Another key component of the Elastic Agent architecture is the policy. A policy, in this context, is a set of inputs and configurations that outline the specific data an agent is tasked with collecting. It's important to note that each Elastic Agent can only be enrolled in one policy. Within such a policy, you will find various configurations for individual integrations. These configurations, specific to each type of input, determine the settings and parameters for data collection. It's also important to mention that, at the time of writing, it's not possible to define a hierarchy or inheritance mechanism between multiple policies.

Figure 4.6 shows the different components, their roles, and interactions:

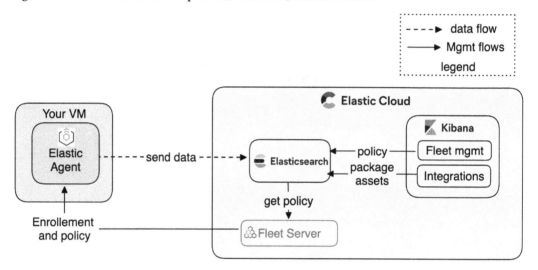

Figure 4.6 – Elastic Agent enrollment and data flow

The interaction between all the components can be summed up as follows:

- Upon the creation of a new agent policy, the Fleet UI records the policy in a Fleet index within Elasticsearch.

- Elastic Agents initiate a request to Fleet Server seeking to enroll in the policy using a specific enrollment key.

- Fleet Server continuously observes the Fleet indices, distributing the policy to all Elastic Agents registered under that policy. Elastic Agent then utilizes the policy's configuration details to gather and transmit data to Elasticsearch.

- Elastic Agent regularly communicates with Fleet Server to stay updated, maintaining a consistent connection.

- Whenever there is an update to a policy, Fleet Server acquires the revised policy from Elasticsearch, prompting each connected Elastic Agent to adjust in response.

There's more...

In this recipe, the Elastic Agent we have deployed is sending data directly to Elasticsearch. This is called the **default output**. But you can also configure Logstash or Kafka as an alternative output for the agent to send data. It's a suitable choice for scenarios requiring advanced data parsing and transformation before the data reaches Elasticsearch or other endpoints.

As with the other components of the Elastic Stack, Elastic Agent can also be deployed through the ECK operator if you plan to run it in a Kubernetes environment.

See also

- If you're interested in running Elastic Agent without Fleet (aka standalone), check out the *Deploying standalone Elastic Agent* recipe in this chapter
- To deploy Elastic Agent and Fleet Server in ECK, check out the following manifests in our official ECK repository: `https://github.com/elastic/cloud-on-k8s/tree/vcluster/config/recipes/elastic-agent`
- To learn more on how to scale your deployment of Elastic agents, you can refer to this official Elastic documentation: `https://www.elastic.co/guide/en/fleet/current/fleet-server-scalability.html`

Monitoring Apache HTTP logs and metrics using the Apache integration

In the previous recipe, we covered setting up and deploying Elastic Agent on a host, managed via Fleet. This recipe will present a practical example of using Elastic Agent's out-of-the-box integrations for monitoring a widely-used service such as Apache. Integrations play a key role in data ingestion strategies, particularly for timestamped events. They offer an easy and standardized way to gather data from a broad spectrum of applications and services. These integrations are housed in the **Elastic Package Registry** and can be accessed through Kibana.

> **Important note**
>
> It's worth mentioning that Kibana requires access to the public Elastic Package Registry for integration discovery in the Integrations app. For environments where network access is restricted, you can still deploy your own self-hosted Elastic Package Registry. Refer to the following documentation page for more information: `https://www.elastic.co/guide/en/fleet/current/air-gapped.html#air-gapped-diy-epr`.

Getting ready

To follow along with this recipe, ensure you have completed the previous one: *Deploying Elastic Agent with Fleet*. Otherwise, you'll need to have a running Elastic Stack and a Linux host with Elastic Agent deployed. You can follow *step 2* of this tutorial to set up a VM with Apache HTTP installed in Google Cloud: `https://www.cloudskillsboost.google/focuses/56596?parent=catalog`.

If you already have a VM or a host up and running, you can install an Apache HTTP Server on it by following the instructions provided in the official Apache documentation: `https://httpd.apache.org/docs/2.4/install.html`.

Make sure to identify the policy used by the agent as we will use it to add the **Apache integration**.

How to do it...

The goal of this recipe is to set up the Apache integration on Elastic Agent to start collecting logs and metrics from a running Apache HTTP Server. Let's get started:

1.   In **Kibana**, head to **Management | Integrations**.

> **Important note**
> The integrations displayed are for Elastic Agent and Beats. To make sure you're viewing only the integrations applicable to Elastic Agent, scroll down and choose **Elastic Agent only**, as shown in *Figure 4.7*:

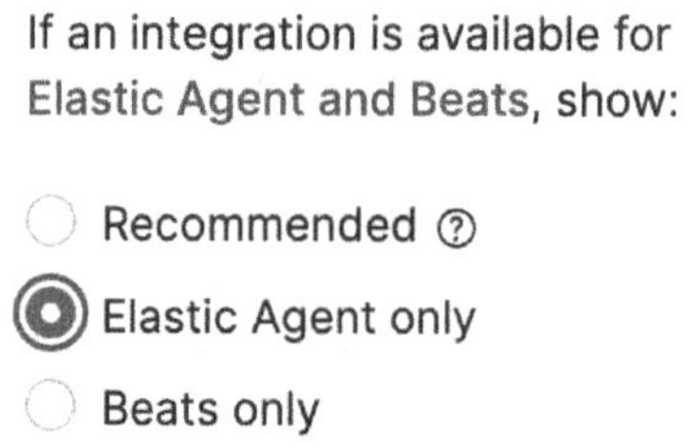

If an integration is available for
Elastic Agent and Beats, show:

○ Recommended ⑦
◉ Elastic Agent only
○ Beats only

Figure 4.7 – Filtering out the integration list

2. In the search bar of the **Integrations** page, type Apache and press *Enter*:

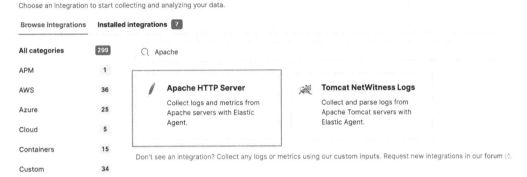

Figure 4.8 – Choosing the Apache HTTP Server integration

3. Choose the Apache HTTP Server integrations. This action will launch the **Integrations** wizard configuration. On this page, you'll find key details about the integration's purpose, its version, compatibility, and the fields it exports. Additionally, you can view screenshots of the dashboards that come with the integration:

Figure 4.9 – Integration details page

4. Click on **Add Apache HTTP Server** at the top right. This will open the integration configuration page. This page contains two main sections: the first section is the **Configure Integration** section. Here, enter the basic information such as the name and description for the integration, and specify how you want to collect logs and metrics from Apache instances. We'll leave the default settings as we're running a default Apache HTTP Server installation for this recipe:

Figure 4.10 – Integration configuration

At the bottom of the configuration page, you'll find the second section: **Where to add this integration?**. Here, you can choose either to add the integration to a new policy or an existing one. For this recipe, we will choose **Existing hosts** and add the integration to the policy we've set up in the previous recipe: *Deploying Elastic Agent with Fleet*.

 Where to add this integration?

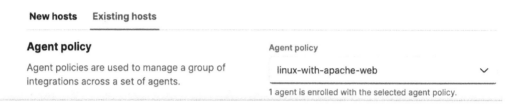

Figure 4.11 – Applying the agent policy

5. Click **Save and continue** at the bottom right of the page to finish the configuration. This action will bring up a popup; click **Save and deploy changes** to proceed.

6. The Apache HTTP Server integration has been added to the policy. To confirm that the integration has been correctly applied and Apache HTTP Server logs and metrics are being processed, head to **Analytics | Dashboard**. Then, among the list of the dashboards, click on **[Logs Apache] Access and error logs**; check whether the dashboard is being populated, as shown in *Figure 4.12*:

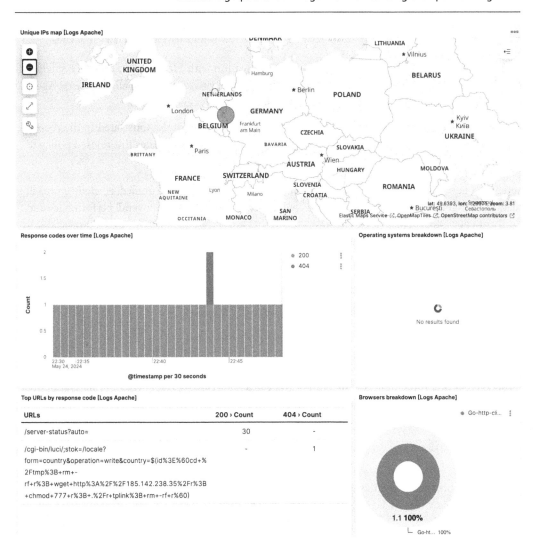

Figure 4.12 – Apache Access and error logs dashboard

How it works...

Elastic Agent integrations offer a cohesive and simple approach to gathering data from well-known applications and services, also maintaining the security of the system. Every integration includes ready-to-use assets catering to various observability requirements, including data intake, storage, transformation guidelines, configuration settings, pre-designed dashboards, and visualizations, along with comprehensive documentation.

Every time you add an integration to an existing policy, a few things happen behind the scenes:

- Elastic Agent consistently engages in communication with Fleet Server, akin to a satellite periodically checking in with its ground station. This regular interaction allows Elastic Agent to stay updated with any changes or modifications in its assigned policy.

- When a policy is modified with the addition of integration, as demonstrated in the recipe, Fleet Server obtains the updated policy and subsequently disseminates this revised policy to all connected agents.

Overall, this new integration approach with Elastic Agent represents a more unified, efficient approach to data ingestion and management, reducing complexity and enhancing the overall performance and security of your Elastic Stack deployment.

There's more...

In this recipe, we just touch upon adding a new integration into an existing policy, but you can also perform many management tasks from the Fleet application. Check out the official documentation to learn more.

On the **Integration configuration** page, there are some advanced options that we didn't cover in this recipe. For example, you can add Elastic Agent **processors** to reduce the number of fields or enhance the collected events with additional metadata. You can also use **tags** to label your events as they are indexed in Elasticsearch. Those give you more flexibility and customization on top of the simplified onboarding experience that integration and Elastic Agent provide.

See also

To see the full breadth of available integrations, you can also visit this page: `https://www.elastic.co/integrations/data-integrations`.

Deploying standalone Elastic Agent

For advanced users who want to configure manually and manage upgrades of the agents themselves, deploying Elastic Agent in standalone mode is the way to go. In this recipe, we'll go through the required steps to have Elastic Agent up and running without Fleet.

> **Note**
> Elastic recommends using Fleet-managed agents whenever possible as running a standalone Elastic Agent is more demanding in terms of operational workload.

Getting ready

To deploy Elastic Agent in standalone mode, a VM is required. If you have followed the *Deploy Elastic Agent with Fleet* recipe, you will need a separate VM for this recipe. You can follow this quick tutorial to set up a Linux VM on Google Cloud: `https://www.cloudskillsboost.google/focuses/56596?parent=catalog`.

Make sure to have an Elastic Stack cluster ready, either self-managed or on Elastic Cloud.

The snippets of this recipe are available at the following link: `https://github.com/PacktPublishing/Elastic-Stack-8.x-Cookbook/blob/main/Chapter4/snippets.md#deploying-standalone-elastic-agent`.

How to do it...

First, let's start by creating a policy for the standalone agent. To get started quickly, we will leverage Kibana to create our policy with the Apache integration:

1. Go to **Home | Add integrations**.

2. In the search bar, enter `Apache server` and select the **Apache HTTP Server** integration.

3. Enter the following information for the integration in the **Configure** section:

    A. **Integration name**: `apache-standalone`

    B. **Description**: `Apache integration for standalone elastic agent`

4. Under **Where to add this integration?**, select the **New hosts** tab to create a new policy. Name your policy `standalone-policy`.

5. Click **Save and continue** at the bottom right. A popup window will show up; click **Add Elastic Agent to your hosts** to open the **Add agent** flyout:

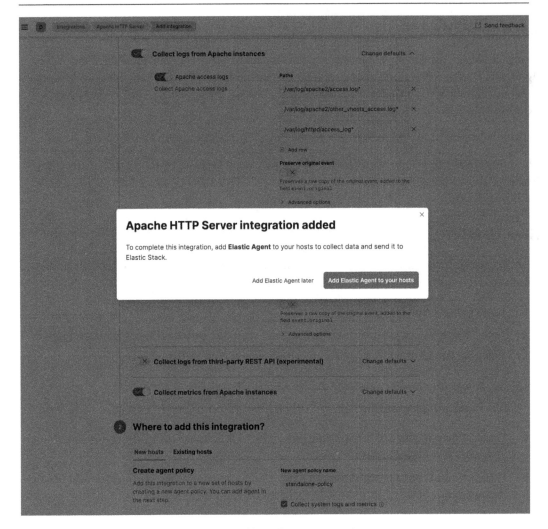

Figure 4.13 – Adding Elastic Agent to hosts

6.   On the flyout page, select the **Run standalone** tab, as shown in *Figure 4.14*:

Add agent

Add Elastic Agents to your hosts to collect data and send it to the Elastic Stack.

Run an Elastic Agent standalone to configure and update the agent manually on the host where the agent is installed.

 Enroll in Fleet?

○ **Enroll in Fleet (recommended)** – Enroll in Elastic Agent in Fleet to automatically deploy updates and centrally manage the agent.

◉ **Run standalone** – Run an Elastic Agent standalone to configure and update the agent manually on the host where the agent is installed.

Figure 4.14 – Selecting Run standalone for Elastic Agent installation

7. Click on **Download Policy** to save it locally:

1 Configure the agent

Copy this policy to the `elastic-agent.yml` on the host where the Elastic Agent is installed. Modify `ES_USERNAME` and `ES_PASSWORD` in the `outputs` section of `elastic-agent.yml` to use your Elasticsearch credentials.

> 📋 Copy to clipboard   ⬇ Download Policy

```
revision: 2
outputs:
  default:
    type: elasticsearch
    hosts:
      - >-

https://
west3.gcp.cloud.es.io:443
      username: '${ES_USERNAME}'
      password: '${ES_PASSWORD}'
output_permissions:
  default:
    _elastic_agent_monitoring:
      indices:
        - names:
            - logs-elastic_agent.apm_server-default
```

Figure 4.15 – Copying the agent policy

Alternatively, we provide a sample `elastic-agent.yml` configuration file, which you can find at `https://github.com/PacktPublishing/Elastic-Stack-8.x-Cookbook/blob/main/Chapter4/elastic-agent-standalone/elastic-agent.yml`.

Modify the credentials (ES_USERNAME and ES_PASSWORD) in the downloaded policy to use your own Elasticsearch credentials.

Now that we have our policy, we can proceed with the installation of the standalone agent. Make sure to copy the downloaded policy onto the target host; for instance, you can place it in the /tmp directory. This will be necessary for the subsequent steps.

8. Connect to the VM where the agent will be installed and follow these steps:

    I. Download and extract the Elastic Agent package with the following commands:

    ```
    $  curl -L -O https://artifacts.elastic.co/downloads/beats/
    elastic-agent/elastic-agent-8.12.2-linux-x86_64.tar.gz
    $  tar xzvf elastic-agent-8.12.2-linux-x86_64.tar.gz
    ```

    II. Copy the downloaded elastic-agent.yml file into the folder where the agent was extracted. In our setup, we used the following command:

    ```
    $ cp /tmp/elastic-agent.yml /home/admin/elastic-agent-8.12.2-
    linux-x86_64
    ```

    III. Go to the agent directory and run the following command:

    ```
    $ sudo ./elastic-agent install
    ```

    IV. Type n when prompted to enroll this agent into Fleet, as illustrated in the following figure:

    ```
    admin@ip-172-31-41-103:~/elastic-agent-8.12.2-linux-x86_64$ sudo ./elastic-agent install
    Elastic Agent will be installed at /opt/Elastic/Agent and will run as a service. Do you want to continue? [Y/n]:
    Do you want to enroll this Agent into Fleet? [Y/n]:n
    ```

Figure 4.16 – Opting out of enrolling the agent into Fleet

    V. When the installation is complete, start the agent:

    ```
    $ sudo systemctl start elastic-agent.service
    ```

    VI. Finally, to confirm the standalone Elastic Agent is running and shipping data, head to **Dashboard | [Logs Apache] Access and error logs**. You should see the dashboard populated with data from the Apache integration.

    VII. One last thing: to confirm that our Elastic Agent is running in standalone mode, in Kibana, go to **Management | Fleet | Agent policies**. Check the standalone-policy entry and you will see that **0** agents are associated with this policy, although the policy is being used by the standalone agent:

Fleet

Centralized management for Elastic Agents.

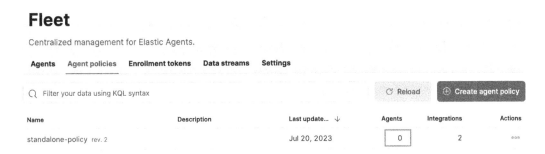

Figure 4.17 – Verifying the Agent policies usage

How it works...

When running in standalone mode, Elastic Agent functions like Beats. There is no Fleet enrollment, you only need to define the authentication mechanism and credentials that the agent will use to send data to Elasticsearch. In this recipe, we've used a username/password as a means of authentication, but you can also use an API key:

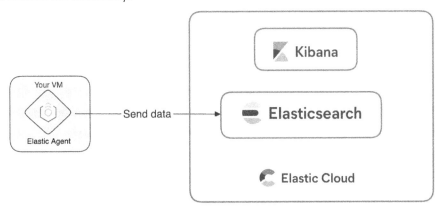

Figure 4.18 – Standalone Elastic Agent data flow

Standalone mode is ideal for users seeking direct control over each agent's configuration, especially in smaller or more specialized environments where centralized management is less critical. However, this approach lacks the scalability and ease of centralized policy management and updates provided by Fleet. On the other hand, Fleet-managed mode offers a streamlined, centralized platform for managing and updating agents across a large and diverse environment, enhancing operational efficiency and consistency. The trade-off here is the potential complexity and overhead in setting up and maintaining Fleet Server, especially in simpler setups where such a level of management might be unnecessary. Therefore, each mode has its specific context of suitability, balancing control and convenience against scalability and management overhead.

There's more...

For this recipe, we've used Kibana to generate the standalone configuration. While it's definitely the easiest way, it also presents a particular advantage: integration assets are automatically set up. You do not need to install them.

The standalone mode is also supported by the ECK operator if you want to deploy your agents in a Kubernetes environment.

See also

To learn how to install integration assets if you decide to generate the policy configuration manually for your agent, see `https://www.elastic.co/guide/en/fleet/current/install-uninstall-integration-assets.html#install-integration-assets`.

You can find the sample configurations for a standalone agent with ECK operator here: `https://github.com/elastic/cloud-on-k8s/blob/2.8/config/recipes/elastic-agent/system-integration.yaml`.

Adding data using Beats

In the previous recipes in this chapter, we learned how to ingest timestamped data using Elastic Agent and its integrations. Sometimes, you may not find an available Elastic Agent integration for your data source, but there might be a Beats module that supports it. In such cases, using Beats becomes the natural alternative. Beats is a suite of lightweight data shippers that can efficiently gather and transport data to Elasticsearch for analysis and visualization. In this recipe, we will learn how to configure and use Beats to collect data from a data source and send it to Elasticsearch.

By following the steps in the upcoming *How to do it...* section, you will learn how to set up **Metricbeat** to monitor **Apache Tomcat** metrics and stream this data to Elasticsearch. Metricbeat, one of the many Beats shippers supported by Elasticsearch, collects metrics from your systems and services and sends them to Elasticsearch or Logstash for indexing. We have chosen Apache Tomcat metrics as an example to demonstrate how Beats can be used as an alternative when Elastic Agent integration is still missing (At the time this book was written, there was no Elastic Agent integration available for Tomcat metrics, although the Metricbeat Tomcat module was available).

Getting ready

Prior to ingesting data, ensure that you have a running Apache Tomcat instance with the supported version, Apache Tomcat 9.x. To fetch metrics, we will use Jolokia, which means that both Apache Tomcat and Jolokia need to be installed. Let's first look into the installation instructions for Apache Tomcat and Jolokia.

Installing Apache Tomcat 9

There are multiple ways to set up an Apache Tomcat instance. For this recipe, we will use the **Google Cloud Platform** (**GCP**) Marketplace solution (`https://console.cloud.google.com/marketplace/browse(cameo:product/cloud-infrastructure-services/tomcat-ubuntu`) to provision an **Ubuntu** VM on the GCP Compute Engine, bundled with a preconfigured Apache Tomcat 9 instance.

> **Note**
>
> If you prefer other cloud providers or a self-managed environment for your Apache Tomcat instances, the instructions in this recipe will still be applicable. For a self-managed installation, the official Tomcat setup guide can be found here: `https://tomcat.apache.org/tomcat-9.0-doc/setup.html`.

Jolokia

Jolokia (`https://jolokia.org/`) is an open source Java library and agent that streamlines the management and monitoring of Java applications by providing a user-friendly HTTP interface to interact with JMX MBeans. The Metricbeat Tomcat module requires Jolokia to fetch JMX metrics; therefore, Jolokia must be installed in the Tomcat application. Detailed instructions are available at `https://jolokia.org/tutorial.html`.

The snippets of the recipe can be found at `https://github.com/PacktPublishing/Elastic-Stack-8.x-Cookbook/blob/main/Chapter4/snippets.md#adding-data-from-beats`.

How to do it...

In this recipe, we detail the installation and configuration of Metricbeat on Ubuntu, aimed at integrating it with Elastic Cloud and capturing metrics through the Tomcat and Jolokia modules. We will also demonstrate how to set up Metricbeat to stream data and use Kibana for visualization. Moreover, we will illustrate how to monitor system health via Metricbeat's System module within Kibana's Observability UI. Let's get started:

1. Start by installing Metricbeat with the appropriate operating system and Elastic Stack version; in our example, we deploy Metricbeat with version 8.12.2 on Ubuntu (deb):

   ```
   $ curl -L -O   https://artifacts.elastic.co/downloads/beats/
   metricbeat/metricbeat-8.12.2-amd64.deb
   $ sudo dpkg -i metricbeat-8.12.2-amd64.deb
   ```

2. Once installed, you need to configure Metricbeat to connect to your Elastic Cloud deployment. The Metricbeat configuration file is located at `/etc/metrcibeat/metricbeat.yml`.

3. In the configuration file, you can provide the `cloud.id` details of your Elasticsearch service and set `cloud.auth` to a user who has the necessary authorization to set up Metricbeat, as shown here:

```
cloud.id: "YOUR_CLOUD_ID"
cloud.auth: "metricbeat_setup:YOUR_PASSWORD"
```

Next, we will configure the Metricbeat Tomcat module.

4. To enable the Tomcat module, you can use the following command:

```
$ sudo metricbeat modules enable tomcat
```

5. The configuration file is located at `/etc/metricbeat/modules.d /tomcat.yml`. For the Tomcat metrics in our example, we don't need to modify the default configuration as the configuration is set in the Jolokia module in the next steps.

6. To retrieve the Tomcat metrics, configure the Metricbeat Jolokia module:

```
$ sudo metricbeat modules enable jolokia
```

The configuration file is located at `/etc/metricbeat/modules.d/jolokia.yml`.

The most important part of the configuration is the `jmx.mapping` attribute, where we map the JMX MBean attributes to **Elastic Common Schema** (**ECS**) attributes. We omit the full mapping here, but the complete configuration file example can be found at `https://github.com/PacktPublishing/Elastic-Stack-8.x-Cookbook/blob/main/Chapter4/metricbeat/jolokia.yml`:

```
- module: jolokia
  metricsets: ["jmx"]
  period: 10s
  hosts: ["localhost"]
  namespace: "metrics"
  path: "/jolokia/?ignoreErrors=true&canonicalNaming=false"
  jmx.mappings:
    ...
    - mbean: 'java.lang:type=Memory'
      attributes:
        - attr: HeapMemoryUsage
          field: memory.heap_usage
        - attr: NonHeapMemoryUsage
          field: memory.non_heap_usage
    ...
  jmx.application:
  jmx.instance:
```

7. Once the configuration is finalized, start Metricbeat to begin collecting Apache Tomcat metrics. Use the following command to set up the related assets (visualizations, dashboards, and index templates):

```
$ sudo metricbeat setup -e
```

8. Then, start Metricbeat with the following command:

```
$ sudo service metricbeat start
```

9. At this stage, Metricbeat will begin streaming metrics to your deployment; you can verify the metrics on the default Metricbeat Tomcat dashboard. In Kibana, go to **Analytics | Dashboards**, and type tomcat in the search bar to find the **[Metricbeat Tomcat] Overview** dashboard, as shown in *Figure 4.19*:

Dashboards

Figure 4.19 – Selecting the Metricbeat Tomcat dashboard

10. Click on the dashboard to verify the metrics collected by the Metricbeat Tomcat and Jolokia modules, such as memory, threading, and global requests, as shown in *Figure 4.20*:

Figure 4.20 – Metricbeat Tomcat dashboard

11. The **System** module is enabled by default, allowing you to also verify metrics in the **Observability | Infrastructure** UI. In Kibana, head to **Observability | Infrastructure | Hosts** to inspect the system metrics of the Ubuntu VM, as shown in *Figure 4.21*:

Figure 4.21 – Elastic Observability Infrastructure view

How it works...

Historically, Beats was the default choice to ingest timestamped data from various sources. Although Elastic Agent has since become the preferred way to ingest timestamped data, Beats are still good alternatives when no Elastic Agent integration exists for the given data source.

In our example, we leveraged Metricbeat's Tomcat and Jolokia modules to collect Apache Tomcat metrics. The modules helped to process the data at source and format the data into Elastic Common Schema, which is a standardized schema for structured logging and event data in the Elastic Stack.

Metricbeat then sends the processed metrics to Elasticsearch in a data stream named `metricbeat-8-12-2` (the name may change if you can use a different version of Metricbeat). In the next recipe, we will learn how to set up a data stream manually and explore the concept behind it and the differences between a data stream and a regular index.

You can also visually check the created data stream in Kibana. Go to **Analytics | Discover**, choose the `metricbeat-*` data view (which includes the `metricbeat-8-12-2` data stream), and select **tomcat.cache.mbean** in the **Break down by** field in the default histogram visualization, as shown in *Figure 4.22* (we will see more in *Chapter 6*):

Figure 4.22 – Checking the data stream in Discover

> **Note**
>
> From version 8.0, all Beats shippers now store events in data streams in Elasticsearch instead of indices.

There's more...

In this recipe, we explored how to install and configure Beats with a concrete sample on Metricbeat. There are seven officially supported Beats in the Beats family: Auditbeat, Filebeat, Functionbeat, Heartbeat, Metricbeat, Packetbeat, and Winlogbeat. The detailed documentation for different Beats and their modules can be found here: `https://www.elastic.co/guide/en/beats/libbeat/current/beats-reference.html`.

See also

If you have used Beats with earlier versions of Elastic Stack, you may consider migrating from Beats to Elastic Agent. The migration guide can be found here:

`https://www.elastic.co/guide/en/fleet/current/migrate-beats-to-agent.html`.

Setting up a data stream manually

Data streams are a new feature introduced in Elasticsearch 7.9 and solidified in version 8. They allow for more efficient management and reduced overhead of time series data. As you learned in the previous recipes of this chapter, both Elastic Agent and Beats create timestamped data automatically in data streams. What about custom data sources where we cannot easily install Elastic Agent or Beats as data shippers? By following the steps in this recipe's *How to do it...* section, you'll be able to create a data stream and start ingesting and analyzing time-based data manually in Elasticsearch.

Dataset

In this and the following recipe, we will use a free dataset provided by European public sector datasets (`https://data.europa.eu/`) – specifically, a real-time traffic status dataset for the city of Rennes, one of the biggest cities in France (`https://data.europa.eu/data/datasets/5caaf5ee9ce2e75d0c8c381a?locale=en`).

> **Dataset citation**
>
> Real-time traffic status (updated 2024-01-08, accessed 2024-03-25). [Dataset]. Rennes Métropole en accès libre. `http://data.europa.eu/88u/dataset/5caaf5ee9ce2e75d0c8c381a` (Original work published 2019).

This dataset offers rich metadata, lending itself to learning about various aspects of timestamped data ingestion, transformation, and visualization data exploration with geodata. We will use this dataset in *Chapters 4* to *8*.

The snippets of this recipe can be found at `https://github.com/PacktPublishing/Elastic-Stack-8.x-Cookbook/blob/main/Chapter4/snippets.md#setting-up-data-stream-manually`.

Getting ready

You need an up-and-running Elastic deployment, as we described in *Chapter 1*, and you will need Kibana access, as most of the commands will be executed in Kibana's **Dev Tools** console.

You will also need an installed Elasticsearch Python client that has successfully connected to your Elasticsearch cluster (please refer to the *Adding data from the Elasticsearch client* recipe in *Chapter 2*).

How to do it...

This recipe will guide you through setting up an **Index Lifecycle Management** (ILM) policy for Rennes traffic data and creating the necessary index templates in Elasticsearch. We'll start with a simple ILM policy in Kibana, and then define mappings and settings for our data. Finally, we'll ingest real traffic data with a Python script and confirm the ingestion in Elasticsearch via Kibana's management UI.

To ingest our Rennes traffic data into a data stream, we need to first prepare the index template with the appropriate mapping, setting, and ILM policy:

1. Let's start by defining the **ILM policy** by executing the following command in the Kibana **Dev Tools** console. The approach at this stage is kept simple, with a policy triggering rollover when the primary shard reaches 50 GB and deleting data after 30 days. (For more information on rollover, please refer to the *Setting up an index lifecycle policy* recipe in *Chapter 12*.) The snippet is available at `https://github.com/PacktPublishing/Elastic-Stack-8.x-Cookbook/blob/main/Chapter4/snippets.md#component-template-for-rennes-traffic-index-lifecyle-policy`:

```
PUT _ilm/policy/rennes_traffic-lifecycle-policy
{
  "policy": {
    "phases": {
      "hot": {
        "actions": {
          "rollover": {
            "max_primary_shard_size": "50gb"
          }
        }
      },
      "delete": {
        "min_age": "30d",
        "actions": {
```

```
              "delete": {}
            }
          }
        }
      }
    }
```

2. From *Chapter 2*, we learned about the concept of index templates and component templates. To prepare the index template, we create the first component template, `rennes_traffic-mapping`, to define the mapping of Rennes traffic data fields (the snippet is available at `https://github.com/PacktPublishing/Elastic-Stack-8.x-Cookbook/blob/main/Chapter4/snippets.md#component-template-for-rennes-traffic-mappings`):

```
PUT _component_template/rennes_traffic-mappings
{
  "template": {
    "mappings": {
      "properties": {
        "@timestamp": {
          "type": "date",
          "format": "date_optional_time||epoch_millis"
        },
        "traffic_status": {"type": "keyword"},
        "location_reference": {"type": "keyword"},
        ...
        "traveltime" : {
          "subobjects" : false,
          "properties" : {
            "reliability" : {
              "type" : "short"
            },
            "duration" : {
              "type" : "short"
            }
          }
        }
        ,
        ...
        "location": {"type": "geo_point"},
        "data_stream": {
          "properties": {
            "namespace": {
              "type": "constant_keyword"
```

```
          },
          "type": {
            "type": "constant_keyword"
          },
          "dataset": {
            "type": "constant_keyword"
          }
        }
      }
    }
  }
},
"_meta": {
  "description": "Mappings for rennes traffic data fields"
}
}
```

3. Next, we create the second component template, rennes_traffic-settings, to host index setting, including the reference to the previously defined rennes_traffic-lifecycle-policy (the snippet is available at https://github.com/PacktPublishing/Elastic-Stack-8.x-Cookbook/blob/main/Chapter4/snippets.md#component-template-for-index-settings):

```
PUT _component_template/rennes_traffic-settings
{
  "template": {
    "settings": {
      "index.lifecycle.name": "rennes_traffic-lifecycle-policy"
    }
  },
  "_meta": {
    "description": "Settings for ILM"
  }
}
```

Finally, we create our index template composed of the two component templates that we just created (the snippet is available at https://github.com/PacktPublishing/Elastic-Stack-8.x-Cookbook/blob/main/Chapter4/snippets.md#rennes-traffic-index-template):

```
PUT _index_template/rennes_traffic-index-template
{
  "index_patterns": ["generic-rennes_traffic-*"],
  "data_stream": { },
  "composed_of": [ "rennes_traffic-mappings", "rennes_traffic-
```

```
settings" ],
    "priority": 500,
    "_meta": {
      "description": "Template for Rennes traffic data"
    }
  }
}
```

4. We can now create a test document for the generic-rennes_traffic-default data stream (the full snippet can be found at https://github.com/PacktPublishing/Elastic-Stack-8.x-Cookbook/blob/main/Chapter4/snippets.md#component-template-for-index-settings):

```
POST generic-rennes_traffic-default/_doc
{
  "@timestamp": "2024-01-17T23:07:00+02:00",
    "traffic_status": "heavy",
    "location_reference": "10273_D",
  . . .
    "maxs_peed": "70",
    "average_vehicle_speed": "46",
    "location": {
      "lat": 48.04479275590756,
      "lon": -1.6502152435538264
    }
}
```

After submitting this document, we confirm that it has been correctly created, as shown in *Figure 4.23*:

```
                                          201 - Created    53 ms
 1 - {
 2     "_index": ".ds-generic-rennes_traffic-default-2023.07.21-000001",
 3     "_id": "PaoGeokBqK4lISUDLRoi",
 4     "_version": 1,
 5     "result": "created",
 6 -   "_shards": {
 7       "total": 2,
 8       "successful": 2,
 9       "failed": 0
10 -   },
11     "_seq_no": 2859,
12     "_primary_term": 1
13 - }
```

Figure 4.23 – Confirmation of the document ingestion to the data stream

5. Now, let's ingest real Rennes traffic data using the Python script. The script can be accessed at `https://github.com/PacktPublishing/Elastic-Stack-8.x-Cookbook/blob/main/Chapter4/python-client-sample/datastream.py`. Adjust the connection parameters in the `.env` file and execute the script with the following commands:

```
$ pip install -r requirements.txt
$ python datastream.py
```

We can then verify that the documents are successfully indexed in the log messages, as shown in *Figure 4.24*:

```
Loading dataset...
Indexing documents...
100%|█████████████| 2859/2859 [00:00<00:00, 4382.03docs/s]
Indexed 2859/2859 documents
```

Figure 4.24 – Index log messages

6. We can verify the `generic-rennes_traffic-default` data stream in Kibana Dev Tools with the following command:

```
GET generic-rennes_traffic-default/_search
```

We can then make sure that it returns the response with the Rennes traffic data we just ingested, as shown in *Figure 4.25*:

```
                                              200 - OK    179 ms
 1 ▾ {
 2       "took": 28,
 3       "timed_out": false,
 4 ▾     "_shards": {
 5           "total": 1,
 6           "successful": 1,
 7           "skipped": 0,
 8           "failed": 0
 9 ▾     },
10 ▾     "hits": {
11 ▾         "total": {
12               "value": 2860,
13               "relation": "eq"
14 ▾         },
15           "max_score": 1,
16 ▾         "hits": [
17 ▾             {
18                   "_index": ".ds-generic-rennes_traffic-default-2024.03.25
                          -000001",
19                   "_id": "bi6Yd44BfC5SFPqL-wAB",
20                   "_score": 1,
21 ▾                 "_source": {
22                       "@timestamp": "2024-03-25T22:47:00+01:00",
23                       "traffic_status": "unknown",
24                       "location_reference": "10273_D",
25                       "denomination": "Route départementale 34",
26                       "hierarchie": "Réseau d'armature",
27                       "hierarchie_dv": "Réseau de transit",
28                       "insee": "35206".
```

Figure 4.25 – Verifying the data stream from Dev Tools

7. In Kibana, we can also navigate to **Stack Management | Index Management | Data Streams** to verify that the `generic-rennes_traffic-default` data stream has been correctly created, as shown in *Figure 4.26*:

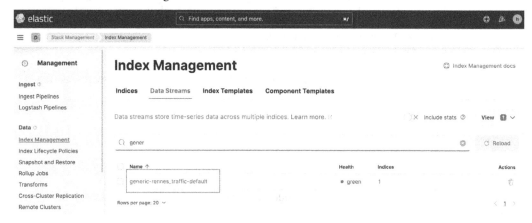

Figure 4.26 – Verifying the data stream from index management

How it works...

We'll now see how Elasticsearch has redefined the management of time series data from version 8.0. We'll also highlight the benefits of using data streams over aliases for indexing and querying.

Why data streams?

Historically, Elasticsearch provides the **alias** concept, which is a secondary name to an index. In time series data indexing, utilizing an alias enables consistent querying of the most current data, even as underlying indices change on a daily basis.

Starting from version 8.0, data streams have become the default way to manage timestamped data. A data stream consists of a collection of automatically generated hidden indices that store timestamped data, which automatically roll over according to conditions defined in the ILM policy. As you can see in *Figure 4.27*, while using a data stream, index operations only take place on the latest index by using the **append-only** concept while the search/query operations address the entire hidden collection of indices.

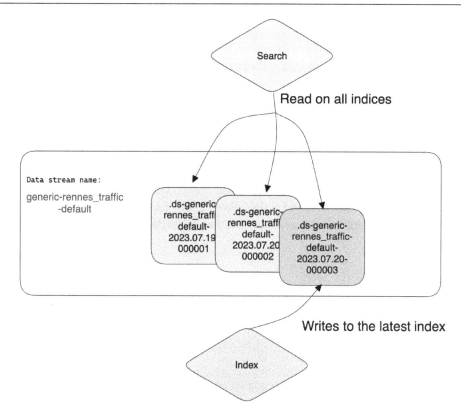

Figure 4.27 – How a data stream works

In this recipe, we successively created an ILM policy, two component templates, and, finally, the index template for the Rennes traffic data. The links between them are explained in *Figure 4.28*:

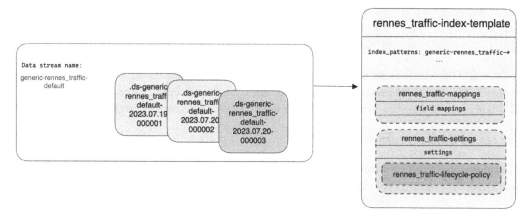

Figure 4.28 – Data stream index template logic

We used a Python script to download the real-time Rennes traffic data and bulk ingest the documents to Elasticsearch. Given the immutable and *append-only* nature of data streams, the `streaming_bulk` API requires the specification of `create` as the operation type for bulk ingestion, as data streams support only the `create` action. We accomplished this by defining the operation type in our script as follows:

```
doc = {
"_op_type": "create",
...
}
```

The detailed documentation can be found at the following link: `https://www.elastic.co/guide/en/elasticsearch/reference/current/docs-bulk.html#docs-bulk-api-prereqs`.

There's more...

In addition to the explanation provided for our recipe, there are several essential elements of data streams as discussed in the following sub-sections.

Naming scheme

When Elastic introduced data streams, a naming scheme was also presented. Consistent naming of your data streams is particularly useful when dealing with multiple streams, exploring data in the Elastic Observability app, or correlating different datasets.

Here is the recommended naming convention: `{data_stream.type}-{data_stream.dataset}-{data_stream.namespace}`.

This is why we named our data stream `generic-rennes_traffic-default`.

To find the official definition of the naming schema, please refer to the official documentation: `https://www.elastic.co/guide/en/ecs/current/ecs-data_stream.html`.

To fully benefit from Elastic's naming convention, documents need to contain three key fields – `data_stream.type`, `data_stream.dataset`, and `data_stream.namespace` – as defined in the ECS. These are set as constant keyword fields, optimizing query efficiency by targeting fewer shards.

Time series data stream

We just created our first data stream. In the next recipe, we will see how to proceed if we need to create a **time series data stream (TSDS)**.

See also

To see more on the benefits of the Elastic data stream naming scheme, please refer to this blog post: `https://www.elastic.co/blog/an-introduction-to-the-elastic-data-stream-naming-scheme`.

In our example, we indexed our documents with fields that contain dots/periods in their names, such as `traveltime.reliability` and `data_stream.type`. Elastic automatically expands them to their corresponding subobject structure. This is a new feature starting from Elasticsearch 8.3. You can refer to the documentation for more details: `https://www.elastic.co/guide/en/elasticsearch/reference/current/subobjects.html`.

Setting up a time series data stream manually

TSDSs are built on top of Elasticsearch data streams and are specifically designed to optimize the storage and retrieval of time series data. Starting from Elasticsearch 8.7, they are generally available as an officially supported feature. In this recipe, we will explore how to manually create a TSDS for the real-time traffic data of Rennes.

Getting ready

You will need to finish the previous recipe, *Setting up a data stream manually*, as we are going to reuse some of the objects that we created in the previous recipe.

The full code snippets for this recipe can be found at `https://github.com/PacktPublishing/Elastic-Stack-8.x-Cookbook/blob/main/Chapter4/snippets.md#setting-up-time-series-data-stream-tsds-manually`.

How to do it...

Continuing from our previous efforts, we'll now focus on setting up time series data streams for Rennes traffic analysis. We'll begin by creating component templates and index templates, which will help structure our data and improve query efficiency. Next, we will practice what we've learned by ingesting real traffic data using a Python script and then verifying our results in Kibana. Let's get started:

1. We will reuse the ILM policy that we created in the previous recipe. We can start by directly creating the `metrics-rennes_traffic-mappings` component template for metrics mapping with the following command (the snippet is available at `https://github.com/PacktPublishing/Elastic-Stack-8.x-Cookbook/blob/main/Chapter4/snippets.md#creates-a-component-template-for-tsds-mappings`):

   ```
   PUT _component_template/metrics-rennes_traffic-mappings
   {
     "template": {
       "mappings": {
   ```

```
      "properties": {
        "@timestamp": {
          "type": "date",
          "format": "date_optional_time||epoch_millis"
        },
        "traffic_status": {
          "type": "keyword"
        },
        "location_reference": {
          "type": "keyword",
          "time_series_dimension": true
        }
        ...
        "vehicles":{
          "type":"long",
          "time_series_metric": "gauge"
        },
        "average_vehicle_speed":{
          "type":"float",
          "time_series_metric": "gauge"
        },
        "location": {"type": "geo_point"},
        "data_stream": {
          ...
        }
      }
    }
  },
  "_meta": {
    "description": "Mappings Rennes traffic metrics",
    "data_stream": {
      "dataset": "rennes_traffic",
      "namespace": "default",
      "type": "metrics"
    }
  }
}
```

2. We can now create our `metrics-rennes_traffic-index-template` index template, which is composed of two component templates. We just created one of them, and we created the other one, `rennes_traffic-settings`, in the previous recipe (the snippet is available

at https://github.com/PacktPublishing/Elastic-Stack-8.x-Cookbook/
blob/main/Chapter4/snippets.md#create-index-template-for-tsds):

```
PUT _index_template/metrics-rennes_traffic-index-template
{
  "index_patterns": ["metrics-rennes_traffic-*"],
  "data_stream": {
  },
  "template": {
    "settings": {
      "index.mode": "time_series"
    }
  },
  "composed_of": [ "metrics-rennes_traffic-mappings", "rennes_
traffic-settings" ],
  "priority": 500,
  "_meta": {
    "description": "Template for Rennes traffic metrics data"
  }
}
```

3. We can now create a test document for the metrics-rennes_traffic-default data
    stream. Before doing that, open your terminal and run the following command to retrieve
    the current time in the correct format. We need to set the current time as the @timestamp
    value of the document we are ingesting because, by default, TSDSs only accept new documents
    that fall within the interval from the current time minus 2 hours to the current time plus 30
    minutes (for more details, please refer to the Elasticsearch documentation on TSDS index
    settings at https://www.elastic.co/guide/en/elasticsearch/reference/
    current/tsds-index-settings.html#tsds-index-settings):

    ```
    date -u +"%Y-%m-%dT%H:%M:%SZ"
    ```

4. Once you get the current time in the terminal, copy and use it to replace the @timestamp value
    in the following snippet. Execute the command in Kibana Dev Tools (the snippet is available
    at https://github.com/PacktPublishing/Elastic-Stack-8.x-Cookbook/
    blob/main/Chapter4/snippets.md#ingest-sample-document-into-tsds):

    ```
    POST /metrics-rennes_traffic-default/_doc
    {
      "@timestamp": "<your_current_time_here>",
      "traffic_status": "heavy",
      "location_reference": "10273_D",
      "traveltime.reliability": "60",
      "traveltime.duration": "16",
      ...
    ```

```
    "maxspeed": "70",
    "average_vehicle_speed": "46",
    "location": {
      "lat": 48.04479275590756,
      "lon": -1.6502152435538264
    }
    ...
  }
```

5. Now, let's create the real Rennes traffic data with the Python script. Check out the script at
    `https://github.com/PacktPublishing/Elastic-Stack-8.x-Cookbook/blob/main/Chapter4/python-client-sample/tsds.py`. Adjust the connection
    parameters in the `.env` file and execute the script with the following command and verify that
    the documents are successfully indexed:

    ```
    $ python tsds.py
    ```

6. We can verify the `metrics-rennes_traffic-default` data stream in Kibana Dev
    Tools with the following command:

    ```
    GET metrics-rennes_traffic-default/_search
    ```

    We can then make sure that it returns the response with the Rennes traffic data we just ingested,
    as shown in *Figure 4.29*:

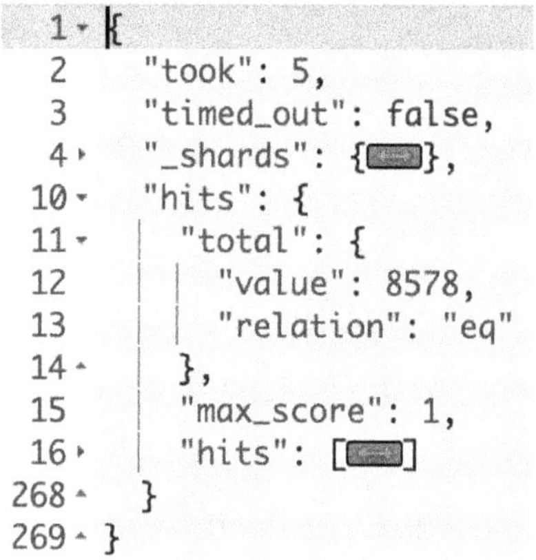

Figure 4.29 – Verifying TSDS

7. We can run a `terms` aggregation on the set of time series dimensions (`_tsid`) to view a date histogram on a fixed interval of one day to verify the dimensions field and the `average_vehicle_speed` metric field (the snippet is available at `https://github.com/PacktPublishing/Elastic-Stack-8.x-Cookbook/blob/main/Chapter4/snippets.md#test-tsds`):

```
GET metrics-rennes_traffic-default/_search
{
  "size": 0,
  "aggs": {
    "tsid": {
      "terms": {
        "field": "_tsid"
      },
      "aggs": {
        "over_time": {
          "date_histogram": {
            "field": "@timestamp",
            "fixed_interval": "1d"
          },
          "aggs": {
            "min": {
              "min": {
                "field": "average_vehicle_speed"
              }
            },
            "max": {
              "max": {
                "field": "average_vehicle_speed"
              }
            },
            "avg": {
              "avg": {
                "field": "average_vehicle_speed"
              }
            }
          }
        }
      }
    }
  }
}
```

```
       200 - OK    69 ms
 1 ▾ [
 2     "took": 7,
 3     "timed_out": false,
 4 ▾   "_shards": {
 5     |   "total": 2,
 6     |   "successful": 2,
 7     |   "skipped": 0,
 8     |   "failed": 0
 9 ▴   },
10 ▸   "hits": {▨},
18 ▾   "aggregations": {
19 ▾   |   "tsid": {
20         |   "doc_count_error_upper_bound": 1,
21         |   "sum_other_doc_count": 2849,
22 ▾       |   "buckets": [
23 ▾       |   |   {
24 ▾       |   |   |   "key": {
25         |   |   |   |   "denomination": "Route départementale 34",
26         |   |   |   |   "hierarchie": "Réseau d'armature",
27         |   |   |   |   "hierarchie_dv": "Réseau de transit",
28         |   |   |   |   "insee": "35206",
29         |   |   |   |   "location_reference": "10273_D"
30 ▴       |   |   |   },
31         |   |   |   "doc_count": 2,
32 ▾       |   |   |   "over_time": {
33 ▾       |   |   |   |   "buckets": [
34 ▾       |   |   |   |   |   {
35             |   |   |   |   |   "key_as_string": "2024-01-17T00:00:00.000Z",
36             |   |   |   |   |   "key": 1705449600000,
37             |   |   |   |   |   "doc_count": 2,
38 ▾           |   |   |   |   |   "avg": {
39             |   |   |   |   |   |   "value": 58
40 ▴           |   |   |   |   |   },
41 ▾           |   |   |   |   |   "min": {
42             |   |   |   |   |   |   "value": 46
43 ▴           |   |   |   |   |   },
44 ▾           |   |   |   |   |   "max": {
45             |   |   |   |   |   |   "value": 70
46 ▴           |   |   |   |   |   }
47 ▴           |   |   |   |   }
48 ▴           |   |   |   |   ]
49 ▴           |   |   |   }
50 ▴           |   |   },
```

Figure 4.30 – Verifying TSDS aggregation

How it works...

Figure 4.31 provides an overview of a TSDS, and the link between the index template, TSDS indices, and documents:

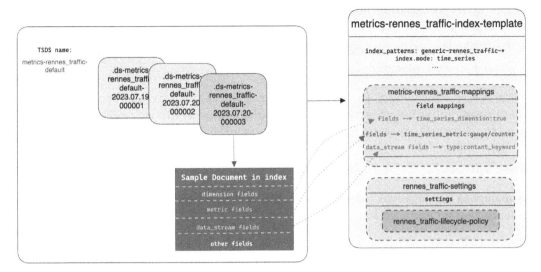

Figure 4.31 – TSDS index template logic

From a document standpoint, a TSDS metric document has different types of fields such as dimension fields and metric fields. Additionally, the same as with a regular data stream, it's recommended to have `data_stream` fields for the TSDS metric document. We defined specific mappings for dimension fields and metric fields in our `metrics-rennes_traffic-mappings` component template, and these field mappings have specific attributes such as `time_series_metric` and `time_series_dimension`.

At least one dimension field is required because dimension fields define the dimension of the entity that you are measuring, which is particularly important for aggregations and downsampling. In our example, we chose to use `location_reference`, `denomination`, `hierarchie`, `hierarchie_dv`, and `insee` as dimension fields because they represent the aspects of the entity *a road or a route in the city of Rennes* that we are measuring.

Metric fields are fields that represent numeric measurements; a TSDS would logically contain at least one metric field. A metric field mapping should have `time_series_metric` set to `counter` or `gauge`:

- `counter` is a cumulative metric that represents a single, continuously increasing value, which can only go up over time
- `gauge` is a metric that represents a single numerical value capable of fluctuating both upward and downward without any constraints

In our recipe, we chose `vehicles`, `traveltime.reliability`, `traveltime.duration`, and `average_vehicle_speed` as time series metrics, as these metrics are capable of fluctuating upward and downward. We set the metric type to `gauge` for all those fields.

There's more...

Downsampling is an important concept of TSDS. In *Chapter 5*, we will see a concrete example in the *Downsampling your time series data* recipe.

To ingest the Rennes traffic data in real-time, we will need to find a more consistent way to ingest continuously rather than using a single Python script. We will see how to do it in *Chapter 5*.

See also

There are a couple of blog post series digging into how to use Elasticsearch and time series data streams for observability metrics. You can find the links here:

- `https://www.elastic.co/blog/elasticsearch-time-series-data-streams-observability-metrics`
- `https://www.elastic.co/blog/nginx-metrics-elastic-time-series-data-streams`

Detailed documentation for TSDS can be found at the following link: `https://www.elastic.co/guide/en/elasticsearch/reference/master/tsds.html`.

5

Transform Data

In this chapter, we'll cover the process of **data transformation** within the Elastic Stack. Data transformation is a pivotal step that transforms the vast influx of information in your systems into actionable insights. It converts raw data into meaningful metrics, and the Elastic Stack provides an extensive array of tools for this purpose.

The transformation of data within the Elastic Stack involves a variety of techniques. These include structuring your data with ingest pipelines, enriching it with contextual details ranging from geolocation to user profiles, reorganizing data using transforms, and downsampling, which is particularly essential for efficiently managing large volumes of time series data.

We will explore the numerous capabilities that the Elastic Stack provides for handling the types of use cases mentioned here. As you will see, the functionality extends across several components including **Elastic Agents**, **Beats**, **Logstash**, and **Elasticsearch** itself.

We're going to cover the following main recipes in this chapter:

- Creating an ingest pipeline
- Enriching data with a custom ingest pipeline for an existing Elastic Agent integration
- Using a processor to enrich your data before ingesting with Elastic Agent
- Installing self-managed Logstash
- Creating a Logstash pipeline
- Setting up pivot data transform
- Setting up the latest data transform
- Downsampling your time series data

Technical requirements

To follow the different recipes in this chapter, you will need an Elastic Stack deployment that includes the following:

- Elasticsearch for searching and storing the data
- Kibana for data visualization and Stack Management (a Kibana user with full privileges for Fleet and integrations is required)
- Integrations Server (included by default in Elastic Cloud deployment)

To quickly spin up a deployment on Elastic Cloud, you can follow the *Deploying Elastic Stack on Elastic Cloud* recipe in *Chapter 1*.

The snippets of this chapter are available at `https://github.com/PacktPublishing/Elastic-Stack-8.x-Cookbook/blob/main/Chapter5/snippets.md`.

Creating an ingest pipeline

In this recipe, we'll dive into one of the lesser-known yet incredibly powerful features of Elasticsearch: **ingest pipelines**. These pipelines enable seamless transformation, enrichment, and manipulation of data as it enters Elasticsearch. Whether you're managing log files, sensor data, or any other structured or unstructured data, ingest pipelines can significantly ease the workload for data engineers and analysts.

What sets Elasticsearch ingest pipelines apart is their simplicity combined with effectiveness. You'll discover how easy it is to create a pipeline tailored to your specific needs. Additionally, we'll demonstrate how to simulate the pipeline to test and validate your data processing logic against sample documents, ensuring data is handled exactly as intended.

Getting ready

You'll need an up-and-running Elastic deployment with Elasticsearch, Kibana, and Fleet Server. To spin up a cluster on Elastic Cloud, refer to the *Deploying Elastic Stack on Elastic Cloud* recipe in *Chapter 1*.

Also, you'll need some sample data; we're going to leverage the Apache integration we've implemented in the *Monitoring Apache HTTP logs and metrics using the Apache integration* recipe in *Chapter 4*.

The snippets of this recipe are available at `https://github.com/PacktPublishing/Elastic-Stack-8.x-Cookbook/blob/main/Chapter5/snippets.md#creating-an-ingest-pipeline`.

How to do it...

We're going to create an ingest pipeline to generate a fingerprint from the values in the host field produced by the Apache integration:

1. In Kibana, navigate to **Stack Management | Ingest Pipelines**, and click on **Create pipeline**. Then, select **New pipeline**, as shown in *Figure 5.1*:

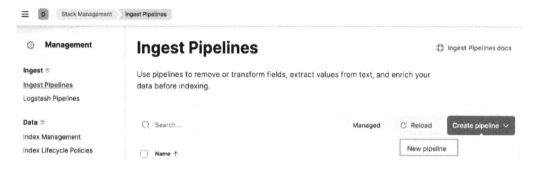

Figure 5.1 – The Ingest Pipelines menu in Stack Management

Name the pipeline, for example, `apache-logs-custom`, and provide a brief description, as shown in *Figure 5.2*:

Figure 5.2 – Create an ingest pipeline configuration page

2. On the **Processors** canvas, click **Add a processor**. This action will open the flyout for processor configuration. From the list of processors, select **Set** and input the following values: for the **Field** name, type `description`, and for **Value**, input `computed hash for Apache access documents`. After entering these details, click on **Add processor** at the bottom of the screen. The following figure displays the configuration you should expect:

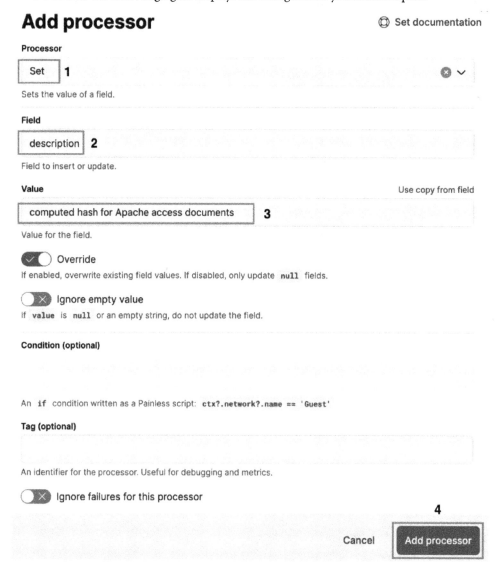

Figure 5.3 – Adding the Set processor

This completes the setup of our first processor.

3. Next, let's incorporate the fingerprint processor to enhance our data's integrity and uniqueness. Click on **Add a processor** once more, and from the available list, choose **Fingerprint**. Configure it with the following parameters:

    • **Fields**: `host`

    • **Method**: Select `SHA-512` from the drop-down list

    After setting these values, your configuration should look like *Figure 5.4*:

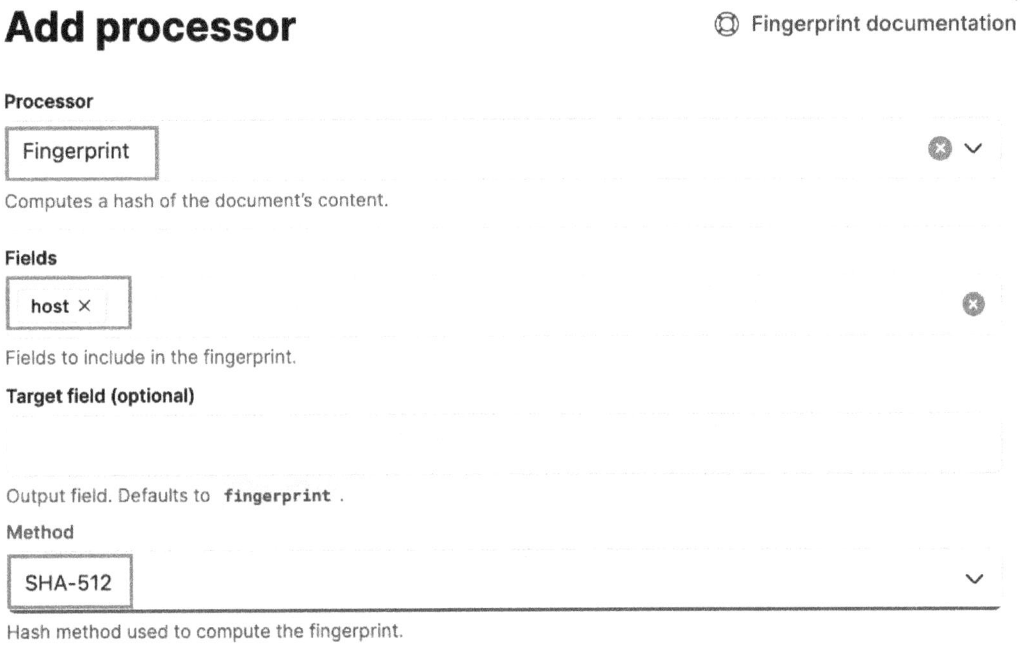

Add processor

⊙ Fingerprint documentation

Processor

    Fingerprint   ⊗ ∨

Computes a hash of the document's content.

Fields

    host ✕   ⊗

Fields to include in the fingerprint.

Target field (optional)

Output field. Defaults to `fingerprint` .

Method

    SHA-512   ∨

Hash method used to compute the fingerprint.

Figure 5.4 – Configuring the Fingerprint processor

4. Click on **Add processor**. This processor will compute a SHA-512 hash for the host field of each document, creating a unique identifier or fingerprint based on the host's value. It's particularly useful for data deduplication, integrity checks, and sensitive data anonymization.

    Your final pipeline configuration should look like the following figure; click **Create pipeline**:

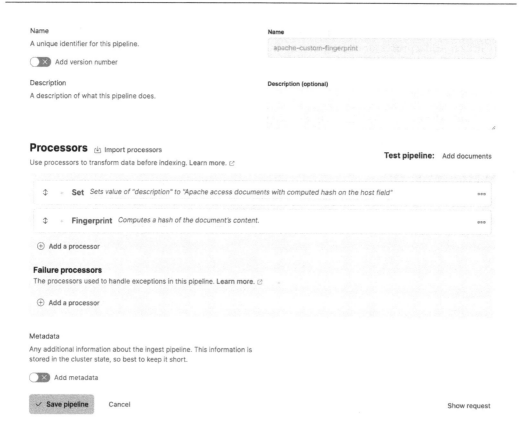

Figure 5.5 – Ingest pipeline full configuration

We've successfully created our ingest pipeline with two processors: **Set** and **Fingerprint**. Now, it's time to test and validate its functionality to ensure it works as expected.

The **Ingest Pipelines** interface in Kibana provides a convenient way to test our pipeline directly. To do this, we'll need a sample document, which can be easily obtained from the **Discover** section of Kibana. Once we have a document to work with, we can input it into the pipeline test interface to see how our pipeline processes the data. This step is crucial for verifying that our **Set** and **Fingerprint** processors are correctly adding a description and computing a hash for the host field, respectively, ensuring our pipeline is ready for real data processing:

5.   In a new tab, navigate to **Analytics | Discover** within Kibana:

  I.   Select the `logs-*` data view.

  II.   Add a filter for `event.dataset: apache.access` to focus specifically on Apache access logs.

6.   From the list of documents that appear, expand the first document in the table to review its details.

7. On the document's flyout page, switch to the **JSON** tab. Here, you'll find various fields related to the document. Carefully copy the values of the `_index` and `_id` fields. These values will be essential for testing our ingest pipeline, as they allow you to reference a specific document within your dataset. *Figure 5.6* shows you where to find those values:

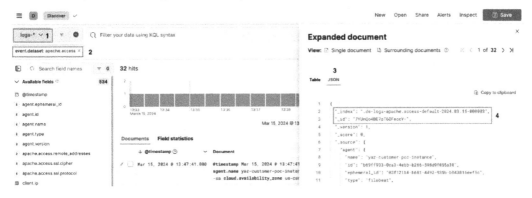

Figure 5.6 – Finding _index and _id from document in Discover

8. Return to the **Ingest Pipelines** section by navigating to **Ingest | Ingest Pipelines** and open the pipeline we previously created, named `apache-logs-custom`. Then, proceed to test the pipeline:

I.   Click on **Add documents**, next to **Test pipeline**, as shown in *Figure 5.7*:

Figure 5.7 – Add documents to test ingest pipelines

II. In the **Index** field, paste the value you previously copied from the `_index` field.

III. In the **Document ID** field, paste the value from the `_id` field, and then click on **Add document**.

Once the document is added, it will be displayed in the **Documents** canvas. Click on **Run the pipeline** to test it:

Figure 5.8 – Testing the apache-logs-custom pipeline with a sample document

In the output data, you should observe the two fields we've added: `description` and `fingerprint`, as displayed in *Figure 5.9*:

Test pipeline

```
{
  "docs": [
    {
      "doc": {
        "_index": ".ds-logs-apache.access-default-
2024.03.15-000003",
        "_version": "-3",
        "_id": "7YUnQo4BE7pT6OFeccV-",
        "_source": {
          "agent": {
            "name": "yaz-customer-poc-instance",
            "id": "b69ff933-0ca3-4ebb-b266-598d9f055a38",
            "ephemeral_id": "03f121b4-b681-4492-935b-
b843016eef5c",
            "type": "filebeat",
            "version": "8.12.2"
          },
          "log": {
            "offset": 73868,
            "file": {
              "path": "/var/log/apache2/access.log"
            }
          },
          "elastic_agent": {
            "id": "b69ff933-0ca3-4ebb-b266-598d9f055a38",
            "version": "8.12.2",
            "snapshot": false
          },
          "description": "computed hash for Apache access
documents",
          ],
          "architecture": "x86_64"
        },
        "fingerprint":
"gvwwsw1Q1IZg+cc96R6Dwng+Rq0wUFE3z/4xu2TL/4y2xBrDfyGTDRETHilTbd0I
```

Figure 5.9 – Visualizing output after testing the apache-logs-custom pipeline

This validation confirms that our pipeline functions as expected and is now prepared for application on incoming data. In the next recipe, we will explore how to integrate this pipeline with an existing Elastic Agent integration, enabling you to apply these transformations to your data streams in real time.

How it works...

Ingest pipelines in Elasticsearch are powerful tools designed to perform common transformations on data before it gets indexed. Essentially, a pipeline is a sequence of configurable tasks, known as processors, each one designed to carry out a specific modification or transformation on incoming documents. As data arrives, it passes through each processor in the order specified, undergoing alterations such as field addition or removal, data formatting, and content enrichment. After the series of processors has applied their transformations, Elasticsearch then adds these processed documents to the designated data stream or index.

The following figure depicts the operating model of ingest pipelines and processors:

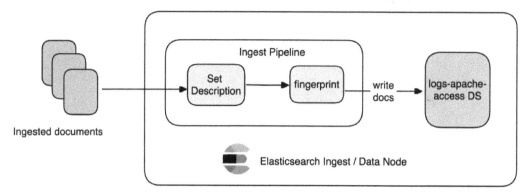

Figure 5.10 – Ingest pipeline flow

Ingest pipelines in Elasticsearch can be created and managed using two primary methods: leveraging the ingest pipelines feature within Kibana, or directly utilizing the ingest APIs. For those requiring programmatic access, the ingest APIs provide flexibility to manage pipelines via code, which facilitates automation and integration with external systems.

Elasticsearch stores these pipelines in the cluster state, ensuring that they are available across the cluster and can be applied to data processed by any node. This centralized storage means that once a pipeline is defined, it can be used by any ingestion process within the cluster, maintaining consistency in data processing, regardless of the entry point.

The following is the code for our pipeline (the snippet is available at https://github.com/PacktPublishing/Elastic-Stack-8.x-Cookbook/blob/main/Chapter5/snippets.md#dev-tools-snippet-to-create-a-custom-apache-log-ingest-pipeline):

```
PUT _ingest/pipeline/apache-logs-custom
{
  "processors": [
    {
      "set": {
        "field": "description",
        "value": "Apache access documents with computed hash on the
host field"
      }
    },
    {
      "fingerprint": {
        "fields": [
          "host"
        ],
```

```
            "method": "SHA-512"
        }
      }
  ]
}
```

Nodes assigned the `ingest` role are responsible for pipeline processing. By default, ingest pipelines are executed on "data hot" nodes. However, for intense ingest workloads, it may be beneficial to designate dedicated ingest nodes.

There's more...

The **enrich processor** in Elasticsearch is a valuable asset for dynamically augmenting documents with pertinent data from another index during the ingest process. This functionality facilitates a range of practical applications, such as the following:

- Identifying web services or vendors from IP addresses

- Adding detailed product information to retail orders using product IDs

- Enriching records with contact details using email addresses

- Appending postal codes to datasets based on user coordinates

The processor employs an enrich index, which is a specially optimized version of source indices designed to expedite the lookup process. These enrich indices are treated as system indices, meaning they are internally managed by Elasticsearch and exclusively used by enrich processors.

Correctly handling exceptions in Elasticsearch ingest pipelines is critical for preserving data integrity and ensuring continuous processing. The `on_failure` parameter within a processor's configuration offers a robust mechanism for error management. Specifying one or more processors to execute upon failure allows for appropriate error logging or notifications, thus maintaining operational oversight and enabling quick resolution.

Finally, most Elastic Agent integrations include dedicated ingest pipelines responsible for parsing and normalizing data before it is indexed into Elasticsearch.

See also

- For a comprehensive list of available processors in ingest pipelines, refer to the official Elasticsearch documentation: `https://www.elastic.co/guide/en/elasticsearch/reference/current/ingest.html`

- For a practical example of how to handle failures in a pipeline, consult the following link: `https://www.elastic.co/guide/en/elasticsearch/reference/current/ingest.html#handling-pipeline-failures`

Enriching data with a custom ingest pipeline for an existing Elastic Agent integration

As we saw in the previous recipe, ingest pipelines are instrumental for data transformation and enrichment. They can be used independently, or you can define a pipeline in the settings of an index so that each incoming document to this index passes through the specified pipeline. In this recipe, we'll examine how to leverage a custom pipeline to enrich documents generated by an integration.

Getting ready

You'll need an up-and-running Elastic deployment with Elasticsearch, Kibana, and Fleet Server.

Make sure to have implemented the previous recipe: *Creating an ingest pipeline*.

How to do it...

As mentioned, we'll use the ingest pipeline created in the previous recipe:

1. Head to **Kibana | Fleet | Agent Policies**.
2. Select the policy running the Apache HTTP Server integration.
3. Click on **Edit Integration** on the right side of the integration.
4. Under the Apache access logs, click on the drop-down arrow to reveal the **Advanced options**. You'll need this to access the pipeline configuration:

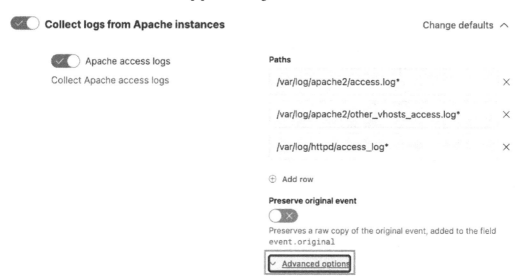

Figure 5.11 – Advanced section in Apache integration configuration

5. Scroll down to the ingest pipelines and click on **Add custom pipeline**:

Figure 5.12 – Adding a custom ingest pipeline to Apache integration

> **Note**
>
> Each integration in Elasticsearch, such as `logs-apache.access@custom`, comes with a predefined custom pipeline for customization purposes. It is considered a best practice to use these custom pipelines as they allow for tailor-made data processing without modifying the default settings. Importantly, it ensures that your custom settings are preserved during integration updates, as standard pipelines may change, but custom pipelines will not.

6. On the pipeline creation menu, click on **Add a processor**.

7. In the processor configuration menu, select **Pipeline processor**.

8. Enter the name of the pipeline that we created in the previous recipe: `apache-logs-custom`.

9. Save the pipeline.

Now, head back to **Discover**. If everything is set up correctly, you should notice that all incoming documents now have both the `description` and `fingerprint` fields populated.

How it works...

What we've accomplished in this recipe essentially boils down to two things:

*   First, we've incorporated a custom data processing layer into an existing Elastic integration by leveraging the `@custom` pipeline, which comes prebuilt with all Elastic integrations. This pipeline is already included in the package but is initially empty. It is advisable to modify only the custom ingest pipeline to prevent disrupting integration policies or losing your modifications in the event.

*   Within the custom pipeline, we've utilized the pipeline processor. The pipeline processor in Elasticsearch executes another pipeline, allowing for nesting pipelines. The name of the pipeline you wish to execute is specified in the name parameter. In our case, we've entered the name of the ingest pipeline that we created in the first recipe of the chapter.

There's more...

Custom integrations are defined at the dataset level. In this case, the integration applies only to the Apache access logs data stream and its associated backing indices. This means that if you need to process error logs, a custom pipeline would have to be added within the error logs configuration. This methodology provides significant flexibility, enabling the development of distinct parsing strategies tailored to your integration.

Using a processor to enrich your data before ingesting with Elastic Agent

In previous recipes, we learned how to use ingest pipelines to transform data during the Elasticsearch ingestion process. In this recipe, we'll explore a different approach, which involves parsing data at the edge – before the documents are sent to Elasticsearch – using Elastic Agent processors.

In this recipe, we'll see how we can enrich data right at the source with custom host metadata.

Getting ready

You will need an up-and-running Elastic deployment with Elasticsearch, Kibana, and Fleet Server.

Additionally, we will leverage what we have implemented in the *Creating an ingest pipeline* recipe in this chapter.

The snippets for the recipe are available at `https://github.com/PacktPublishing/Elastic-Stack-8.x-Cookbook/blob/main/Chapter5/snippets.md#using-processor-to-enrich-your-data-before-ingesting-with-elastic-agent`.

How to do it...

1. In Kibana, head to **Management | Fleet | Agent Policies**.
2. Locate the policy running the Apache integration and click **View Policy**.
3. Click on **Edit integration** to the right of the Apache integration. This action will open the integration settings page.
4. Under the **Collect logs from Apache instances**, click on the **Apache access logs** toggle, collapse the **Advanced options** option, and you will see the **Processors** field. Add the following code in the **Processors** field as shown in *Figure 5.13* (the code snippet is available at this address: `https://github.com/PacktPublishing/Elastic-Stack-8.x-Cookbook/blob/main/Chapter5/snippets.md#custom-processor-for-apache-access-logs`); remember to adapt these values to your environment and host metadata:

```
- add_host_metadata:
  geo:
    name: iowa-dc
```

```
location: 41.8780, 93.0977
continent_name: North America
country_iso_code: US
region_name: Iowa
region_iso_code: IA
city_name: Council Bluffs
```

Collect logs from Apache instances   Change defaults ∧

       Apache access logs

       Collect Apache access logs

Paths

/var/log/apache2/access.log*   ✕

/var/log/apache2/other_vhosts_access.log*   ✕

/var/log/httpd/access_log*   ✕

⊕ Add row

Preserve original event

  ✕

Preserves a raw copy of the original event, added to the field
`event.original`

∨ Advanced options

Tags

apache-access

⊕ Add row

Processors   Optional

```
- add_host_metadata:
    geo:
      name: iowa-dc
      location: 41.8780, 93.0977
      continent_name: North America
      country_iso_code: US
      region_name: Iowa
      region_iso_code: IA
      city_name: Council Bluffs
```

Figure 5.13 – Editing the Processors field

5.   Click **Save integration** at the bottom right, and wait until the changes are applied to the running
Elastic Agents.

6. Now, let's return to **Discover** and look at the documents pushed by the Agent after adding the processor. Select the `logs-*` data view and add the following filter: `event.dataset: apache.access`.

7. Expand the first document in the table, and you will see the additional fields added by the Agent processors.

> **Important note**
>
> You'll likely notice that the fields added by the Agent processor don't have a proper type, indicated by the **?** mark to the left of the field names. This occurs because no explicit mappings have been defined for those fields. You can refer to the *Defining index mappings* recipe in *Chapter 2* to adjust this.

How it works...

Elastic Agent processors are powerful, modular components designed to manipulate and enhance data at its source. They take in an event, apply a specified action – such as parsing, filtering, transforming, or enriching – and then pass the event forward. When multiple processors are defined in a list, they execute in the specified order. This sequential execution allows for a flexible, step-by-step data transformation and enrichment process that ensures data is optimized for analysis even before it reaches the ingest phase.

> **Note**
>
> Elastic Agent processors operate at an early stage in the data processing workflow, executing before any ingest pipelines or Logstash processing takes place. Therefore, processor configurations set within the Elastic Agent cannot depend on or refer to fields that are created or modified by later stages, such as ingest pipelines or Logstash.

There's more...

Elastic Agent processors are a way to modify or filter data before sending it to Elasticsearch. However, they do have limitations:

- They do not have the ability to enrich data from Elasticsearch or external data sources
- They cannot process data after it's normalized to Elastic Common Schema since this happens in the ingest pipelines
- Caution should be exercised with integration-based ingest pipelines to avoid removing or modifying fields expected by these pipelines, which could cause them to break
- As mentioned in the recipe, you'll need to define mappings in the `*-@custom` component template to ensure they are applied correctly

Despite these limitations, processors are a valuable tool in the data processing toolkit of the Elastic Stack, particularly for filtering out events at the edge. Not all integrations support processors, so always consult the Elastic official documentation for confirmation.

If you require capabilities beyond those provided by available processors, consider the `Script` processor. This processor offers flexibility by allowing you to embed Java directly into your configuration.

See also

To see the full list of available processors, refer to the latest version of the Stack documentation here: `https://www.elastic.co/guide/en/fleet/current/elastic-agent-processor-configuration.html`.

Installing self-managed Logstash

Logstash is an open-source tool that can collect, process, and forward data from various sources in real time. It uses a pipeline architecture where data flows through different plugins that can filter, transform, enrich, or output the data. Logstash can integrate with Elasticsearch, Kibana, Beats, and other products in the Elastic Stack to create powerful data analysis and visualization solutions.

In this recipe, we will see how to set up Logstash on a **virtual machine** (**VM**). As an example, we will use a Debian-based machine. Let's get started!

Getting ready

To install Logstash on a Debian machine, you need to have the following requirements:

- **A Debian-based system with root or sudo privileges**: You can follow this tutorial to set up a VM in Google Cloud: `https://www.cloudskillsboost.google/focuses/56596?parent=catalog`

- Make sure your VM packages are up to date; on a Debian system, you can apply the following command to update and apply patches:

```
$ sudo sh -c 'apt update && apt upgrade'
```

- A stable internet connection

- A text editor of your choice

- The console commands of this recipe can be found at this address: `https://github.com/PacktPublishing/Elastic-Stack-8.x-Cookbook/blob/main/Chapter5/snippets.md#console-commands`

How to do it...

> **Important note**
>
> At the time of writing, the latest version of Logstash is 8.12.2. The code samples provided in this recipe will use this version. You will need to adapt the commands according to the version you wish to install and always check the release notes and official documentation.

To install Logstash on a Debian-based machine, follow these steps:

1. Download and install the public signing key for the Elastic repository:

    ```
    $ wget -qO - https://artifacts.elastic.co/GPG-KEY-elasticsearch
    | sudo gpg --dearmor -o /usr/share/keyrings/elastic-keyring.gpg
    ```

> **Note on GNU Privacy Guard (GPG)**
>
> If you encounter a **gpg: command not found** error message, try to install it with the following command: `sudo apt install gnupg2`.

2. Install the `apt-transport-https` package if it is not already installed:

    ```
    $ sudo apt-get install apt-transport-https
    ```

3. Add the Elastic repository to your sources list:

    ```
    $ echo "deb [signed-by=/usr/share/keyrings/elastic-keyring.gpg]
    https://artifacts.elastic.co/packages/8.x/apt stable main" |
    sudo tee -a /etc/apt/sources.list.d/elastic-8.x.list
    ```

4. Update your package index and install Logstash:

    ```
    $ sudo apt-get update && sudo apt-get install logstash
    ```

5. Verify that Logstash is installed by checking its version:

    ```
    $ /usr/share/logstash/bin/logstash --version
    ```

    You should see the following installed Logstash version:

    ```
    $ logstash 8.12.2
    ```

How it works...

Logstash is installed as a service that runs in the background and can be controlled by the `systemctl` command. You can start, stop, restart, or check the status of Logstash using the following commands:

- `$ sudo systemctl start logstash.service`
- `$ sudo systemctl stop logstash.service`
- `$ sudo systemctl restart logstash.service`
- `$ sudo systemctl status logstash.service`

Logstash reads its configuration from the `/etc/logstash/conf.d` directory, where you can create and edit pipeline files that define how Logstash processes your data. You can also use the `-f` option to specify a custom configuration file or directory when running Logstash manually.

Logstash writes its logs to the `/var/log/logstash` directory, where you can find useful information for troubleshooting and debugging.

While Logstash is not required in your ingest architecture, integrating it offers several benefits:

- **Highly advanced data processing pipelines**: Logstash supports a variety of architectural patterns to combine multiple pipelines
- **Resiliency**: In your ingest pipeline with the persistent queue mechanism that allows Logstash to buffer events on disks
- **Diversity of data sources**: Logstash can ingest from many different sources and can also emit to multiple destinations, making it the perfect tool for data pipelines

There's more...

The method used in this recipe is one of the easiest ways to install Logstash on a Debian machine, but it is not the only one. There are other methods that you can use, depending on your preferences and needs, which include the following:

- Downloading and unpacking the Logstash installation file for your host environment (TAR. GZ, DEB, ZIP, or RPM) and running Logstash manually or as a service
- Setting up and running Logstash using Docker or the **Elastic Cloud on Kubernetes** (**ECK**) operator
- Configuring Logstash using environment variables, command-line flags, or the `logstash. yml` file
- Storing sensitive settings such as passwords or API keys in a secret keystore

For more information about these alternative setup methods, please refer to the Logstash documentation. You can also check out some Logstash configuration examples to see how you can use different plugins and conditionals to process your data.

See also

For a comprehensive list of Logstash inputs plugins, refer to `https://www.elastic.co/guide/en/logstash/current/input-plugins.html`.

Creating a Logstash pipeline

Creating a Logstash pipeline involves defining a series of configurations that specify how data should be collected, parsed, filtered, and where it should be sent for further analysis or storage. This process enables you to harmonize data from various sources, making it ready for visualization, search, and analysis through Elasticsearch and Kibana, as well as third-party destinations.

In this recipe, we will walk you through the steps to create a Logstash pipeline, from configuring input sources to defining filters, specifying output destinations, and running the pipeline. Our example is based on `Rennes Traffic Data`, which we introduced in *Chapter 4*.

Getting ready

You will need to have completed the previous *Installing self-managed Logstash* recipe and the *Setting up time series data stream (TSDS) manually* recipe in *Chapter 4* as we are going to reuse the objects that we created in this recipe such as the index life cycle policy, mapping, setting, and the index template.

The dataset we will use is the one introduced in *Chapter 4*, `Rennes Traffic Data`. This time, however, instead of importing the latest traffic status of Rennes just once, we will create a new data stream to continuously receive real-time traffic data. This means that every 10 minutes, we will import the latest traffic data from the public data API and append it to the data stream.

The snippets of this recipe are available at `https://github.com/PacktPublishing/Elastic-Stack-8.x-Cookbook/blob/main/Chapter5/snippets.md#creating-a-logstash-pipeline`.

How to do it...

You will need to first create the Logstash configuration file in the Logstash configuration folder (`/etc/logstash/conf.d/` for Debian). The reference configuration file can be found at this address: `https://github.com/PacktPublishing/Elastic-Stack-8.x-Cookbook/blob/main/Chapter5/logstash-conf/rennes_traffic-default.conf`. If you are not deploying Logstash on Debian, the details of the Logstash directory layout for different installation packages can be found here: `https://www.elastic.co/guide/en/logstash/current/dir-layout.html`.

Now let's go through the process of setting up a Logstash configuration to fetch and process real-time traffic data from Rennes. We'll outline each section of the configuration file, beginning with the input plugin and concluding with the output configuration. By following these instructions, you'll be able to ingest and later visualize traffic data.

1. Let us start by configuring the input plugin, which specifies where Logstash should fetch data from. Here, we are using the `http_poller` input plugin to fetch the Rennes traffic data every 10 minutes in CSV format:

```
input {
    http_poller {
        urls => {
            rennes_data_url => "https://data.rennesmetropole.
fr/explore/dataset/etat-du-trafic-en-temps-reel/
download?format=csv&timezone=Europe/Paris&use_labels_for_
header=false"
        }
        request_timeout => 60
        schedule => { every => "10m" }
        codec => "line"
    }
}
```

2. Next, let us prepare the filter section to format the data correctly before ingesting it into our data stream. We are using different filters in this order:

A. The `csv` filter is used to parse the data from different columns and remove unwanted fields.

B. Fetch and parse the `datetime` field and mutate it to index the `@timestamp` field.

C. Convert the `Sens unique` column to a Boolean field that we name `oneway`.

D. Rename the fields according to our field mapping for the data stream.

E. Remove the unnecessary fields that have been generated from the `csv` filter:

```
filter{
    csv {
        separator => ";"
        skip_header => "true"
        columns => [
            "datetime","predefinedlocationreference",
            "averagevehiclespeed","traveltime",
            "traveltimereliability","trafficstatus",
            "vehicleprobemeasurement","geo_point_2d",
            "geo_shape","gml_id","id_rva_troncon_fcd_v1_1",
            "hierarchie","hierarchie_dv","denomination",
            "insee","sens_circule","vitesse_maxi"
```

```
        ]
        remove_field => [
            "geo_shape","gml_id",
            "id_rva_troncon_fcd_v1_1"
        ]
    }

date {
    match => ["datetime", "UNIX"]
    target => "@timestamp"
}

if [sens_circule] == "Sens unique" {
    mutate {
        add_field => { "oneway" => "true" }
    }
}
else {
    mutate {
        add_field => { "oneway" => "false" }
    }
}

mutate {
    rename => {"traveltime" => "traveltime.duration"}
    ...
}
mutate {
    remove_field => [
        "datetime","message","path","host","@version",
        "original","event.original","tags","sens_circule"
    ]
}
}
```

3. Now, let's finish preparing the Logstash configuration file. We will focus on the **output configuration**. Here, we will set two outputs, one to Elastic Cloud with specific settings for a metrics data stream and the second for STDOUT to display the output for debugging purposes in the shell running Logstash:

```
output {
    elasticsearch {
        cloud_id => "CLOUD_ID"
        cloud_auth => "user:password"
```

```
                    data_stream => true
                    data_stream_type => "metrics"
                    data_stream_dataset => "rennes_traffic"
                    data_stream_namespace => "default"
            }

        stdout { codec => rubydebug }
    }
```

4. Once we finish the configuration file, we can start Logstash with the following command (for Debian):

```
$ sudo systemctl start logstash.service
```

We can now check the data ingestion by creating a data view in Kibana. Go to Kibana, **Stack Management | Data Views | Create Data View**, and then set the name and index pattern with `metrics-rennes_traffic-default`:

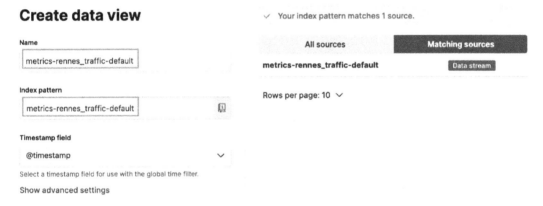

Figure 5.14 – Creating the metrics-rennes_traffic-default data view

5. We can then go to **Kibana | Analytics | Discover** in Kibana and choose `metrics-rennes_traffic-default` to verify that the data has been successfully ingested, as shown in *Figure 5.15*:

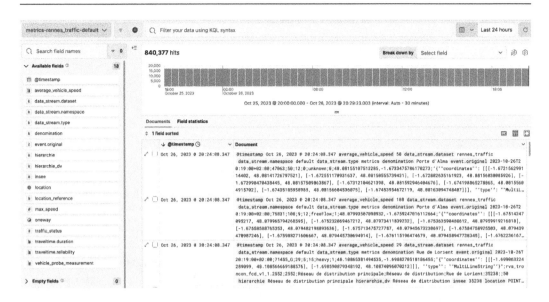

Figure 5.15 – Visualizing the ingested data stream

How it works...

As you can see in *Figure 5.16*, each Logstash instance can host multiple pipelines:

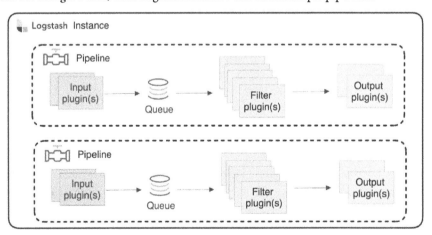

Figure 5.16 – Logstash pipelines

In each pipeline, we define the following:

- **Input plugins**: We define these to interface with different input sources. Each plugin is designed to handle a specific type of input source. For our example, we used the CSV plugin to read Rennes Traffic Data from the public API endpoint.

- **Queue**: This is Logstash's mechanism to buffer events between various stages of the pipeline, such as inputs and pipeline workers. In our example, we did not explicitly define queues as, by default, Logstash uses **in-memory bounded queues**. **Persistent** queues can be employed not only to safeguard against message loss during a standard shutdown but also to handle sudden surges of events without the necessity for external buffering mechanisms. Additionally, there exists a specialized queue known as the **dead-letter queue** (**DLQ**), which serves as a temporary holding area for events that cannot be processed immediately.

- **Filter plugins**: We define these to enable you to process and manipulate the data. Filters can include parsing plugins (as in our case for CSV format parsing), conditional statements (as used for our `oneway` field), and transformation plugins (e.g., `date formatting`).

- **Output plugins**: We define these to send data to various destinations at the same time. For example, there are plugins for Elasticsearch, `STDOUT` to display the output for debugging purposes in our example, and there are also output plugins for Amazon S3, Kafka, and more.

There's more...

As mentioned earlier in this chapter, there are two possible ways to create data transformation pipelines: using ingest pipelines or Logstash pipelines. Both have pros and cons. Here is a comparison of both methods:

Logstash	Ingest Node
☺ Better suited for expressing complex logic with nested conditionals and tagging (for custom logs)	☺ Features a pipeline builder user interface
☺ Offers a full selection of transformation options	☺ Typically delivers faster performance
☺ Capable of handling multiple inputs and outputs	☺ Simplifies testing processes
☹ Adds one extra component to deploy and learn	☺ Makes composing pipelines straightforward
☹ Challenging to offer as a software as a service (SaaS) solution	☺ Allows enrichment using indexed data
☹ Composability can be resource intensive (when using pipeline-to-pipeline communication)	☺ Generally safer to use
	☹ Requires a data shipper
	☹ Difficulty increases when trying to express complex logic
	☹ May impact the performance of the cluster

Table 5.1 – Logstash versus ingest node pros and cons

See also

In this recipe, we covered the basics of setting up and running a single pipeline on a single node Logstash instance:

- To learn more about Logstash production deployment and scaling, please refer to this documentation: `https://www.elastic.co/guide/en/logstash/current/deploying-and-scaling.html`

- Elastic Stack version 8.x introduced a lot of new monitoring capabilities on Logstash; please refer to this documentation: `https://www.elastic.co/guide/en/logstash/current/configuring-logstash.html`

Setting up pivot data transform

Elasticsearch's **data transform** is a capability that allows you to create and execute data transforms on your Elasticsearch indices. It enables you to reshape and aggregate data from one index and store the results in another index, making it easier to analyze and visualize your data in diverse ways. This feature is particularly useful for summarizing and aggregating large datasets, creating denormalized views of your data, and preparing data for analytics or reporting.

For this recipe, we will continue with the dataset for Rennes traffic that we ingested with Logstash from the previous recipe.

Getting ready

To set up pivot data transform, you need to do the following:

- Complete the previous recipe of this chapter, *Creating a Logstash pipeline*.

- Ensure that at least one Elasticsearch node of your cluster is set to the `transform` role. (Note that if you are using Elastic Cloud, this is configured automatically.) To configure transform nodes on your self-managed cluster, you can refer to the *Creating and setting up additional Elasticsearch nodes* recipe in *Chapter 1*.

How to do it...

In this recipe, we will use pivot data transform to demonstrate how to convert existing indices into summarized indices. Our goal is to pivot our Rennes traffic data into an entity-centric index to provide summarized insight into the road/route entities:

1. First, let's go to **Stack Management | Transforms** and click on **Create your first transform**. Then, select the `metrics-rennes_traffic-default` data view that we created in the previous recipe.

New transform / Choose a source

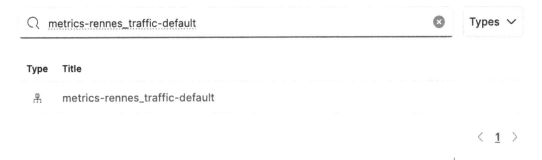

Figure 5.17 – Choosing the source for the data transform

2. On the **Configuration** page, we will keep the default selection as a **Pivot** transform and choose **Use full data** for the time range:

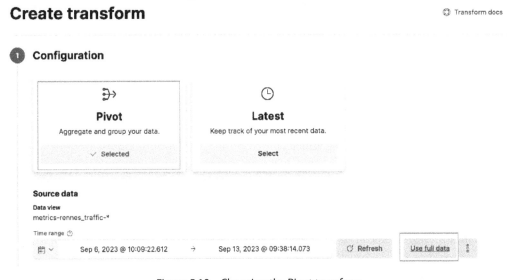

Figure 5.18 – Choosing the Pivot transform

3. On the **Transform configuration** page, define our pivot entity in the **Group by** section with the following fields: denomination, hierarchie, hierarchie_dv, and location_reference. These fields enable us to characterize the road/route entities for traffic status analysis.

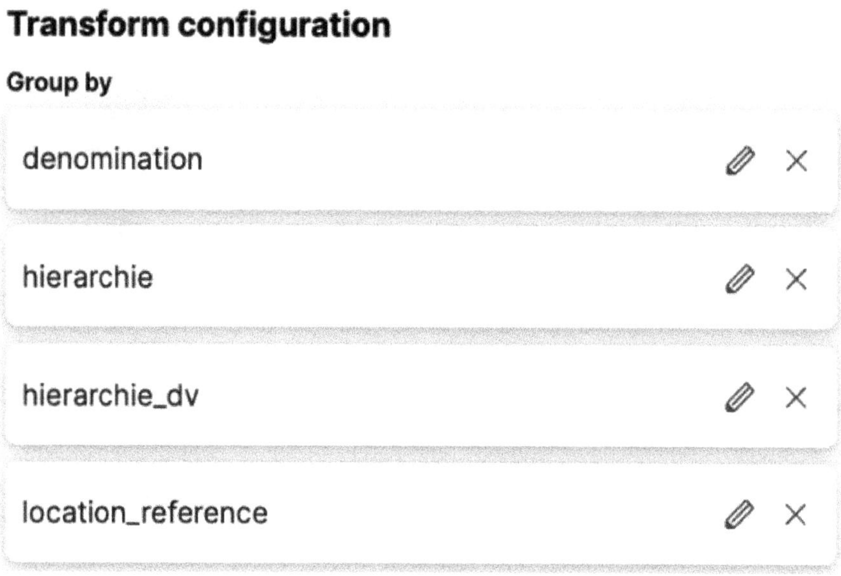

Transform configuration

Group by

denomination	🖉 ✕

hierarchie	🖉 ✕

hierarchie_dv	🖉 ✕

location_reference	🖉 ✕

Figure 5.19 – Defining the pivot transform Group by fields

4. Now, let's calculate the data we wish to analyze for each road/route entity:

    *   The average travel duration

    *   The average travel reliability

    *   The average number of probe vehicles

    We accomplish this by using the `avg` aggregation on fields such as `average_vehicle_speed`, `traveltime.duration`, `traveltime.reliability`, and `vehicle_probe_measurement`. Additionally, we calculate the percentage of the average speed versus the maximum speed to determine whether certain roads are particularly affected by speeding. For this, we will use the `max` aggregation on `max_speed` to prepare for the scripted aggregation in the next step:

Aggregations

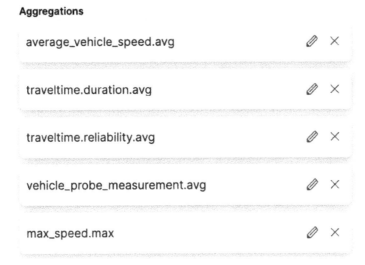

average_vehicle_speed.avg   ✎ ✕

traveltime.duration.avg   ✎ ✕

traveltime.reliability.avg   ✎ ✕

vehicle_probe_measurement.avg   ✎ ✕

max_speed.max   ✎ ✕

Figure 5.20 – Configuring the aggregations for the pivot transform

5. After creating the initial aggregations, we can create one final aggregation, autorized_ speed_percentage, using a bucket script. We can click on the **Edit JSON config** option and add the following script to the aggregation bracket (the code snippet can be found at this address: https://github.com/PacktPublishing/Elastic-Stack-8.x-Cookbook/blob/main/Chapter5/snippets.md#bucket-script-aggregation):

```
"autorized_speed_percentage": {
    "bucket_script": {
        "buckets_path": {
            "avg_speed": "average_vehicle_speed.avg.value",
            "maximum_speed": "max_speed.max.value"
        },
        "script": "(
            params.avg_speed / params.maximum_speed) * 100"
    }
}
```

6. We can now preview the result directly on the configuration screen, as shown in *Figure 5.21*:

Preview

⬡ Columns   ⇅ Sort fields

hierarchie	hierarchie_dv	location_reference	autorized_speed_per...	average_vehicle_spee...	max_speed.max	traveltime.duration.avg	traveltime.reliability.a...	vehicle_probe_mea:
entale 34 Réseau d'armature	Réseau de transit	10273_D	103.13854998423209	72.19558498896247	70	10.061368653421833	38.87064017660044	0.93951434878
entale 34 Réseau d'armature	Réseau de transit	10273_G	103.01916530071871	72.11341571050309	70	10.3693733451015	40.20653133274492	0.98367166813
entale 34 Réseau d'armature	Réseau de transit	10274_D	104.6601941747573	73.2621359223301	70	8.846425419240953	38.51897618946161	0.97371270943
entale 34 Réseau d'armature	Réseau de transit	10274_G	105.69348127600556	73.98543689320388	70	8.841571050308914	39.937334510150045	0.9699911738
entale 34 Réseau d'armature	Réseau de transit	10275_D	104.49501954356325	73.14651368049427	70	14.219770520741395	38.59179170344219	0.93115622241

Rows per page: 5 ⌄   ‹ 1 2 3 4 5 … 20 ›

Figure 5.21 – Transform preview

7. Next, define the destination index, create the data view for the destination index, and set the transform to run in continuous mode, as seen in *Figure 5.22*. Additionally, you may configure the retention policy to manage out-of-date documents in the destination index.

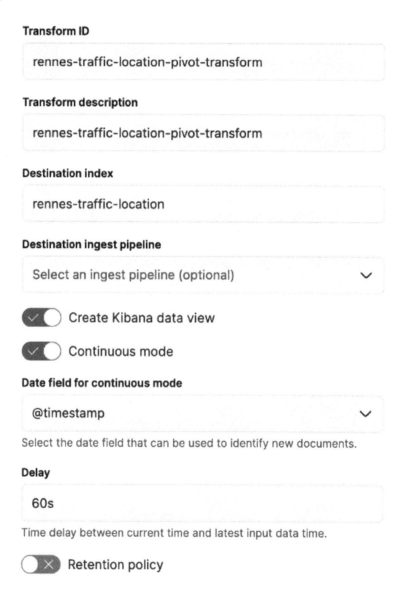

Figure 5.22 – Configuration of transform job details

8. Before creating and starting the transform job, we can copy the transform job creation script to the clipboard via the **Copy to clipboard** button. This is useful if you want to create the transform job in the test environment before deploying it to the production environment. The reference script can be found at this address: `https://github.com/PacktPublishing/Elastic-Stack-8.x-Cookbook/blob/main/Chapter5/snippets.md#dev-tools-snippet-to-create-pivot-data-transform`. We can then click on **Create and Start**, then go to **Analytics | Discover**, and select the `rennes-traffic-location` data view to see the data transform results, as shown in *Figure 5.23*:

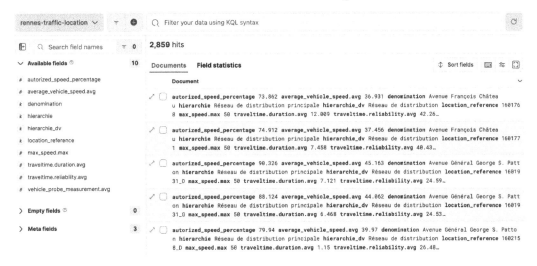

Figure 5.23 – Visualizing the transformed location data

How it works...

In our example, we just created a transform job to convert original data from our source index (`metrics-rennes_traffic-default`) to our new destination index (`rennes-traffic-location`), pivoting on fields that define the location and adding aggregations that will help us to have better analysis on speeding, data quality, and traffic flow.

We started our job in continuous mode; the transform job creates checkpoints by checking changes to the source index. The periodic timer is defined in the `frequency` property of the transform job and, by default, the frequency is 1 minute and the max value is 1 hour. It collects the new raw documents and updates the aggregations to the destination index.

As shown in the following diagram, this allows us to efficiently manage the process of data transformation by performing incremental updates, including only new documents. The checkpoint also ensures the consistency and reliability of transformed data, especially in scenarios where the source data is continuously changing.

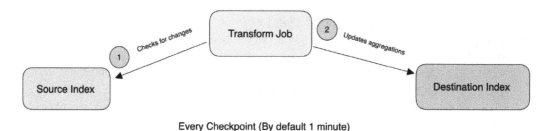

Figure 5.24 – Data transform flow

There's more...

Once you created the data transform job, you can monitor the job status, health, and checkpoint status directly in Kibana by going to the **Stack Management | Transform** tab, as shown in *Figure 5.25*.

Figure 5.25 – Data transform monitoring

Elasticsearch aggregations are powerful and flexible features that allow you to summarize and retrieve complex insights about your data (we will see more examples in *Chapter 6*). A data transform is useful in the following situations:

- You need a complete feature index rather than a top-N set of items
- You need to sort aggregation results by pipeline aggregation
- You want to create summary tables to optimize queries

See also

In the next recipe, we will see another example of a data transform to find the latest document among all the documents that have a certain unique key.

Elastic Stack 8 introduces the capability to create alerting rules to check the health of continuous transforms; please refer to the following documentation: `https://www.elastic.co/guide/en/elasticsearch/reference/current/transform-alerts.html`.

Performance is one of the key considerations while using an Elastic data transform; please refer to this documentation to understand the impact of the transform configuration options on the performance: `https://www.elastic.co/guide/en/elasticsearch/reference/current/transform-scale.html`.

Setting up the latest data transform

While pivot data transforms are valuable for aggregating data and providing summaries for distinct entities, the latest data transform has a different application. It specifically replicates the most current document associated with each entity into the destination index. In this recipe, we will set up the latest data transform for our Rennes traffic data. The result of a data transform will help us to answer this question: For every route location, what was the last time the traffic jam happened?

Getting ready

To set up the latest data transform, you need to do and have the following:

- Complete the previous recipe of this chapter, *Creating a Logstash pipeline*
- At least one Elasticsearch node of your cluster set to the `transform` role

How to do it...

To further analyze the traffic data from Rennes and gain insights into traffic congestion, we can make use of Elastic Stack's latest data transform feature to focus specifically on the latest traffic jams per route location. Here's how we can set up the latest data transform:

1. Let's first go to **Stack Management | Transforms**, click on the **Create a transform** button, and then select the `metrics-rennes_traffic-default` data view. This step is the same as our previous recipe for the pivot data transform.

2. On the configuration page, we will select the **Latest** transform and choose **Use full data** for the time range.

Create transform

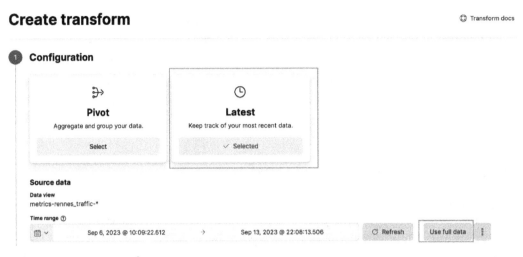

Figure 5.26 – Choosing the Latest transform

3. To prepare the data to understand the traffic jam, we used a search filter to filter out unnecessary data to keep the documents about traffic jams. In the **Search filter**, type the query to filter out all the data with the freeFlow or unknown traffic status, as shown in *Figure 5.27*

Search filter

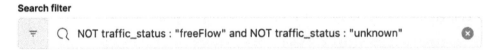

Figure 5.27 – Defining the search filter

For the transform configuration, we just need location_reference for our unique key and we use @timestamp for our **Sort** field to identify the latest document, and then click on the **Next** button:

Figure 5.28 – Configuring the unique keys and sort field

4. We can now define the destination index, create the data view for the destination index, and define the transform in continuous mode, as shown in the following screenshot.

Figure 5.29 – Transform details configuration

Before creating and starting the transform job, we can copy the transform job creation script to the clipboard via the **Copy to clipboard** button. The reference script can be found at this URL: `https://github.com/PacktPublishing/Elastic-Stack-8.x-Cookbook/blob/main/Chapter5/snippets.md#dev-tools-snippet-to-create-latest-data-transform`. We can then click on **Create and Start**, go to **Analytics | Discover**, and select the `rennes-traffic-latest-trafficjam-location` data view to see the data transform results.

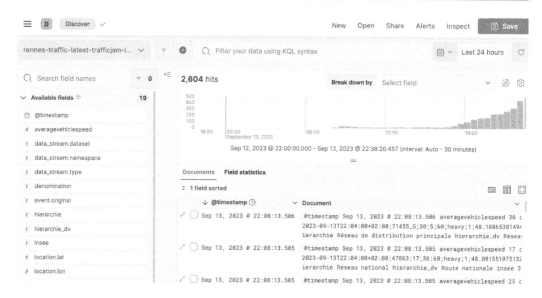

Figure 5.30 – Visualizing the transformed data for the latest traffic jam per location

How it works...

In summary, we've leveraged the option of the **Latest** transform to allow us to track the latest traffic jam for each location in a dataset. How it works is straightforward:

- **Define unique keys**: You decide which field or fields uniquely identify each entity in your dataset. This could be a user ID, a device ID, a combination of fields, or any other identifier that makes each document unique within a grouping. In our recipe, we used the `location_reference` field.

- **Choose a sort field**: Select a timestamp or any other field that indicates recency, which the transform can use to sort documents within each group. In our case, we logically choose the `@timestamp` field.

- **Configure and start the transform**: Once you've defined your unique keys and `sort` field, you start the transform. Elasticsearch will process your source index as follows:

  - It groups documents based on the unique keys you've specified

  - Within each group, it identifies the most recent document using the sort field

  - It indexes this latest document into the destination index

The result is a new index where, for each unique key, only the most recent document according to the sort field is present. The latest transform is particularly powerful for maintaining current snapshots of time-sensitive data, such as user profiles, sensor readings, or the most recent transaction for each customer.

There's more...

While setting up the latest data transform addresses our need to understand the most recent traffic jams, it also opens the door to a variety of other applications as shown in *Table 5.2*:

Use case	Example
Real-time monitoring	By maintaining an index of the latest records, we can create dashboards for real-time monitoring of various systems or processes. For instance, tracking the most current status of Internet of Things (IoT) devices.
Profile enrichment	Use the latest transform to continuously update user or entity profiles with the most recent interactions or transactions. This ensures that downstream processes, such as recommendation systems or customer service, have access to the current state of user data.
Data cleanup	The latest transform can help keep your indexes clean and compact by only retaining the most recent document per unique key.
Event resolution	In cases where multiple events are reported for an entity in quick succession, the latest transform helps in determining the most up-to-date event, simplifying the resolution of event chains.

Table 5.2 – Latest transform use cases

See also

A data transform is also a very common process for data science, and it can be helpful for behavior analytics. We will see some more examples in *Chapter 8*.

Downsampling your time series data

Downsampling, introduced in Elastic 8.7, reduces the granularity of your data by utilizing less detailed time series data. This feature offers flexibility in managing historical metrics within your budget and improves visualization performance in Lens, Timelion, or Time Series Visual Builder (TSVB) when dealing with large datasets.

In this recipe, we will configure downsampling for our Rennes traffic TSDS.

Getting ready

To perform downsampling on your time series data, you need to do the following:

- Complete the previous recipe of this chapter, *Creating a Logstash pipeline*
- Make sure the Rennes traffic data is correctly ingested to the `metrics-rennes_traffic-default` TSDS

How to do it...

We need to configure an index lifecycle policy to control how our data ages and when it transitions between different phases. Let's locate and adjust the lifecycle policy for the `metrics-rennes_traffic-default` data stream.

1. Let's find the index life cycle policy for our TSDS. Go to **Stack Management | Index Lifecycle Policies**. Then, type `rennes_traffic` in the search bar and click on `rennes_traffic-lifecycle-policy`.

2. We can then adjust the **Advanced setting** section for the **Hot** phase, by setting **Maximum age** to `10` minutes (uncheck **Use recommended defaults** first) for **Rollover**, activating **Downsample**, and setting **Downsampling interval** to `1` hour.

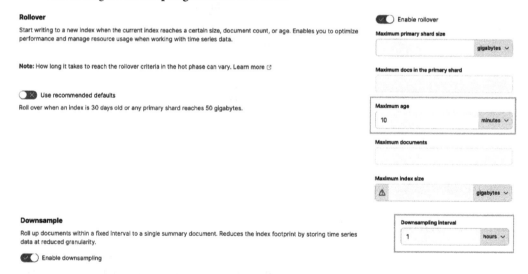

Figure 5.31 – Setting up downsampling in the index life cycle policy

3. After setting the rollover max age to 10 minutes, wait for over 10 minutes, then go to **Stack Management | Index Management**, and stay on the **Indices** tab. We can check the downsampled

indices for our data stream by filtering `data_stream="metrics-rennes_traffic-default"` in the search bar and by activating **Include hidden indices**, as shown in the following screenshot:

Index Management

🌐 Index Management docs

Indices   **Data Streams**   **Index Templates**   **Component Templates**

Update your Elasticsearch indices individually or in bulk. Learn more. ☑

🔲✕ Include rollup indices   ✓⚪ Include hidden indices

🔍 data_stream="metrics-rennes_traffic-default"   ⊗   Lifecycle status ∨   Lifecycle phase ∨   ⟳ Reload indices

☐	Name	Health	Status	Primaries	Replicas	Docs count	Storage size	Data stream
☐	.ds-metrics-rennes_traffic-default-20 23.09.14-000002	● green	open	1	1	0	494b	metrics-rennes_traffic -default
☐	downsample-3pgj-.ds-metrics-rennes _traffic-default-2023.09.06-000001	● green	open	1	1	91488	34.34mb	metrics-rennes_traffic -default

Figure 5.32 – Verifying the downsampled index

4. The downsampling took place successfully and we can go to Kibana Discover to check the `metrics-rennes_traffic-default` data view and then add the `average_vehicle_speed`, `traveltime.duration`, and `vehicle_probe_measurement` metrics field as columns. Ensure that your data is older than 10 minutes, as we have set the rollover period to 10 minutes. We can see the metrics fields correctly downsampled, as shown in *Figure 5.33*.

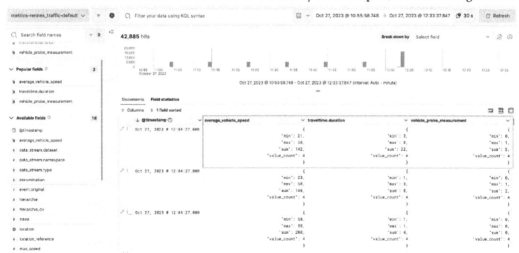

Figure 5.33 – Viewing the downsampling results

How it works...

The following figure illustrates the data flow from the actual metrics to the downsampled data:

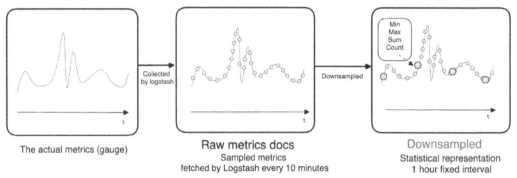

Figure 5.34 – From raw data to downsampled data

Let us understand the figure:

1. The actual metrics are represented by the left box in *Figure 5.34*. They represent real-time traffic flow, occurring every second.

2. Logstash collects these metrics every 10 minutes using the public API and the `http_poller` input plugin. The raw documents corresponding to these metrics are highlighted in the middle box of *Figure 5.34*.

3. We then configured downsampling within the index life cycle policy to aggregate metrics for fixed intervals every 1 hour. The right box shows the results of the downsampled documents that have been stored in our TSDS. We had previously defined the mappings in our component template to choose the dimension fields and the metric fields, which guides the downsampling process in understanding how to perform aggregations and which metric fields to include when generating aggregations for `min`, `max`, `sum`, and `count`.

There's more...

We have just configured the downsampling within an **index lifecycle management** (ILM) policy. It is also possible to set up downsampling manually; please refer to the following documentation: `https://www.elastic.co/guide/en/elasticsearch/reference/current/downsampling-manual.html`.

To see the result quickly, we have set the rollover max age to 10 minutes and the fixed interval for downsampling to 1 hour. In general, a reasonable default rollover max age could be 1 day and a fixed interval for downsampling could be 2 hours. We will see some more details on this subject in the *Optimizing time series data stream with downsampling* recipe in *Chapter 12*.

In this chapter, we have seen two methods for reducing the storage need for timestamped data: pivot data transform and downsampling – let's see what the key differences between each of them are:

- **Pivot data transform**: The pivot transform allows you to summarize and transform your data by pivoting on a specified key. This creates a new entity-centric index from your existing data. It takes a detailed, often denormalized, dataset, and aggregates it into a summarized, more structured format.

- **Downsampling**: Downsampling, on the other hand, is a technique used to reduce the granularity of your data. This is commonly used in time series data when you want to store less data over extended periods. For instance, you might keep high-resolution data (such as per-second records) for a week, then downsample that data to one-minute averages for anything older than a week.

In summary, pivot data transform is used to restructure and summarize data, often for analysis that focuses on certain facets of the data (such as users, devices, or locations). Downsampling, however, is more about conserving resources and maintaining longer-term historical data by compromising the granularity of the dataset.

See also

The list of the current limitations and restrictions of downsampling can be found in the following documentation: `https://www.elastic.co/guide/en/elasticsearch/reference/current/downsampling.html#downsampling-restrictions`

6

Visualize and Explore Data

In this chapter, we'll dive into the process of turning data into visualizations and uncovering its hidden values and insights. So far, we have managed to ingest, onboard, and transform our data. This chapter will show you how to create useful charts, maps, and dashboards that help you understand and use your data effectively.

In this chapter, we're going to cover the following recipes:

- Exploring your data in Discover

- Exploring your data with ES|QL

- Creating visualizations with Kibana Lens

- Creating visualizations from runtime fields

- Creating Kibana maps

- Creating and using Kibana dashboards

- Creating Canvas workpads

Technical requirements

To follow different recipes in this chapter, you'll need an Elastic Stack deployment that includes the following:

- **Elasticsearch** for searching and storing the data

- **Kibana** for data visualization and stack management (Kibana user with **All** privileges on **Fleet** and **Integrations**)

Exploring your data in Discover

In this recipe, we'll see how you can take advantage of one of the most powerful and popular applications in Kibana: **Discover**. After indexing some data in Elasticsearch, Discover should be your starting point to gain insights into the structure of the data, fields, value distribution, and other critical aspects. It's also an extremely efficient tool for confirming that data is indexing correctly into your Elasticsearch cluster.

With the launch of Elastic Stack version 8, Discover has introduced some exciting new features. In this recipe, we will concentrate on the following:

- Field statistics and document explorer

- Adding columns to the document table with drag and drop

- Breaking down the histogram by value

- Extracting more insights from the field lists

Getting ready

To follow along in this recipe, you'll need an up-and-running Elastic deployment with Elasticsearch and Kibana. To spin up a cluster on Elastic Cloud, refer to the *Installing Elastic Stack on Elastic Cloud* recipe in *Chapter 1*.

Also, ensure that you have a running Logstash instance. If not, turn to the *Installing self-managed Logstash* recipe in *Chapter 5* for guidance.

We'll be working with a dataset that we've become familiar with: `Rennes Traffic Data`. In, the *Creating a Logstash pipeline* recipe in *Chapter 5*, we created a **Time Series Data Stream** (**TSDS**) and used **Logstash** to periodically fetch, transform, and send real-time traffic data to Elasticsearch.

In this chapter, we will create a new data stream named `metrics-rennes_traffic-raw`. Unlike TSDS, this regular data stream will allow us to explore advanced features such as **Elasticsearch Query Language** (**ES|QL**), runtime fields, and so on that are currently unsupported in TSDS.

Make sure to complete the following setup as we will be using this new data stream in this chapter as well as in the *Alerting and anomaly detection* recipe in *Chapter 7*, and the *Advanced data analysis and processing* recipe in *Chapter 8*.

Here are the steps to follow:

1. In Kibana, go to **Management | Dev Tools**, and execute the commands provided here: `https://github.com/PacktPublishing/Elastic-Stack-8.x-Cookbook/blob/main/Chapter6/snippets.md#preparing-regular-data-stream-metrics-rennes_traffic-raw`.

These commands allow us to create the following components for our regular `metrics-rennes_traffic-raw` data stream in the following order:

A. **Index lifecycle policy**: `metrics-rennes_traffic-raw-lifecycle-policy`

B. **Index mapping**: `metrics-rennes_traffic-mappings@raw`

C. **Index settings**: `metrics-rennes_traffic-raw-settings`

D. **Index template**: `metrics-rennes_traffic-raw-index-template`

2. Once you have prepared the mapping, setting, and index template for the `metrics-rennes_traffic-raw` data stream, go to your Logstash server and create the `rennes-traffic-raw.conf` Logstash configuration file in the Logstash configuration folder. For Debian, this folder is located at `/etc/logstash/conf.d/`. You can find the reference configuration file at this URL: `https://github.com/PacktPublishing/Elastic-Stack-8.x-Cookbook/blob/main/Chapter6/logstash-conf/rennes_traffic-raw.conf`. Should you deploy Logstash elsewhere, the directory layout for different Logstash installations is detailed here: `https://www.elastic.co/guide/en/logstash/8.12/dir-layout.html`.

3. Modify the output section of the configuration file with your own Elastic deployment connection details (Refer to the *Creating a Logstash pipeline* recipe in *Chapter 5* for more details).

4. Restart your Logstash server to apply the configuration file with the following command:

```
$ sudo systemctl restart logstash.service
```

At this stage, your Logstash instance has been configured to send the data to our new data stream, `metrics-rennes_traffic-raw`. In the next steps, we will create a data view for the data stream to verify the data ingestion.

How to do it...

You will first need to create a **data view**. Those familiar with Elastic Stack might be wondering what a data view is. Essentially, a data view is the rebranded term for what used to be called an **index pattern**. Both concepts function as a method for directing Kibana to the location of your Elasticsearch data. Essentially, a data view designates which indices, or groups of indices, Kibana should search through when retrieving data. Furthermore, data views offer the flexibility to modify data fields, including the creation of new fields using runtime fields.

Head to **Kibana| Analytics | Discover** and follow these steps:

1. In the upper left, navigate to the **Data view** menu and select **Create a data view**. This action will open the data view creation flyout, as shown in *Figure 6.1*:

Figure 6.1 – Creating a new data view from Discover

2. Enter the following information to configure your data view:

    A. Enter the name: `metrics-rennes_traffic-raw`

    B. In the **Index pattern** field, start typing the name of the data stream. By typing `metrics-rennes_traffic-raw`, you should see, on the right-hand side, the matching source.

Figure 6.2 – Creating the metrics-rennes_traffic-raw data view

    C. Leave **Timestamp field** with the **@timestamp** value.

    D. Click on **Save data view to Kibana** at the top of the screen.

You should now see the Rennes data and the available fields in Discover, as illustrated in *Figure 6.3*:

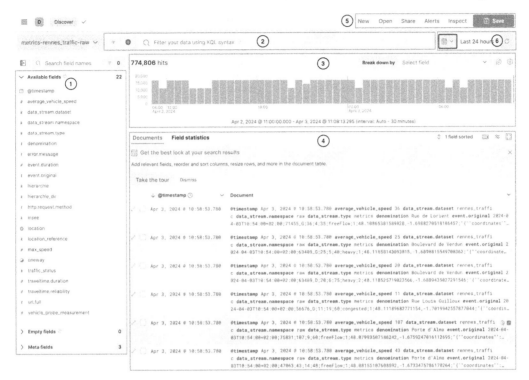

Figure 6.3 – Discover panel breakdown overview

Discover can be broken down into the following sections, as numbered in *Figure 6.3*:

1. **Available fields**: This shows the available fields, empty fields, and meta fields from the documents.

2. **Search bar and filter**: Use these to apply **Kibana Query Language** (**KQL**) or **Lucene queries** for filtering and analyzing documents.

3. **Histogram**: By default, this displays the count of documents broken down by an automatically determined time interval.

4. **Documents table**: This shows the documents that match the criteria entered in the search bar or fall within the selected time range. It allows for customization to include additional columns.

5. **Menu bar**: This provides access to the main actions in Discover.

6. **Time picker**: This enables you to quickly select the time range you wish to apply to the documents.

We will cover each of the sections in detail in the next steps of the recipe.

Now that you have some data in Discover, we can start exploring the indexed documents:

1. On the left of the screen, you can see a list called **Available fields**. Those are the non-empty fields in our dataset. There is a small icon at the left of each field name; this serves as an indication of the field type (keywords, numeric, geo, date, etc.):

Figure 6.4 – Field type icons

2. From the available fields, click on the **average_vehicle_speed** field and you'll see a contextual window showing two columns – **Top values** and **Distribution**:

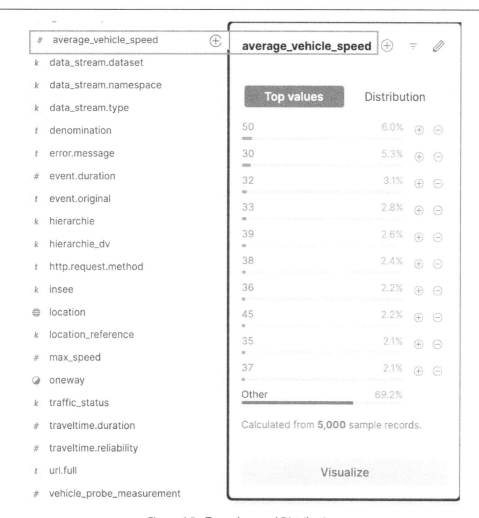

Figure 6.5 – Top values and Distribution

The **Top values** column shows the most common values that appear within a specific field across all documents that match the current search or filter criteria. It provides a quick overview of the predominant values for that field, which is useful for identifying patterns, trends, or anomalies in your data

3. Click on the **Distribution** column to get an idea of the value distribution across the documents. This feature allows you to visualize the frequency of values or ranges of values within a field across the dataset that matches the search or filter criteria. This can be particularly useful for numeric fields or date fields where you're interested in understanding the spread or distribution of values:

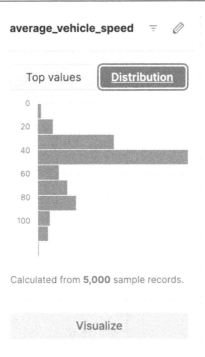

Figure 6.6 – Distribution histogram for average_vehicle_speed

In our case, we see that we have most traffic documents with an average speed of 40 km per hour.

4. Now, let's look at the **Documents** table displayed in the center of the **Discover** application. By default, it has two columns:

    ▪ The first column is for the time field – **@timestamp**, in our case, as we're dealing with a data stream.

    ▪ The second column is **Document**: it displays all the fields for each document

    You can expand a single document and view all the fields either in table or JSON format.

5. Starting with Kibana version 8.1, a valuable machine learning feature, **Field statistics**, was integrated into Discover. Click on **Field statistics** at the top of the table and expand the average_vehicle_speed field to quickly view its top values, cardinality, the percentage of documents in which it is present, and its distribution:

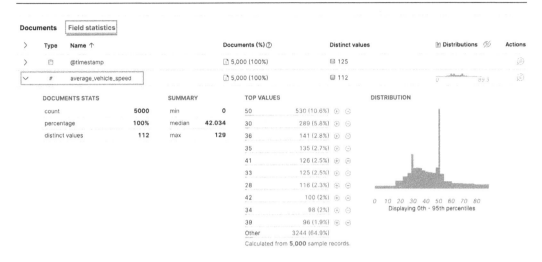

Figure 6.7 – Field statistics

Let's update the table with two new columns.

6. Start by selecting the **denomination** field from the options on your left and drag it onto the **Documents** table to add it as a new column. Then, similarly, add the **average_vehicle_speed** field by dragging it into the table. These changes will enrich your table with data displaying the average speed for each road, alongside the timestamps of these measurements:

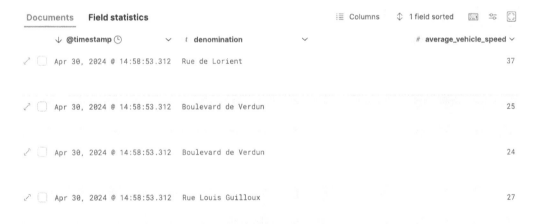

Figure 6.8 – Documents table with a denomination column added

You can sort the table by any of the columns simply by clicking on the column header at the top of the table.

7. At this stage, you have likely noticed the histogram visualization displayed above the **Documents** table. The histogram is one of Discover's hallmark features, and it can be customized to suit your requirements.

8. Above the histogram visualization, look for the **Break down by** field. Click on the drop-down menu next to this text and select **traffic_status**. The histogram will now segment the data by traffic status, providing a clear visualization of the status distribution across the incoming documents, as illustrated in *Figure 6.9*:

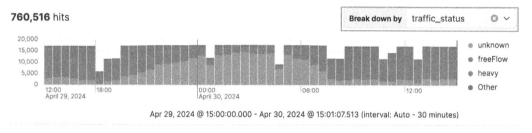

Figure 6.9 – Discover histogram broken down by traffic status

9. One of Discover's standout features is its ability to combine text search and filtering. Suppose we want to filter our documents by a specific traffic status, such as the **congested** one. Click on the + sign to the right of the data view name to open the **Add filter** menu. Configure the following and click on **Add filter**:

    I.   **Field: traffic_status**

    II. **Operator: is**

    III. **Value: congested**

Figure 6.10 – Filter creation in Discover

Once the filter is applied, notice that the **Discover** page updates to display only the documents with the **congested** traffic status.

10. Next, we will utilize the KQL to achieve the same outcome as the filter we just set. Remove the existing filter and click inside the query bar at the top of the histogram:

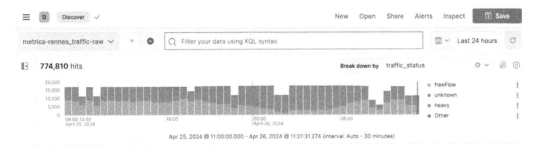

Figure 6.11 – Using KQL within the Discover search bar

11. Enter the following query and hit *Enter*:

```
traffic_status: " congested"
```

Notice how the autocomplete feature suggests field names as you type. We can refine our query by adding another filter for the road type with the following condition:

```
traffic_status : "congested" and not hierarchie : "Réseau
national"
```

The results of this query will be displayed, as shown in *Figure 6.12*:

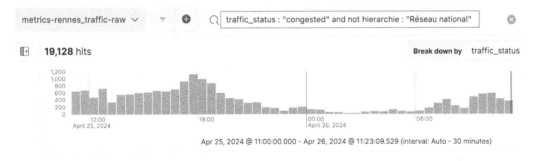

Figure 6.12 – KQL result query applied to the histogram

Let's save our customized query and **Documents** table as a saved search. Saved searches allow users to quickly access frequently used queries and filters, enhancing workflow efficiency. They are also useful when building dashboards.

12. Click **Save** in the top-right corner of the screen and name it—let's call it `congested road` in the popup:

Figure 6.13 – Save search in Discover

13. To access your saved search, click on the **Open** option in the top-right corner of the screen, and select your saved search, as shown in *Figure 6.14*:

Figure 6.14 – Retrieve and open the saved search

14. Finally, let's export our results as an easily shareable CSV report. Click **Share** in the top-right corner, choose **CSV Reports**, and then select **Generate CSV**.

> **Note**
>
> The CSV export feature in Discover within Kibana does have limits. By default, it is configured to export up to a maximum number of rows, which can be adjusted in the Kibana settings (`kibana.yml`). The limit is in place to manage performance and resource utilization on both the Elasticsearch cluster and the Kibana server. If the number of documents matching your search criteria exceeds this limit, only the top documents, up to the specified limit, will be exported. To export larger datasets, adjustments to the `xpack.reporting.csv.maxSizeBytes` settings in `kibana.yml` may be necessary. However, always consider the performance implications of exporting large datasets.

Figure 6.15 – Generating CSV reports from Discover

15. The CSV is generated in the background; once it is completed, a popup at the bottom right of the screen will appear. You can now click on the **Download report** button to retrieve the CSV file:

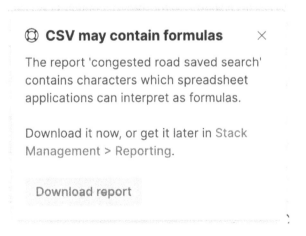

Figure 6.16 – Downloading the CSV report

How it works...

Discover is a powerful feature in Kibana that enables you to search and filter your data, learn about the structure of the fields, and present your findings through visualization. To begin using Discover in Kibana, you need a data view (formerly known as an index pattern), which directs Kibana to the relevant Elasticsearch data. Once set up, you can select the data to work with and specify the time range for which you want to view data.

When you use Discover, Kibana sends a query to Elasticsearch, which then searches through your data and returns the results to Kibana; the results are displayed in table format that can be customized to your needs. By default, Discover's search bar uses the KQL, but you can also select the Lucene syntax if you need more advanced features such as regex or fuzzy queries. From Kibana version 8.11, the **ES|QL** new query language is also available from Discover. We will learn more about this new language in the following recipe: *Exploring your data with ES|QL*.

Discover is an essential tool for data exploration within Kibana because it allows you to instantly search and filter through large amounts of data, making it easier to identify patterns and trends in your data. With Discover, you can also customize and save your searches and place them on a dashboard, which makes it easier to share insights with others. In addition, Discover provides several menus that allow you to interact with your data in diverse ways.

There's more...

In this recipe, we went through an example data exploration flow within Discover, but you can do a lot more with Discover:

- **Create search alert threshold rules**: For example, with our sample dataset, we can have a rule that triggers an alert when the number of documents with the `traffic_status: congested` status has been above 100 for the last 5 minutes (We will explore more in the *Creating alerts in Kibana* recipe in *Chapter 7*):

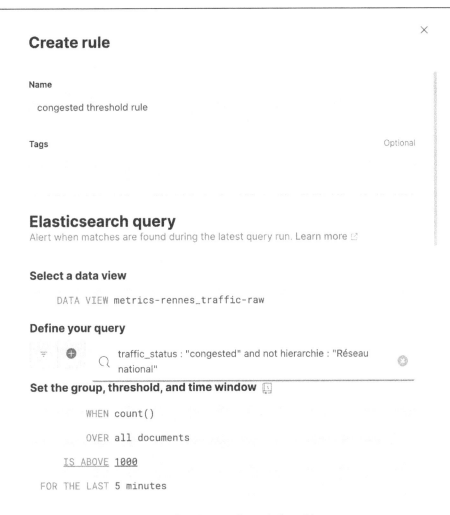

Create rule   ×

Name

congested threshold rule

Tags   Optional

Elasticsearch query

Alert when matches are found during the latest query run. Learn more ☑

Select a data view

DATA VIEW metrics-rennes_traffic-raw

Define your query

traffic_status : "congested" and not hierarchie : "Réseau national"

Set the group, threshold, and time window 🗒

WHEN count()

OVER all documents

IS ABOVE 1000

FOR THE LAST 5 minutes

Figure 6.17 – Creating an alert rule from Discover

- **Add a new field by leveraging the runtime field feature**: You will learn more about that in the *Creating visualizations with Kibana Lens* recipe later in this chapter.

- **Run pattern analysis on text fields**: You can find patterns in unstructured log messages, which makes it easier to examine your data. The value of log pattern analysis is that it allows you to quickly identify patterns and trends in your data, making it easier to troubleshoot issues and identify areas for improvement. By categorizing log messages based on their content, you can quickly identify which messages are important and which ones can be ignored.

- **Visualize (in Lens) the fields in your documents**: Quick start your visualization straight from Discover.

- **Inspect the queries and the response**: This can be very useful to get the Query DSL structure of the search in Discover; you can then pivot to **Dev Tools** or the **Search Profiler**.

See also

- For a global overview of the evolution of Discover, you can refer to the following blog post: `https://www.elastic.co/blog/the-evolution-of-discover-in-kibana`

- If you are interested in solving issues in Kibana Discover, check out this excellent article: `https://www.elastic.co/blog/troubleshooting-guide-common-issues-kibana-discover-load`

- For the official documentation and Discover walk-through, have a look at the following page: `https://www.elastic.co/guide/en/kibana/current/discover.html`

Exploring your data with ES|QL

In this recipe, we will explore **ES|QL**, a relatively new and highly effective query language in Elasticsearch. **ES|QL** is crafted to encompass all the functionalities of existing languages within Elasticsearch through a unified, piped command syntax and data model. It supports many commands and functions, allowing users to execute a variety of data operations, including filtering, aggregation, and time series analysis, among others. This makes **ES|QL** a versatile tool for managing and analyzing data within the Elastic Stack.

You'll get a glimpse of the vast possibilities in terms of data exploration and analysis offered by **ES|QL**. Specifically, we're going to cover the following aspects:

- Chaining processing commands
- ES|QL aggregations
- Lookups

Getting ready

To follow along with this recipe, you'll need an up-and-running Elastic deployment with Elasticsearch and Kibana. Additionally, you should have completed the first recipe of this chapter, *Exploring your data in Discover*.

The queries used in this recipe can be found at the following location in the book's GitHub repository: `https://github.com/PacktPublishing/Elastic-Stack-8.x-Cookbook/blob/main/Chapter6/snippets.md#exploring-your-data-with-esql`.

> **Important note**
> ES|QL is available in the Elastic stack starting from version 8.11 as a technical preview. Make sure to have an environment that meets this minimum version requirement before proceeding with this recipe.

How to do it...

ES|QL is fully integrated into Kibana's Discover application, and this is where we're heading to take advantage of this powerful query language:

1. Go to **Kibana | Discover** and select the **metrics-rennes_traffic-raw** data view.
2. Click on the arrow on the right side of the data view name to expand the menu and select **Try ES|QL**:

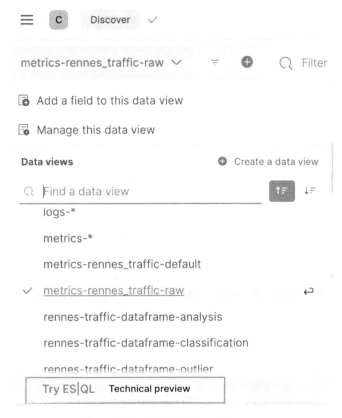

Figure 6.18 – Try ES|QL from Discover

3. Once you have clicked on **Try ES|QL**, you should see the query bar populated with a default query:

Figure 6.19 – ES|QL default query and results table

4. Let us update the query to filter on the `congested` traffic status; our query will look like this:

    ```
    from metrics-rennes_traffic-raw
    | where traffic_status == "congested"
    | limit 50
    ```

    Notice how we use the pipe to chain the processing. With the `limit` processing command, we cap the number of rows to `50`. Inspecting the resulting documents, you should notice that they have all been filtered based on the `congested` traffic status.

5. Now, let's switch gears a bit and add an aggregation to calculate the average time by denomination using the `stats...by` processing command. In addition, we will sort the resulting table in descending order:

    ```
    from metrics-rennes_traffic-raw
    | where traffic_status == "congested"
    | stats avg_traveltime = avg(traveltime.duration) by
    denomination
    | sort avg_traveltime desc
    | limit 50
    ```

By executing the query, you should get the result shown in *Figure 6.20*:

Figure 6.20 – The ES|QL stats...by command results

As you can see, we just computed a new field on the fly and used it to sort the resulting table.

6. We will keep tweaking our query further. The `avg_traveltime` field we just computed is expressed in seconds; let's say we want to convert it into minutes. We can do so by using the EVAL command. Our query will now look like this:

```
from metrics-rennes_traffic-raw
| where traffic_status == "congested"
| stats avg_traveltime = avg(traveltime.duration) by
denomination
| eval avg_traveltime_min = round(avg_traveltime/60)
| sort avg_traveltime_min desc
| keep denomination, avg_traveltime_min
| limit 50
```

You will have certainly noticed that we also introduced the KEEP command. This allows us to specify what columns are returned as well as their order:

Figure 6.21 – Introducing the EVAL and KEEP commands

By now, you can see the game-changing flexibility that ES|QL brings to the table. But it does not stop there.

Let's explore another powerful and handy capability of the new query language: the ENRICH command. Imagine we want to enrich our Rennes traffic dataset with information such as postal code and municipality. Let's see how we can achieve that with ES|QL. To use the ENRICH capability, we need to set up an enrich policy. This is where we define how we combine multiple indices for enrichment:

1. First, we will start by setting up the source index we will use for enrichment. The source index will contain data such as postal code and municipality, as well as an insee code field that we will leverage for the lookup.

> **Notes on the insee code**
>
> The insee code in France refers to a set of numeric codes used by the National Institute of Statistics and Economic Studies for various administrative purposes. These codes are used to identify different geographical and administrative entities within the country.

2. Head to **Kibana | Home | Upload a file**.

3. Use the CSV file in the repository: `https://github.com/PacktPublishing/Elastic-Stack-8.x-Cookbook/blob/main/Chapter6/esql/insee-postal-codes.csv`.

4. On the **Import data** page, click on the **Advanced** tab, name the index `enrich-insee-codes`, and replace the default mappings with the one provided in this file: `https://github.com/PacktPublishing/Elastic-Stack-8.x-Cookbook/blob/main/Chapter6/snippets.md#index-mapping-for-insee-codes`. The **Import data** page should look like *Figure 6.22*:

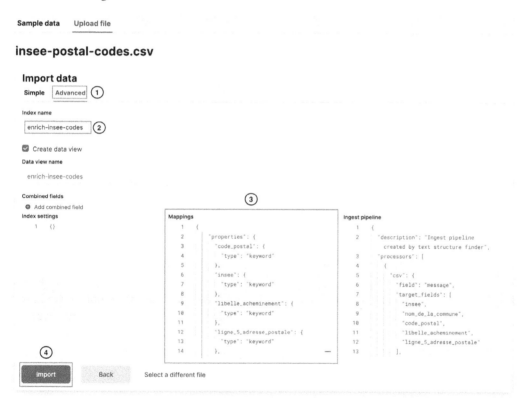

Figure 6.22 – Importing insee code for enrich index

Once the import process is completed, you can view the index in Discover, as shown in *Figure 6.23*:

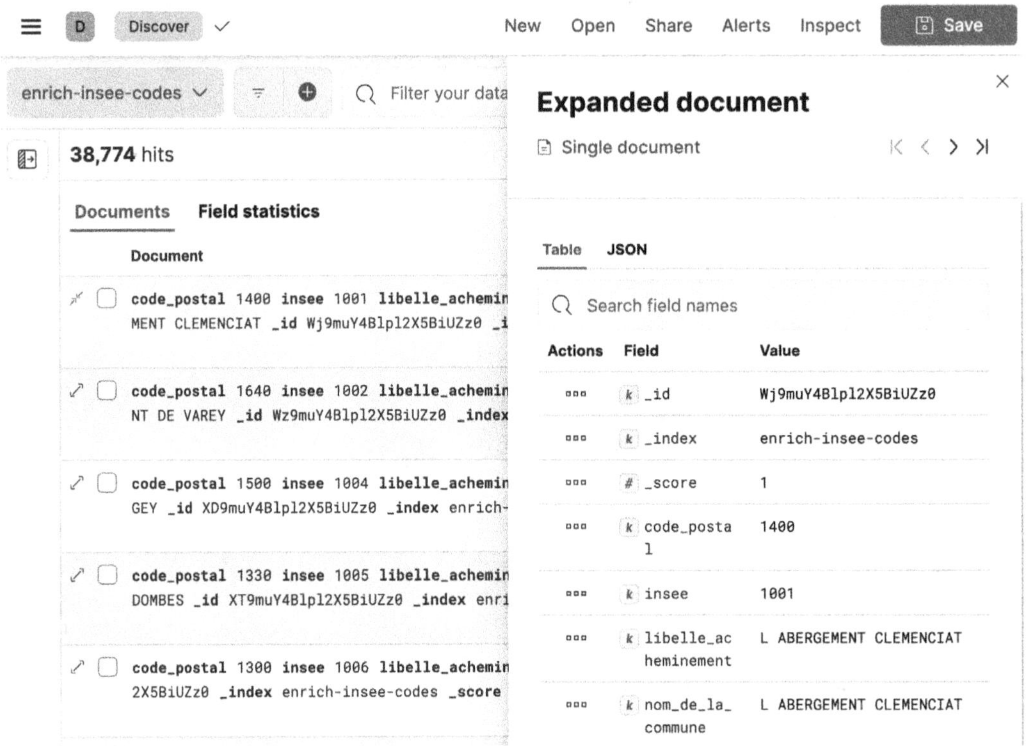

Figure 6.23 – Viewing enrich-insee-codes in Discover

5. Now that we have created our `enrich` index, let's configure the enrich policy. We have provided the Dev Tools request to use in the following file: `https://github.com/PacktPublishing/Elastic-Stack-8.x-Cookbook/blob/main/Chapter6/snippets.md#enrich-policy`. You can also find the corresponding requests here:

```
# Create the enrich policy
PUT _enrich/policy/rennes-data-enrich
{
  "match": {
    "indices": [
      "enrich-insee-codes"
    ],
    "match_field": "insee",
    "enrich_fields": [
      "code_postal",
      "nom_de_la_commune"
```

```
      ]
    }
  }
```

6. Once the enrich policy has been defined, the next step is to execute it with the following command:

```
# execute the enrich policy
PUT _enrich/policy/rennes-data-enrich/_execute
```

7. To validate that you have an enrich policy perfectly set up and ready to be used, go to **Stack Management | Index Management | Enrich Policies**. You should see your enrich policy as shown in *Figure 6.24*:

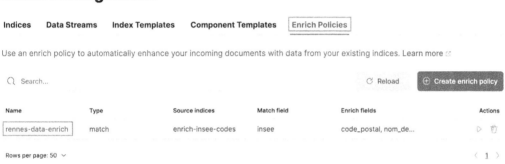

Figure 6.24 – Setting up an enrich policy

Now we have our source index and enrich policy, let's head back to Discover to use it with ES|QL.

8. To do so, we'll use the ENRICH process command. We'll enrich our metrics-rennes dataset with two fields, code_postal and nom_de_la_commune, by doing a lookup on the insee field. The resulting query will be as follows:

```
from metrics-rennes_traffic-raw
| where traffic_status == "congested"
| enrich rennes-data-enrich on insee with code_postal, nom_de_
la_commune
| keep average_vehicle_speed, code_postal, nom_de_la_commune,
denomination
| sort average_vehicle_speed desc
| limit 50
```

By running the preceding query, you get the following result table:

Figure 6.25 – ES|QL query with the enrich policy result

As you can see in the resulting table, we have used the enrich policy to add the following two columns—code_postal and nom_de_la_commune.

9. You can also use the newly added columns for aggregation, as demonstrated in the following query, where we calculate the average travel time and aggregate it by the nom_de_la_commune field:

```
from metrics-rennes_traffic-raw
| where traffic_status == "congested"
| enrich rennes-data-enrich on insee with code_postal, nom_de_
la_commune
| stats avg_traveltime = avg(traveltime.duration) by nom_de_la_
commune
| sort avg_traveltime desc
| limit 50
```

While there is much more to explore in ES|QL, these examples should provide a glimpse into the vast possibilities it offers for data manipulation within the Elastic Stack.

How it works...

An ES|QL query is composed of a series of commands chained together by pipes. There are essentially two types of commands:

- **Source commands**, which retrieve or generate data in the form of tables.

- **Processing commands**, which take a table as input and produce a new table as output. Processing commands can be chained by using the | pipe character, as you saw in this recipe.

Figure 6.26 describes the way it operates:

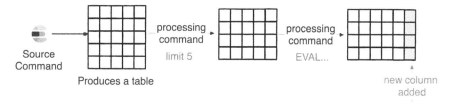

Figure 6.26 – ES|QL operation overview

ES|QL also provides some functions and operators for working with data.

The following table shows the main ES|QL commands classified by type:

Source commands	Processing commands
FROM	DISSECT
ROW	DROP
SHOW	ENRICH
	EVAL
	GROK
	KEEP
	LIMIT
	MV_EXPAND
	RENAME
	SORT
	STATS... BY
	WHERE

Table 6.1 – ES|QL commands and processing commands

In our recipe, we have used the ENRICH processing command. This command merges, during the search, data from one or more original indices with combinations of field values stored in Elasticsearch *enrich* indices. The following figure shows how it works:

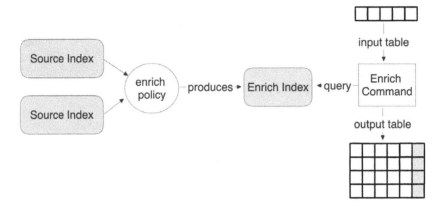

Figure 6.27 – ES|QL and enrich policy data flow

As displayed in the preceding figure, the ENRICH command requires a few pieces:

- **Source index**: There is at least one source index, which contains the Enrich data the ENRICH command will use that'll be used to add data to the input tables.

- **Enrich index**: This is the place where the enrich policy will look up records based on the match field value. The enrich index is a read-only internal Elasticsearch-managed index. It's also force merged for performance aspects.

- **Enrich policy**: This is a set of configurations used to add the right Enrich data to the input table.

There's more...

ES|QL also has another unique feature: its compute engine. To meet both the functional and performance demands of ES|QL, a completely new computing architecture had to be developed. The search, aggregation, and transformation functions of ES|QL are executed directly within Elasticsearch. Instead of translating query expressions into Query DSL for execution, they are run natively. This method enables ES|QL to achieve exceptional speed and flexibility.

ES|QL can also be used for alerting in Discover. The ES|QL query API is exposed under the _query endpoint and accepts an ES|QL query string in the query parameter.

See also

- If you're interested in learning more on the performance side of things about ES|QL, check out the nightly benchmark here: `https://elasticsearch-benchmarks.elastic.co/#tracks/esql/nightly/default/90d`

- For more information on how you can leverage this powerful new query language, read this blog article: `https://www.elastic.co/blog/introduction-to-esql-new-query-language-flexible-iterative-analytics`

Creating visualizations with Kibana Lens

Kibana Lens offers a user-friendly and intuitive approach for anyone to visualize their data within Kibana. With its seamless drag-and-drop interface, one-click data exploration, and capability to provide visualization recommendations, Lens is the fastest way to uncover valuable insights within your Elasticsearch data. In this recipe, you'll discover the power and flexibility of data visualization with Kibana Lens, and we will also explore some useful new features that have been introduced in Elastic 8.

Getting ready

Make sure that you have completed the previous recipes of this chapter, as the same dataset on Rennes traffic will be used in this recipe.

How to do it...

Let's get started with visualizing your data using Kibana Lens.

By the end of this recipe, you will have created the following visualizations:

- Number of unique locations (metric chart)

- Overall average speed gauge (gauge visualization)

- Global traffic status (waffle chart)

- Traffic status by road type (donut chart with sub-bucket)

- Average speed and traffic status comparison (multilayer chart)

- Average speed by road hierarchy (advanced metric chart)

Let's get started!

Number of unique locations (metric visualization)

The metric visualization is the most fundamental type, displaying a numerical value. Let's create such a visualization:

1. Navigate to **Kibana | Analytics |Visualize Library**, click on **Create visualization**, then choose **Lens**.

2. Select the **metrics-rennes_traffic-raw** data view.

3. Pick **Legacy Metric** as the chart type.

4. From the list of fields on the left navigation bar, drag the **location_reference** field to the central workspace, as shown in *Figure 6.28*:

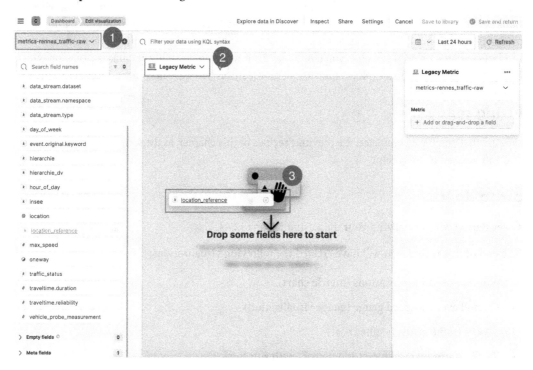

Figure 6.28 – Kibana Lens creating a legacy metric visualization

5. Once you have dropped the field in the central workspace, a count of records will show up; this represents the number of unique locations in our dataset. The **Unique count** function is selected by default. To improve readability, let's change the label. Click on **Unique count of location_reference** in the right panel and enter Locations as the display name in the **Appearance** section.

6. Click **Save** at the top right. As shown in *Figure 6.29*, set the title to [Rennes Traffic] Number of locations and select **None** for the **Add to dashboard** option. Then, add a tag named **cookbook** to the visualization. Finally, click on **Save and add to library** (we will build our dashboard in the *Creating and using Kibana dashboards* recipe later in this chapter):

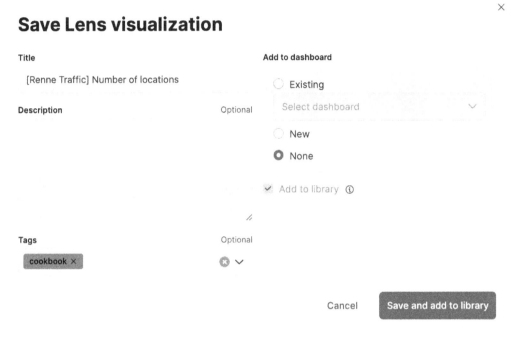

Figure 6.29 – Saving the Lens visualization

The final metric visualization result should appear as shown in *Figure 6.30*:

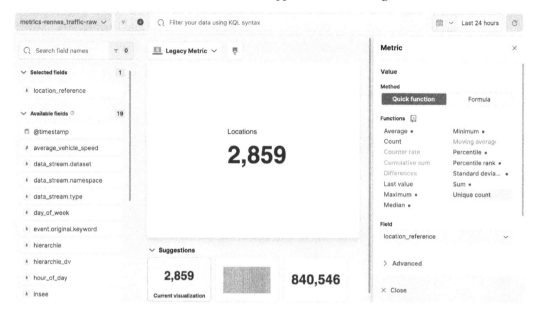

Figure 6.30 – Saved Lens visualization

Overall average speed gauge (gauge visualization)

We will now create another new visualization introduced in Elastic Stack version 8: the gauge chart. A gauge chart is a type of visualization used to display a single value within a given range, showing where that value falls compared to preset thresholds or target values. Follow these steps:

1. Navigate to **Kibana | Analytics | Visualize Library**, click on **Create visualization**, and select **Lens**.

2. Choose the **metrics-rennes_traffic-raw** data view.

3. Pick **Gauge Horizontal** as the chart type.

4. From the list of the fields on the left panel, drag the **average_vehicle_speed** field to the central workspace.

5. Click on **Median of average_vehicle_speed** in the **Metrics** section on the right panel and change **Functions** to **Average**.

6. In the **Appearance** section, change the name to Overall avg speed and toggle **Band colors**. Then, click to edit the color, as shown in *Figure 6.31*:

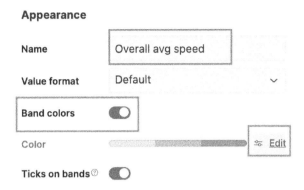

Appearance

Name	Overall avg speed
Value format	Default
Band colors	⬤
Color	Edit
Ticks on bands	⬤

Figure 6.31 – Setting the gauge metrics appearance

7.   This brings out the **Color** configuration flyout. Choose the status color palette and click on **Reverse colors** (we want to set the lower speed range to red (**45**) and the higher speed range to green (**51**)). Then, click on the **Number** tab in the **Value type** section and set the color ranges as shown in *Figure 6.32*. Click on the *back* button after the configuration:

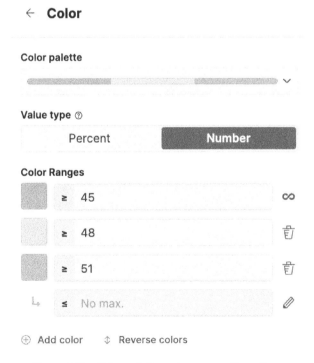

← **Color**

Color palette

Value type ⓘ

Percent	Number

Color Ranges

	≥ 45	∞
	≥ 48	🗑
	≥ 51	🗑
↳	≤ No max.	✎

⊕ Add color   ↕ Reverse colors

Figure 6.32 – Color palette and range configurations

8. Once configured, go to the **Metric** section on the right panel. Set the minimum value to a static value and your gauge visualization should look like the one shown in *Figure 6.33*:

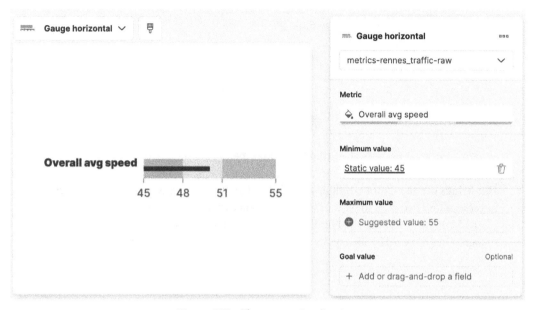

Figure 6.33 – The gauge visualization

9. Click **Save** at the top right, then set the title to [Rennes Traffic] Average speed gauge. Select **None** in the **Add to Dashboard** section and add the **cookbook** tag to the visualization. Finally, click on **Save and add to library**.

Global traffic status (waffle visualization)

We will now create another new visualization introduced in Elastic Stack version 8, the waffle chart. This chart is excellent for showing proportions, including the global distribution of traffic status. Follow these steps:

1. Navigate to **Kibana | Analytics | Visualize Library**, click on **Create visualization**, and select **Lens**.

2. Choose the **metrics-rennes_traffic-raw** data view.

3. Pick **Waffle** as the chart type.

4. Next, click on the **traffic_status** field from the list of fields on the left navigation bar to inspect the distribution. You will notice there are four types of traffic status in the Rennes traffic dataset, as shown in *Figure 6.34*:

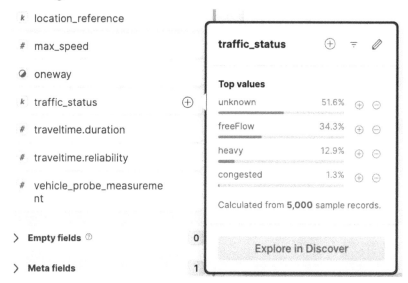

Figure 6.34 – Inspecting the field value distribution in Kibana Lens

5. From the list of the fields on the left panel, drag the **traffic_status** field to the central workspace.

6. Click on **Traffic status** in the **Group by** section on the right panel, and adjust the number of values to 4 to align with the four distinct traffic statuses in our dataset.

7. The count of records represents the number of occurrences of each traffic status in different locations. Let's update the label to reflect this. Click **Count of records** on the right panel and enter `Traffic status distribution` as the display name.

8. Click **Save** at the top right, then set the title to `[Rennes Traffic] Traffic status waffle`. Select **None** in the **Add to Dashboard** section and add the **cookbook** tag to the visualization. Finally, click on **Save and add to library**. The result of the waffle visualization should look as shown in *Figure 6.35*:

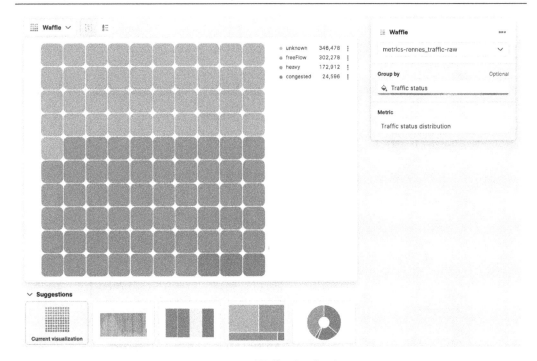

Figure 6.35 – Waffle visualization

Traffic status by road type (donut chart with sub-bucket)

Next, we will create a more advanced visualization based on a donut chart. The goal here is to understand the traffic status and analyze the distribution of road types (defined by the maximum authorized speed) within each traffic status. We will first break down the donut chart by the values of the **traffic_status** field, and then further sub-group the data by breaking down each pie chart using sub-buckets. Follow these steps:

1. Head to **Kibana | Analytics | Visualize Library**, click on **Create visualization**, and choose **Lens**.

2. Select the **metrics-rennes_traffic-raw** data view.

3. Select **Donut** as the chart type.

4. Drag and drop **traffic_status** onto the workspace, then click in **Top 5 values of traffic_status** to adjust it to Top 4 values of traffic_status.

5.   Drag and drop **max_speed** onto the **Slice by** section. Your chart should look like the one shown in *Figure 6.36*. The inner ring shows the percentage of each traffic status. The outer ring breaks down the traffic status into road type categories:

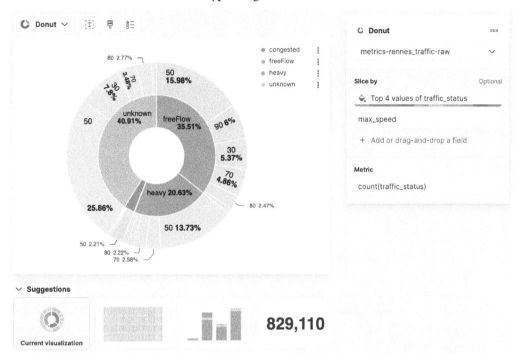

Figure 6.36 – Creating the donut chart

6.   You can customize the labels on this donut chart. Click the first button to the right of the visualization type dropdown (it looks like the letter *T* in a dotted square). Set **Maximum decimal places for percent** to 1, as shown in *Figure 6.37*:

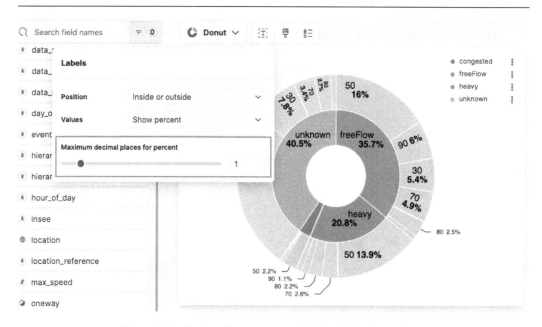

Figure 6.37 – Setting the percentage display in the donut chart

7. By default, the outer ring only shows the values sliced by default intervals, as **max_speed** is a numeric field. You can change the number of values to display in the slices by clicking on **max_speed** in the right panel and adjusting the granularity either with the slider or by creating custom ranges, as shown in *Figure 6.38*:

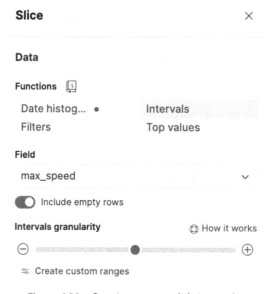

Figure 6.38 – Creating custom slide intervals

8. Click **Save** at the top right, then set the title to `[Rennes Traffic] Traffic status by road type`. Select **None** in the **Add to Dashboard** section and add the **cookbook** tag to the visualization. Finally, click on **Save and add to library**.

Average speed and traffic status comparison (multiple layers)

A single visualization can consist of multiple layers, which is one of the many new features introduced in Elastic 8. This type of visualization is useful for comparing multiple fields or for understanding data more contextually. Follow these steps:

1. Head to **Kibana | Analytics | Visualize Library**, click on **Create visualization**, and choose **Lens**.

2. Select the **metrics-rennes_traffic-raw** data view.

3. Select **Bar vertical percentage** as the chart type.

4. Drag and drop **Records** and **traffic_status** onto the workspace. Click on **Top 3 values of traffic_status** on the right panel and change **Number of values** to 4, set the name to `Traffic status`, and adjust the color palette to **Temperature** as shown in *Figure 6.39*:

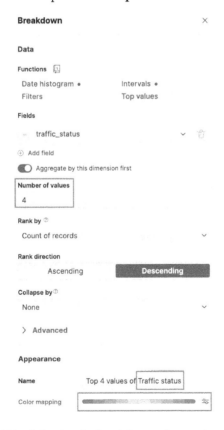

Figure 6.39 – Adjusting the breakdown values and appearance

At this stage, your visualization should look like *Figure 6.40*:

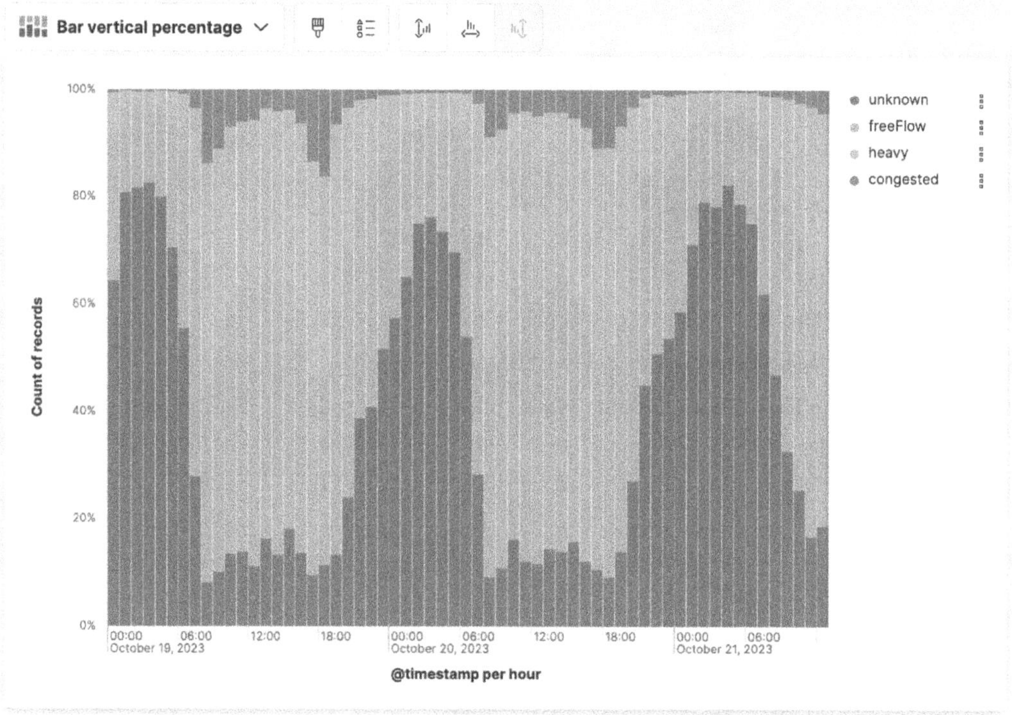

Figure 6.40 – Bar vertical percentage visualization

5. Now, let's add another layer to compare the traffic status with the global average speed. Click on **Add layer** on the right panel then select the **Visualization** layer type.

6. Choose **Line** as the chart type for the new visualization layer, then do the following:

    A. Drag and drop **@timestamp** to the horizontal axis of the new visualization layer.

    B. Drag and drop **average_vehicle_speed** to the vertical axis of the new visualization layer.

    C. Adjust the **Aggregation** function of **average_vehicle_speed** to **Average**; then, in the **Appearance** section, change the name to `Global average speed`, change the series color to make it more visible on our final visualization, and finally, set **Axis side** to **Right**.

7. Click **Save** at the top right, then set the title to `[Rennes Traffic] Average speed & Traffic Status`. Select **None** in the **Add to Dashboard** section and add the **cookbook** tag to the visualization. Finally, click on **Save and add to library**.

We now have a multilayer visualization that displays the impact of traffic status distribution on the city's global average speed. Your completed visualization should look similar to *Figure 6.41*:

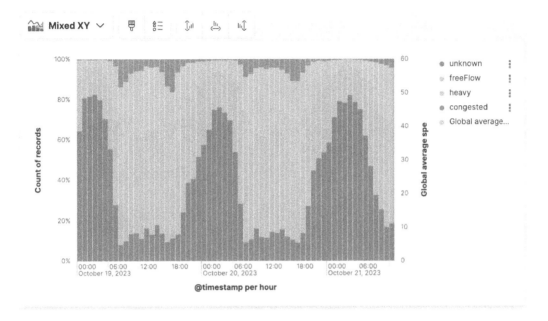

Figure 6.41 – Bar vertical percentage visualization with multiple layers

Average speed by road hierarchy (advanced metric)

Next, we will create a metric visualization, a feature introduced in Elastic 8. This new chart type allows for the addition of a secondary metric, providing more context—especially when looking at data over time. Here, we will create a visualization that enables us to organize multiple average speed metrics in a grid by road hierarchy. Follow these steps:

1. Head to **Kibana | Analytics | Visualize Library**, click on **Create visualization**, and choose **Lens**.
2. Select the **metrics-rennes_traffic-raw** data view.
3. Select **Metric** as the chart type.
4. Drag and drop **average_vehicle_speed** onto **Primary metrics** in the right panel.
5. Drag and drop **hierarchie** onto **Break down by** in the right panel.
6. Click on **Median of average_vehicle_speed** in the right panel, then change the aggregation function to **Average**. In the **Appearance** section, change the name to Average speed and change the color mode to **Dynamic**.

7. Click on the **Edit** button and then click on **Reverse colors**, so that higher speeds are represented by green and lower speeds by red. Lastly, in the **Supporting visualization** section, select the **Line** type.

8. Drag and drop **max_speed** to **Second metrics** in the right panel. Click on **Median of max_speed**. Change the function to **Average**, then change the name to Average authorized speed. You can also change **Value format** to **Number** and set **Decimals** to 2. Your visualization should look like the one shown in *Figure 6.42*:

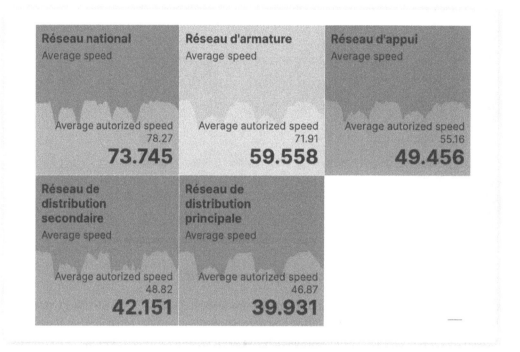

Figure 6.42 – Advanced metric visualization

9. Click **Save to library** at the top right, then set the title to [Rennes Traffic] Speed by road hierarchy. Select **None** in the **Add to Dashboard** section and add the **cookbook** tag to the visualization. Finally, click on **Save and add to library**.

How it works...

As you can see, Kibana Lens offers a very intuitive way to build visualizations without necessarily having data analytics skills. At the beginning of the creation of a Lens visualization, we started with the default view, as shown in *Figure 6.43*:

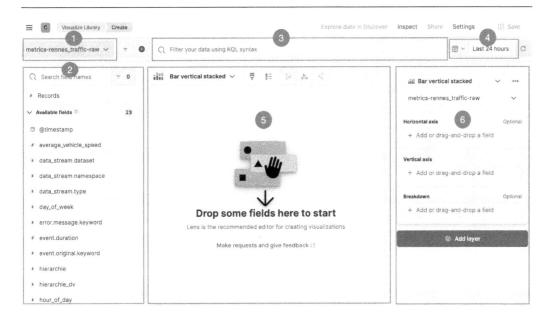

Figure 6.43 – Kibana Lens UI

Let's break down what you see in this view:

1. At the top left, you can select a *data view*. A data view corresponds to one or more indices in Elasticsearch. As you create visualizations, you will only see data from the indices behind the selected data view.

2. Below the **Data view** dropdown, you see a list of the *fields* that exist in your data. Take note of the list of fields presented on the left-hand side. You can search for a specific field or group them by type. For convenience, Lens only displays fields that contain data. You can click on a field for a quick view of its top values or distribution.

3. At the top center, there is a *query bar*, where you can write queries and create filters.

4. At the top right, the *time filter* enables you to set the time range you wish to visualize.

5. In the middle, you have the *workspace* where your visualization will be shown.

6. Finally, on the right, you find the *layer pane*, with the controls to configure your visualization.

We primarily use drag and drop to start visualizing a field. It is then possible to change the chart type on the fly. Below the visualization area, Kibana Lens also offers some very useful suggestions for alternative visualization types, as you can see in *Figure 6.44*:

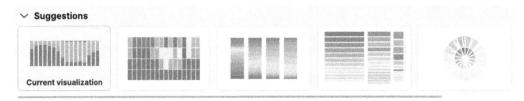

Figure 6.44 – Alternative visualization suggestions

There's more...

In the following subsections, we will delve into some of the latest features of Kibana Lens in Elastic Stack version 8 that enhance data analysis and visualization. These features enable analysts to display data relationships, improve performance, and enhance the interpretability of visual data representations.

Formula

Kibana Lens also has a powerful feature called **Formula**, which can combine aggregation and mathematical functions. We can, for example, create a **Line** visualization that gives a week-over-week average vehicle speed comparison breakdown by traffic status with a formula using the `percentile` function and `shift` parameter, as shown in *Figure 6.45*. You can do this by creating a **Line** chart visualization and selecting **Formula** for **Vertical axis**; once selected, you can try to fill the **Formula** section with the one shown in the example:

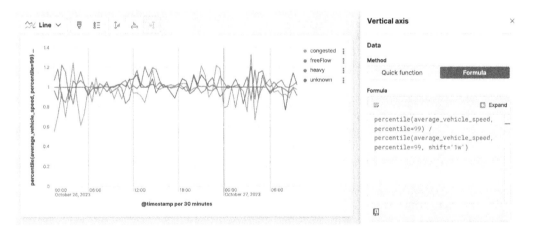

Figure 6.45 – Kibana Lens formulas

Sampling

To improve the visualization loading time, it is possible to use the **Sampling** feature to reduce the number of documents used for the aggregation. This feature is based on random sampler aggregation, which was introduced in Elastic Stack 8.2 (`https://www.elastic.co/guide/en/elasticsearch/reference/current/search-aggregations-random-sampler-aggregation.html`). It is possible to bring up the layer setting of each visualization and adjust the **Sampling** rate, as shown in *Figure 6.46*:

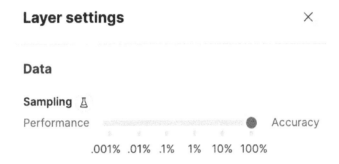

Figure 6.46 – Sampling setting for Kibana Lens

Annotation

Kibana Lens also offers an annotation capability that enables you to highlight data points in your visualizations, emphasizing their significance, such as notable data changes. You can either make annotations with static dates or use a custom query. We can, for example, modify the existing visualization, **[Rennes Traffic] Average speed & Traffic Status**, and add a layer of annotation in the **Annotation query** field to show that when the highway (`max_speed=110`) is congested, these annotations are marked as numbered lines, as shown in *Figure 6.47*:

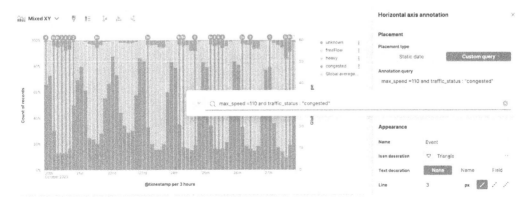

Figure 6.47 – Kibana Lens annotations

See also

- As Kibana Lens becomes the default choice to build visualizations and **Time Series Visual Builder (TSVB)** visualizations are deprecated in Elastic Stack 8.x. There is a new feature to convert the existing TSVB visualizations to Kibana Lens visualizations: `https://www.elastic.co/guide/en/kibana/current/tsvb.html#edit-visualizations-in-lens`

- To learn more about Kibana Formula, please check out this blog post, which gives you concrete examples of the use cases: `https://www.elastic.co/blog/kibana-10-common-questions-formulas-time-series-maps`

Creating visualizations from runtime fields

While building data visualizations or exploring data, you may need to include fields derived from the values of other fields. A **runtime field** is a very useful feature that allows you to define and enrich the index schema at runtime without altering your source data. In this recipe, we will see how to create runtime fields in Kibana and how to build useful visualizations from them.

Getting ready

To begin creating visualizations for runtime fields, ensure that you have completed the previous recipes in this chapter. We will continue to use the same dataset on Rennes traffic for this guide.

You can find the code snippets for this recipe at the following address: `https://github.com/PacktPublishing/Elastic-Stack-8.x-Cookbook/blob/main/Chapter6/snippets.md#creating-visualization-from-runtime-fields`.

How to do it...

For Rennes traffic, let's compare the traffic statuses between weekdays and weekends and between rush hour and nighttime using runtime fields.

First, let's create a runtime field that represents the hour of day of each document.

Navigate to **Kibana | Analytics | Discover** and then follow these steps:

1. Select the **metrics-rennes_traffic-raw** data view.
2. At the bottom of the left panel, click on **Add a field**.

3. Set the name to `hour_of_day`, and leave the field type as **keyword**. Then, toggle **Set value** and set the value with the following Painless snippet and click on the **Save** button (the snippet is available at `https://github.com/PacktPublishing/Elastic-Stack-8.x-Cookbook/blob/main/Chapter6/snippets.md#hour-of-the-day-painless-script`):

```
ZonedDateTime date =  doc['@timestamp'].value;
ZonedDateTime cet = date.withZoneSameInstant(ZoneId.of('Europe/
Paris'));
int hour = cet.getHour();
if (hour < 10) {
    emit ('0' + String.valueOf(hour));
} else {
    emit (String.valueOf(hour));
}
```

4. Head to **Kibana | Analytics | Visualize Library**, click on **Create visualization**, and choose **Lens**.

5. Select the **metrics-rennes_traffic-raw** data view.

6. Choose **Bar vertical percentage** as the chart type.

7. Drag and drop **Records** and **traffic_status** onto the workspace. Click on **Top 3 values of traffic status** in the right panel, change the number of values to 4, change the name to `Traffic Status`, and adjust the color palette to **Temperature**.

8. In the right panel, remove **@timestamp** from the **Horizontal axis** field and drag and drop the **hour_of_day** runtime field into its place. Click on **Top 3 values of hour_of_day**, change the value to `24`, and rank it by **Alphabetical**. Your visualization should resemble *Figure 6.48*:

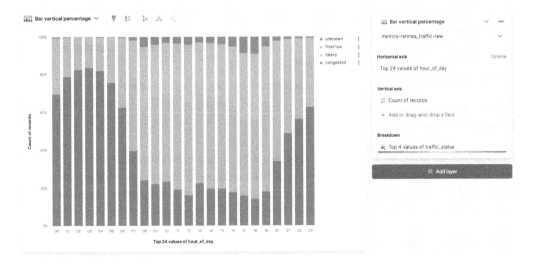

Figure 6.48 – Kibana Lens visualization with the hour_of_day runtime field

9. Click the **Save** button at the top right, then set the title to `[Rennes Traffic] Traffic status by hour`. Select **None** in the **Add to Dashboard** section and add the **cookbook** tag to the visualization. Finally, click on **Save and add to library**.

Now, try creating a second runtime field called `day_of_week` to represent the day of the week for each document. Use the following snippet for the value of the field (the snippet can be found at `https://github.com/PacktPublishing/Elastic-Stack-8.x-Cookbook/blob/main/Chapter6/snippets.md#day-of-the-week-painless-script`):

```
emit(doc['@timestamp'].value.dayOfWeekEnum.getDisplayName(TextStyle.
FULL, Locale.ROOT))
```

Following a similar process, create a visualization that depicts the distribution of traffic statuses according to the day of the week, as shown in *Figure 6.49*:

Figure 6.49 – Kibana Lens visualization with the day_of_week runtime field

How it works...

In our example, we first created a runtime field in a Kibana data view within the Discover UI. We defined the field's name and type directly in the Kibana runtime field creation interface. To set the value, we used a Painless script and employed the `emit()` function to assign the value to our field. Once created, the runtime field appears just like any other field; during searches, Kibana automatically adds a `runtime_mappings` clause.

We created two runtime fields, `hour_of_day` and `day_of_week`, derived from the `@timestamp` field. This approach enabled us to create these fields at runtime without reindexing. Furthermore, runtime fields can be suitable for various use cases, such as the following:

- Retrieving a disabled field from `_source`
- Exploring unstructured data
- Overriding the value or the type of an indexed field at query time

Painless script

While defining the value of the field, we used a scripting language called Painless. In our `hour_of_day` example, it facilitated extracting the value from `@timestamp`, adjusting the time zone, and formatting it properly as Strings. Painless was explicitly designed for Elasticsearch and provides a user-friendly, Groovy-like syntax, allowing direct usage of Java APIs, as we demonstrated in our examples.

There's more...

In our guide, we created the runtime field directly in our Kibana data view. However, there are two other methods to create a runtime field:

- **Defining the runtime field as part of a search**: To do this, you can add a `runtime_mappings` clause to your search, name the field, indicate its type, write the script to determine the value, and use `emit()` to assign that value to the field. Detailed documentation is available here: `https://www.elastic.co/guide/en/elasticsearch/reference/current/runtime-search-request.html`.

- **Adding the runtime field to an index mapping**: Rather than adding a runtime field to each search, you can include a runtime clause in the index mapping. As in other cases, the runtime field will operate the same as a regular field. Documentation can be found here: `https://www.elastic.co/guide/en/elasticsearch/reference/current/runtime-mapping-fields.html`.

Performance considerations

Since runtime fields are computed for each document every time you run a search, it is generally more efficient to create the field within an ingest pipeline where the script runs only once per document. When you do create a runtime field, be aware of the time and space complexity of your script and strive to optimize it. Inefficient runtime fields can slow down the Elasticsearch cluster or even cause a node to crash.

When to use runtime fields

Runtime fields are good candidates for the following situations:

- You need to control the mapping explosion

- You need to save space by not indexing a field

- You require a field calculated from data available at query time

- You are prototyping but plan to convert it into an indexed field later

- You are exploring unstructured data, such as testing Grok processing without an ingest pipeline

See also

For additional insights into how to use the Painless scripting language, please consult the official documentation: `https://www.elastic.co/guide/en/elasticsearch/reference/current/modules-scripting-painless.html`.

Creating Kibana maps

Kibana Maps is a place where you can create dynamic, informative maps from your data. In this recipe, you will learn how to use Kibana Maps to visualize and understand data in a spatial context.

Getting ready

To create map visualizations, ensure that you have completed the previous recipes in this chapter, as we will use the same dataset on Rennes traffic for this recipe.

How to do it...

For the Rennes traffic, we would like to understand the traffic fluidity by using a geographic map. Follow these steps:

1. Head to **Kibana | Analytics | Maps** and then click on **Create map**.

2. You'll arrive at an empty map from which you can start adding layers. First, make sure that the time filter in the top-right corner is set appropriately for your data.

3. Click **Add layer**. Here, you will see different types of layers available for your map. Select **Documents** (the term for individual records in the Elastic Stack). This layer allows you to plot individual documents for road portion locations in the Rennes traffic data.

4. From the **Data view** dropdown, choose **metrics-rennes_traffic-raw**, and leave **Geospatial Field** set to **location**. Then, click on **Add and continue**.

5. In the **Layer settings** panel, provide a name for the layer, such as `traffic-fluidity`.

6. Scroll down to the **Layer style** panel and set the following, as shown in *Figure 6.50*, then click on **Add and continue**:

    A. In the **Fill color** section, choose **By value** and select **average_vehicle_speed**.

    B. Alter the color range in the **As number** field to a full rainbow and activate the **Reverse colors** toggle.

    C. Set the symbol size to **fixed** and change the pixel size to 4 px.

Figure 6.50 – Adjusting the map layer style

7. Click on **Keep changes**, then use the **Fit to data bounds** button on the left toolbar to zoom into the Rennes traffic data. The resulting map displays locations, with the color of each document indicating the average vehicle speed from that location:

Figure 6.51 – First layer of the Rennes traffic map

It's possible to add multiple layers to one map. We can now add another layer by clicking on **Add layer** and selecting **Clusters**.

8. Select **metrics-rennes_traffic-raw** from the **Data view** dropdown.

9. Keep the Cluster field value as **location** and, in the **Show as** section, keep **Clusters** selected. Then, click on **Add and continue**.

10. On the next screen, in the Layer settings panel, name the layer, for example, `maximum permissible speed`. Then, set the opacity to **50%**. This setting makes the map more visible when it contains multiple layers.

11. In the **Metrics** section, select **max** in the **Aggregation** field and **max_speed** in the **Field** field.

12. In the **Layer style** flyout, verify that the following values are pre-filled, as shown in *Figure 6.52*:

    A. The **Fill color | By value** field should show **max max_speed**.

    B. The **Symbol size | By value** field should also show **max max_speed**.

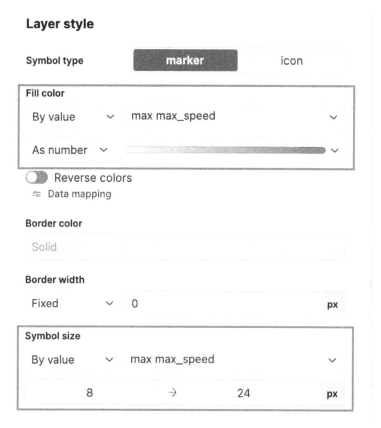

Figure 6.52 – Adjusting the second map layer style

13. Click on **Keep changes**. The map now displays the additional layer representing maximum permissible speeds in a clustered format. We've adjusted the opacity to ensure that both layers are visible. In *Figure 6.53*, you can see a map displaying both average speed metrics and zones indicating maximum allowable speeds:

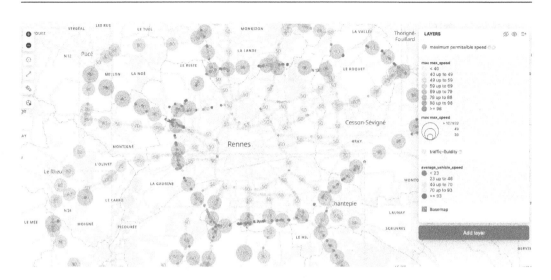

Figure 6.53 – Rennes traffic map with two layers

Finally, click **Save** at the top right, then set the title to [Rennes Traffic] Traffic fluidity. Select **None** in the **Add to Dashboard** section and add the **cookbook** tag to the visualization. Finally, click on **Save and add to library**.

How it works...

As you can see, Kibana Maps lets us display highly useful geographical data. In our example, we constructed a multilayer map that represents geographic points and clusters of aggregated documents. We leveraged the styling options available to customize our map according to the specifics of our dataset and visualization objectives.

There's more...

We only explored a subset of the layer types— **Documents** and **Clusters**—in our recipe. Kibana Maps offers an extensive array of layer types that we encourage you to try out and discover the full potential of Kibana Maps, including **Choropleth**, **Create index**, **EMS Boundaries**, **Point to point**, **Top hits per entity**, **Tracks**, and **Upload file** and more. We encourage you to explore these additional types to unlock the full capabilities of Kibana Maps. For in-depth information, you can refer to the full documentation here: https://www.elastic.co/guide/en/kibana/current/vector-layer.html.

With our example, we constructed a map with multiple layers. Generally, Kibana Maps enables you to do the following:

- Create maps with multiple layers and indices

- Upload custom GeoJSON files

- Embed your map in a dashboard (we will see this in the next recipe, *Creating and using Kibana dashboards*)

- Embed your map in a Canvas workpad (we will see this in the last recipe of this chapter, *Creating Canvas workpads*)

Furthermore, you can also leverage the **Elastic Maps Service** (**EMS**), a free hosting service that provides you with more than 60 vector layers built for Kibana to cover all your geo-visualization needs for any geography. By default, Kibana Maps is configured to use EMS, which connects to the following domains:

- `tiles.maps.elastic.co`

- `vector.maps.elastic.co`

See also

The comprehensive documentation for EMS is available at this link: `https://www.elastic.co/elastic-maps-service`.

If you cannot access EMS from Kibana due to licensing or connectivity issues, you have the option to self-host the service on your infrastructure; the full documentation can be found here: `https://www.elastic.co/guide/en/kibana/current/maps-connect-to-ems.html#elastic-maps-server`.

Creating and using Kibana dashboards

In this guide, we will integrate all previously created visualizations into a comprehensive dashboard consisting of multiple panels. Additionally, we will explore how to enhance user interaction using control-based drilldowns.

Getting ready

Make sure to complete the following recipes from this chapter:

- *Creating visualizations with Kibana Lens*

- *Creating visualizations from runtime fields*

- *Creating Kibana maps*

At the end of this recipe, you will have dashboards composed of the various visualizations and elements built into the aforementioned recipes.

How to do it...

Building dashboards is very straightforward in Kibana, especially if you've already created some visualizations. Follow these steps:

1. Go to **Kibana | Analytics | Dashboard** and click on **Create dashboard**.

    This will bring you to a blank canvas, where you can start adding some visualizations.

2. We will start by adding a nice image! You can be creative, but we provided a sample picture:

    A. Click on **Add panel | Image**.
    B. Select the **Use link** tab and set **Link to image** with the following URL: `https://upload.wikimedia.org/wikipedia/commons/6/60/Ville_de_RENNES_Noir.svg`. Then, click on **Save**:

Figure 6.54 – Adding an image for a logo

The logo will be added to the panel. Including a picture is a great way to add some personalization and branding to your dashboards. Let's add some proper visualizations from the ones we've built in the last three recipes.

3. Click on **Add from library** and select the **[Rennes Traffic] Number of locations** visualization. Make sure to align it to the right with the image panel.

4. Let's add another visualization; this time, we'll pick **[Rennes Traffic] Average speed gauge**.

At this stage, your dashboard should look like the one shown in *Figure 6.55*:

Figure 6.55 – Rennes traffic dashboard – first step

You can easily rearrange the position of the different panels by clicking on the title section and moving the panel with your mouse anywhere you want on the canvas. To adjust the size and fit of the panel, position your mouse on the small arrow at the bottom right of the panel.

Let's keep adding more panels to our dashboard.

5. Click on **Add from library** and add the following visualizations in the respective order:

    I.   **[Rennes Traffic] Traffic status waffle**

    II.   **[Rennes Traffic] Speed by road hierarchy**

    III. **[Rennes Traffic] Average speed & Traffic Status**

    IV.   **[Rennes Traffic] Traffic status by hour**

6. Finally, let's add a **Map** visualization for a real-time view of the traffic; select the one named **[Rennes Traffic] Traffic fluidity**.

By now, your dashboard should look like the one shown in *Figure 6.56*:

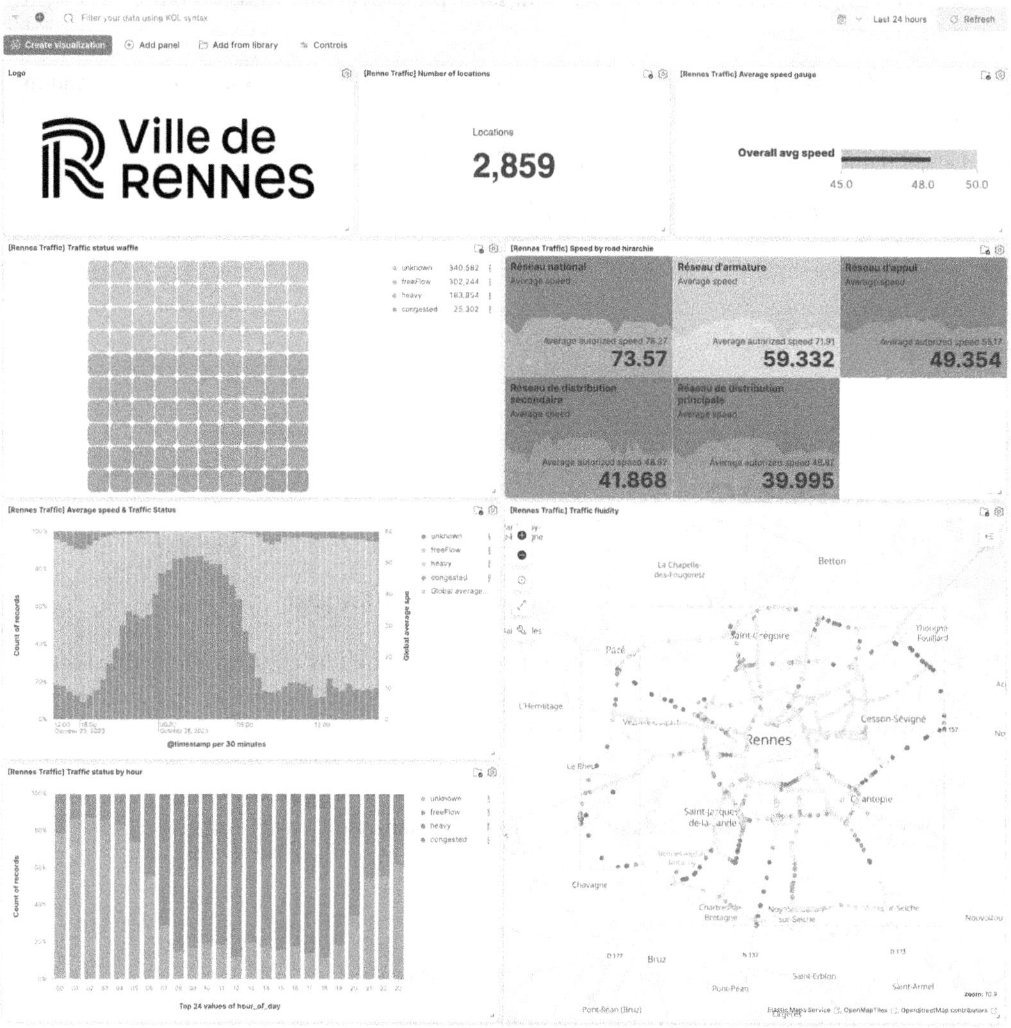

Figure 6.56 – Rennes traffic dashboard – more visualizations

You can start playing around with the dashboard to see the built-in interactivity of the panels. For example, clicking on a specific road hierarchy will automatically apply the filter to the entire dashboard.

You can also have dedicated panels to filter and display only the data you are interested in with **Controls**. Let's add some to our dashboard.

7.   On the dashboard toolbar, click on **Controls**:

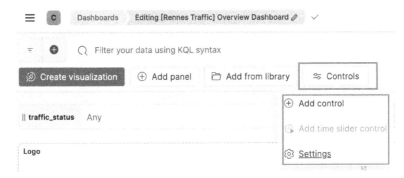

Figure 6.57 – Adding controls to the dashboard

8. From the drop-down list, select **Add control**; the **Create control** flyout will appear on the right of the screen.

9. Select the **traffic_status** field and click on **Save and close**.

10. Back to the dashboard, you now have a new panel on top of the visualization named **traffic_status**. By clicking on it, you will see a drop-down list where you can select the values associated with the status of the traffic you want to filter, as shown in *Figure 6.58*. Select **congested** as an example:

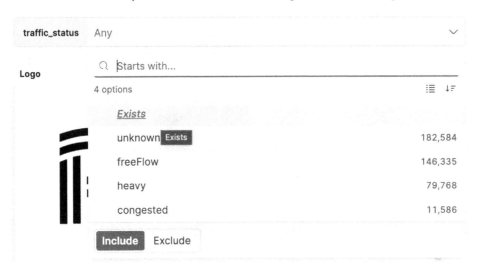

Figure 6.58 – Using controls in the dashboard

11. You can see on your dashboard that all the panels have been updated according to the value selected in the **traffic_status** control.

    Imagine you want to filter your traffic data to analyze it within a specific time range, such as early in the morning or late in the afternoon, to better understand traffic patterns. This is where the *time slider* control proves to be incredibly useful.

12. Go to the **Controls** menu again in the dashboard toolbar and select **Add time slider control**. You'll see a new panel to the right of **traffic_status**:

Figure 6.59 – Time slider control

By clicking the *play* icon, you will see your dashboard animate and your data change over the defined time range. You can advance the time range forward as well as backward, which is especially useful when working with time series data.

Your dashboard should now look as shown in *Figure 6.60*, with our two controls:

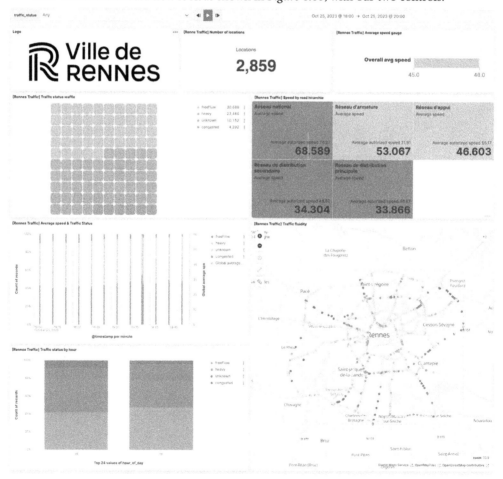

Figure 6.60 – Rennes traffic dashboard with controls

13. Save the dashboard by clicking the **Save** button in the upper-right corner. Name it `[Rennes Traffic] Overview`.

To enhance our dashboard further, consider this: users frequently manage multiple dashboards, and the ability to navigate seamlessly from one to another is crucial, especially when aiming to refine analysis or focus on more detailed panels related to a specific dataset. Dashboard drilldowns are invaluable in this scenario as they allow you to transition between dashboards while maintaining the overall context. Let's explore how to implement and use this feature effectively!

For this exercise, we have already built a drilldown dashboard. Download and save the NDJSON file of the exported dashboard from the following location: `https://github.com/PacktPublishing/Elastic-Stack-8.x-Cookbook/blob/main/Chapter6/kibana-objects/rennes-data-drilldown-dashboard.ndjson`. Then, follow these steps:

1. To import the dashboard, go to **Stack Management | Saved Objects**.

2. Click on **Import** and select the NDJSON file you have previously downloaded from the GitHub repository. Upon completing the import process, you will notice a warning in the flyout about data view conflicts. The reason is straightforward: our saved objects rely on an existing data view. To resolve the conflict, simply click on the drop-down list under the **New data view** column and select **metrics-rennes_traffic-raw**, as shown in *Figure 6.61*, then click on **Confirm all changes** to finalize the import procedure:

Figure 6.61 – Importing saved objects and selecting the right data view

3. Once all the objects have been imported, you will get a recap as shown in the following screenshot:

Figure 6.62 – Saved objects successfully imported from the file

Return to the [**Rennes Traffic**] **Overview** dashboard. Then, open the menu for the [**Rennes Traffic**] **Speed by road hierarchie** panel and select **Create drilldown**:

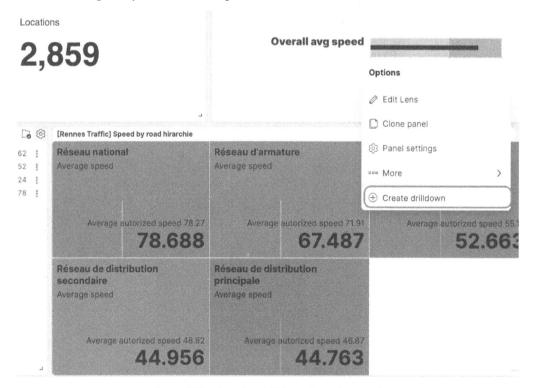

Figure 6.63 – Creating drilldown from the panel

4. Navigate to the drilldowns page and select the **Go to Dashboard** option. Here, you will need to name your drilldown—consider `View Details for Road Hierarchy` as a suggestion. Then, from the **Choose destination dashboard** drop-down menu, select **[Rennes Traffic] Detailed traffic drilldown dashboard**, which you have recently imported. This process sets up a targeted navigation path within your dashboard environment, allowing for a seamless transition between your overview and detailed analysis dashboards:

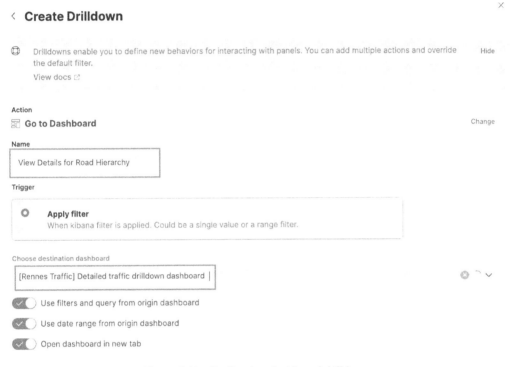

Figure 6.64 – Configuring dashboard drilldown

5. Click on **Create drilldown**. Save the dashboard to test our drilldown, click on one of the five charts in the **[Rennes Traffic] Speed by road hierarchie** panel. You will be redirected to the detailed dashboard filtered on the value you have selected.

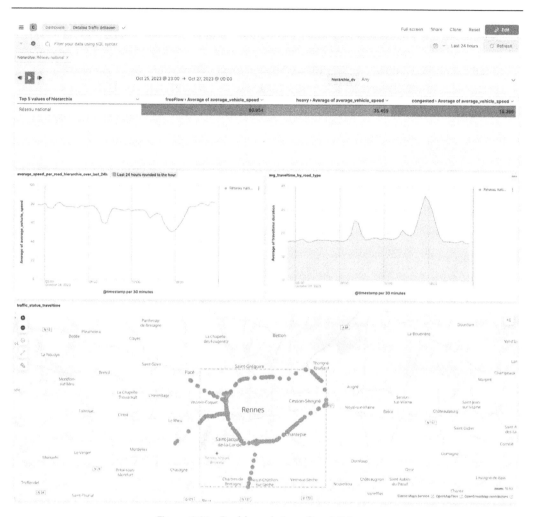

Figure 6.65 – Dashboard view after drilldown

Et voilà! You have just built your first dashboard with a nice touch of interactivity thanks to controls and drilldowns.

How it works...

In Kibana, a dashboard is a collection of visualizations and saved searches that you can arrange and customize to display the data that is most important to you. You can create multiple dashboards for different use cases, and each dashboard can have its own set of visualizations and searches.

Dashboards are a powerful tool for data analysis because they allow you to see multiple visualizations side by side and quickly identify patterns and trends in your data. You can also use dashboards to monitor key metrics in real time, which is especially useful for operational use cases. Kibana provides a wide range of visualization types that you can use to create custom dashboards, including bar charts, line charts, pie charts, tables, and more.

The following table outlines a framework for choosing the right visualization:

Use case	Recommended type of visualization
Comparison and correlation	Many items: Horizontal bar Few items: Vertical bar
Comparison over time	Few periods and categories: Stacked bar Few time periods but many categories: Line graph
Distribution of values	Few numbers of points: Vertical bar histogram Many points: Line histogram
Composition of a whole	Simple compositions with few items: Waffle or Treemap Multiple grouping dimensions for a few bottom-level items: Mosaic Multiple grouping dimensions for many bottom-level items: Treemap
Eye-catching summary	One value: Metric Many values: Table with color styling
Visualizing goals or targets	Vertical bar or Line with reference lines Metric

Table 6.2 – Choosing the right visualization

In addition to visualizations, Kibana dashboards also support saved searches, which allow you to quickly filter your data based on specific criteria. You can save searches that you use frequently and add them to your dashboard for easy access.

Overall, Kibana dashboards are a powerful tool for data analysis and monitoring. They allow you to quickly identify patterns and trends in your data, monitor key metrics in real time, and customize your view of the data to suit your needs.

There's more...

In our recipe, we have used dashboard drilldowns, but you can also create URL and Discover drilldowns. With the former, you can link to data outside of Kibana, and with the latter, you can open Discover from a Lens panel while keeping all the contextual information.

Dashboards are great when used in Kibana, but you can also share them with teams and colleagues outside of Kibana. You have many options that are easily accessible from the **Share** menu in the toolbar when it comes to sharing dashboards: you can interactively embed dashboards as an iFrame, export them as reports in various formats (PNG, CSV, PDF, etc.), and share them as direct links for easy access.

When building dashboards, design thinking is a good practice. Start by asking yourself the following questions:

- What is the outcome or the goal of the dashboard? Is it about understanding high-level behaviors, visually correlating specific metrics at the same time, or finding the root cause of an issue?

- Who is using this dashboard to do their job? If you are building it for a team or someone else, step into their shoes to visualize their perspective when they will need that data.

See also

- Looking for more design tips to elevate your dashboards? Look no further and check out this blog: `https://www.elastic.co/blog/designing-intuitive-kibana-dashboards-as-a-non-designer`

- If you're interested in delving deeper into the topics of creating dashboards more efficiently, be sure to check out this technical blog: `https://www.elastic.co/blog/building-kibana-dashboards-more-efficiently`

- For developers interested in debugging their Kibana dashboard, the following article will be very useful: `https://www.elastic.co/blog/debugging-kibana-dashboards`

Creating Canvas workpads

In this recipe, we'll delve into the art of crafting engaging and informative visual workpads using Kibana's Canvas feature. **Canvas workpads** transform your data into live infographics. Starting from a blank slate, you can incorporate logos, colors, and design elements to present your data uniquely.

Getting ready

To prepare for creating a Canvas workpad, ensure that you have completed the following recipes from this chapter:

- *Creating visualizations with Kibana Lens*
- *Creating visualizations from runtime fields*
- *Creating Kibana maps*

The snippets of this recipe are available at `https://github.com/PacktPublishing/Elastic-Stack-8.x-Cookbook/blob/main/Chapter6/snippets.md#creating-canvas-workpad`.

By the end of the recipe, you will have built a Canvas workpad that should resemble *Figure 6.66*. Refer to this figure as a guide throughout the recipe:

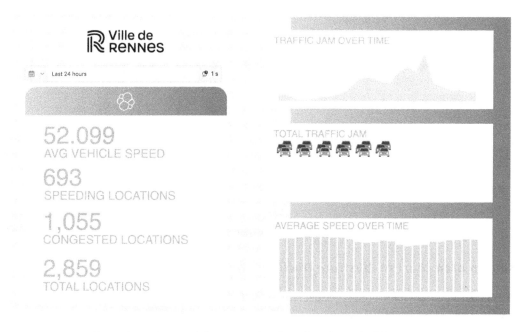

Figure 6.66 – Expected Canvas workpad result at the end of this recipe

How to do it...

You will start with a workpad template. Download the Canvas workpad from the following URL: `https://github.com/PacktPublishing/Elastic-Stack-8.x-Cookbook/blob/main/Chapter6/kibana-objects/canvas-workpad-rennes-traffic-template.json`. Then, follow these steps:

1. Go to **Kibana | Analytics | Canvas**, then drag and drop the JSON file, `canvas-workpad-rennes-traffic-template.json`, that you've downloaded into the **Add your workpad** field. If you previously created other Canvas workpads, click on **Import workpad JSON file** instead and select the JSON file you downloaded.

2. Next, click **View** in the top menu, followed by **Zoom | Fit to window**. You should now see the initial workpad that you imported. We will add various chart elements to the workpad, following the order shown in *Figure 6.67*:

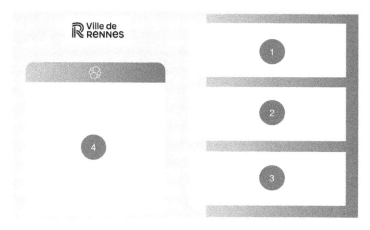

Figure 6.67 – Canvas workpad starting point

Using an area chart to illustrate traffic congestion over time

Let's add our first element to create a visual representation of traffic congestion over time:

1. We will start by clicking **Add element**, then, from the **Chart** menu, select **Area**. Once added, move the area chart into position, making sure it fits within the first white box on the right-hand side.

2. Next, select the **Data** tab on the right. By default, **Demo data** is selected. Change the data source by clicking on **Demo data** and selecting **Elasticsearch SQL**.

3. Once you have selected **Elasticsearch SQL**, update the SQL query to the following:

```
SELECT HOUR_OF_DAY("@timestamp") hour, COUNT(*) locations
FROM "metrics-rennes_traffic-raw"
WHERE traffic_status = 'congested' or traffic_status = 'heavy'
```

```
GROUP BY hour
ORDER BY hour
```

4. You can click **Preview data** to get a quick preview of the data that this SQL statement returns. After reviewing the data, click **Save**.

5. Now, adjust the chart's appearance by navigating to the **Display** tab and modifying the settings according to *Figure 6.68*:

    A. For **X-axis**, select the **hour** value.

    B. For **Y-axis**, select the **locations** value.

    C. Update the color to a specific shade of green, rather than leaving it on auto.

Figure 6.68 – Canvas area chart display setting

6. As shown in *Figure 6.68*, you can adjust the axis configuration by adding **X-axis** and **Y-axis** configurations in the **Chart style** section. Use the toggles to turn off both axes, creating a cleaner chart appearance.

7. Click the + button to the right of **Element style** and choose **Container style**. Change **Opacity** to **50%**, as shown in *Figure 6.68*.

8. To add a title to the chart, insert a new text element and use the Markdown editor to replace the placeholder text with TRAFFIC JAM OVER TIME. Resize and align the text element with the area chart.

Using an image chart to illustrate a total traffic jam

Next, let's add the second element to our Canvas workpad using an **Image Repeat** chart to illustrate the total traffic jam:

1. Clone the TRAFFIC JAM OVER TIME text element by selecting it and choosing **Clone** from the **Edit** menu. Update the text to TOTAL TRAFFIC JAM and place the element in the second box on the right-hand side.

2. Click **Add element** and select the **Image** menu | **Image repeat**. Then, resize the element and place it in the second white box, under the title.

3. On the right-hand side under the **Display** tab, choose the traffic jam image under **Asset**.

4. Switch to the **Data** tab and modify the data source with the following Elasticsearch SQL query:

```
SELECT COUNT(*)/10000 as locations
FROM "metrics-rennes_traffic-raw"
WHERE traffic_status = 'congested' or traffic_status = 'heavy'
```

5. Switch back to the **Display** tab. Delete the existing value under **Measure**. Now, in the **Value** field, select **Value** in the first dropdown and **locations** in the second dropdown, as shown in *Figure 6.69*:

Figure 6.69 – Image repeat chart display setting

6.   Click the + button to the right of **Repeating image** and select **Image size** and **Max count**. Set them to 50 and 33, respectively, as shown in *Figure 6.69*.

Using a bar chart to illustrate average traffic speed over time

Continuing with our visualization, let's use the third white box on the right-hand side to provide insights into the average traffic speed:

1.   Clone the earlier text element as we did for the previous text element, and replace the text with AVERAGE SPEED OVER TIME. Position it in the third white box on the right-hand side.

2.   Add a new element by selecting **Chart | Vertical bar**. Place it in the third white box.

3.   On the **Data** tab, change the data source with the following Elasticsearch SQL query:

```
SELECT
  HOUR_OF_DAY("@timestamp") hour,
  AVG("average_vehicle_speed") speed
FROM "metrics-rennes_traffic-raw"
GROUP BY hour
ORDER BY hour
```

4. On the **Display** tab, make the following changes:

    A. In the **Dimensions & measures** section, set the **X-axis** field to the **hour** value.

    B. Then, set the **Y-axis** field to the **speed** value.

    C. In the same section, click **x** on the color line to remove the color dimension.

    D. In the **Default style** section, change the color from auto to the green color you used before.

    E. Add **Container style** to the **Element style** section and set **Opacity** to **30%**.

    F. Bring out the axis configuration by adding **X-axis** and **Y-axis** configurations in the **Chart style** section. Then, use the toggles to turn them off to make the chart cleaner.

    Your final display setting should look like *Figure 6.70*:

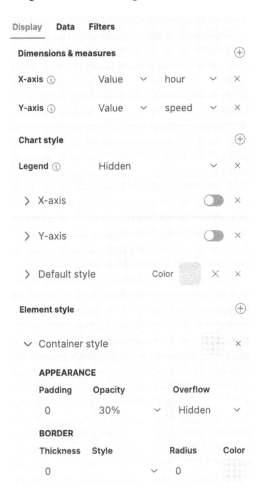

Figure 6.70 – Vertical bar chart display setting

5. At this stage, you can check that your Canvas workpad's right-hand side looks like the screenshot in *Figure 6.71*:

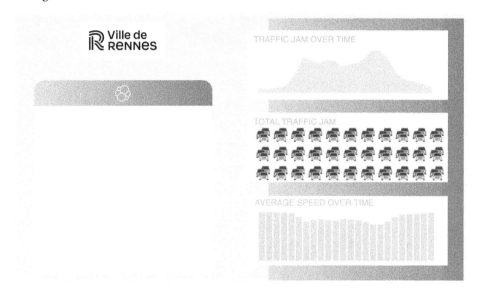

Figure 6.71 – Canvas workpad with three first elements

Using metric charts to illustrate summarized traffic metrics

Now, let's add some summarized traffic metrics to our Canvas workpad for quick insights:

1. Add a metric by clicking **Add element** and selecting **Chart | Metric**. Move the chart and place it in the white box on the left.

2. On the **Data** tab, change the data source with this Elasticsearch SQL query:

    ```
    SELECT AVG(average_vehicle_speed) metric
    FROM "metrics-rennes_traffic-raw"
    ```

3. On the **Display** tab, make the following changes:

    A. Under **Measure**, choose both **Value** and the new metric.

    B. Change the label to AVG VEHICLE SPEED.

    C. Adjust the alignment of both the label and metric text to the left.

    D. Set the font sizes to **60** for the metric and **30** for the label.

    E. Apply the same green color used in prior elements to the font.

4. Clone the metric and place it below the first metric.

5. Change the label to SPEEDING LOCATIONS and the SQL query to the following:

```
SELECT COUNT(DISTINCT location_reference) metric
FROM "metrics-rennes_traffic-raw"
WHERE average_vehicle_speed/max_speed > 1
```

6. Clone the metric again. Position it below the others. Change the label to CONGESTED LOCATIONS and the SQL query to the following:

```
SELECT COUNT(DISTINCT location_reference) as metric
FROM "metrics-rennes_traffic-raw"
WHERE traffic_status = 'congested'
```

7. Clone the metric again. Position it below the others. Change the label to TOTAL LOCATIONS and the SQL query to the following:

```
SELECT COUNT(DISTINCT location_reference) as metric
FROM "metrics-rennes_traffic-raw"
```

To control the timeframe of the data displayed in your metrics, add a **Time filter** element from the **Filter** menu and situate it above the series of metrics.

By following these steps, you have created a Canvas workpad that mirrors the example provided in the *Getting ready* section of this recipe.

How it works...

As you can see, Canvas allows us to add various types of elements according to our preferences, providing extensive options for customizing the look and feel. We utilized SQL syntax to access our data in Elasticsearch. If you come from a relational database background, this syntax should be very familiar to you.

For the complete documentation of Elasticsearch SQL, visit https://www.elastic.co/guide/en/elasticsearch/reference/current/sql-spec.html.

There's more...

In Elastic Stack version 8.x, you can add panels created and saved in the Visualize Library to Canvas workpads. This feature allows for the reuse of existing elements, enabling even non-technical users to create visualizations with Kibana Lens and integrate them into Canvas workpads.

Canvas also supports creating elements on the fly in addition to adding saved ones, such as Lens, maps, machine learning anomaly detection, images, log streams, TSVB, and Vega visualizations.

In our recipe, we've provided a template to help you get started. Elastic also offers a variety of generic templates that can inspire multi-page workpads with diverse visual elements. Explore these at **Analytics | Canvas | Templates**.

See also

- Check out the Canvas demo gallery to draw inspiration from real-world examples: `https://www.elastic.co/demo-gallery/finance-revenue-sales-marketing-workpad`

7

Alerting and
Anomaly Detection

Alerting and **anomaly detection** tools are crucial for proactive decision-making across various domains, from IT operations to business intelligence. In this chapter, we will explore the strategies, techniques, and tools needed to set up effective alerts and detect anomalies in your data. This chapter aims to equip you with the knowledge and skills necessary for the early detection of anomalies, enabling timely responses in your data-driven environment.

In this chapter, we're going to cover the following main topics:

- Creating alerts in Kibana
- Monitoring alert rules
- Investigating data with log rate analysis
- Investigating data with log pattern analysis
- Investigating data with change point detection
- Detecting anomalies in your data with unsupervised machine learning jobs
- Creating anomaly detection jobs from a Lens visualization

Technical requirements

To follow different recipes in this chapter, you'll need an Elastic Stack deployment that includes the following:

- Elasticsearch for searching and storing the data
- Kibana for data visualization and stack management (a Kibana user with **All** privileges set for **Fleet** and **Integrations**)

Creating alerts in Kibana

Alerting is a vital component of the Elastic Stack. You can use the data stored in Elasticsearch to trigger alerts based on specified conditions. Alerting actions can include sending an email or Slack message, writing data to an Elasticsearch index, and invoking an external web service with passed data, among others. In this recipe, we will learn how to create an alerting rule using the Elastic Stack's new alerting framework.

Getting ready

Ensure you have access to the mailbox associated with the email address you used to sign up for Elastic Cloud to receive alerts from your deployment.

Make sure that you finished the *Exploring your data in Discover* recipe in *Chapter 6*.

How to do it...

Let's create a rule that uses the **metrics-rennes_traffic-raw** data stream, which was created in the *Exploring your data in Discover* recipe in *Chapter 6*.

1. Navigate to **Kibana | Management | Stack management**, then select **Rules** under **Alerts and Insights**.

2. Under **Rules**, select **Create your first Rule**. In the flyout, select the **Index Threshold** rule type (located near the bottom). We can begin preparing our rule in the next steps.

3. Set the rule name to `Rennes Traffic congestion average speed` and the rule tag to `cookbook`. Under the **Select indices** field, set **Index** to `metrics-rennes_traffic-raw` and choose **@timestamp** for the time field as shown in *Figure 7.1*:

Figure 7.1 – Setting indices for the rule

4. Under the **Define the condition** section, set WHEN to average(), OF to average_
    vericle_speed, OVER to all documents, IS BELOW to 12, and FOR THE LAST
    to 35 minutes as shown in *Figure 7.2*.

Note

You may need to adjust the values for IS BELOW and FOR THE LAST depending on the
real-time traffic data to ensure the rule can trigger some alerts.

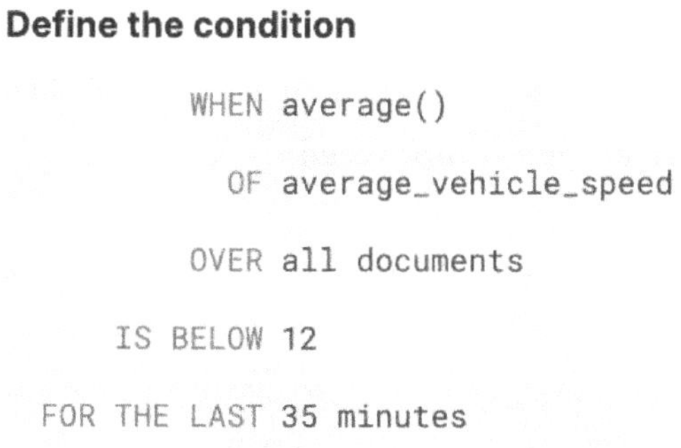

Define the condition

```
WHEN average()

  OF average_vehicle_speed

OVER all documents

IS BELOW 12

FOR THE LAST 35 minutes
```

Figure 7.2 – Defining the rule conditions

5. Next, set **Filter** to `traffic_status: "congested"` and adjust the rule frequency to check every **30 minutes** as shown in *Figure 7.3*:

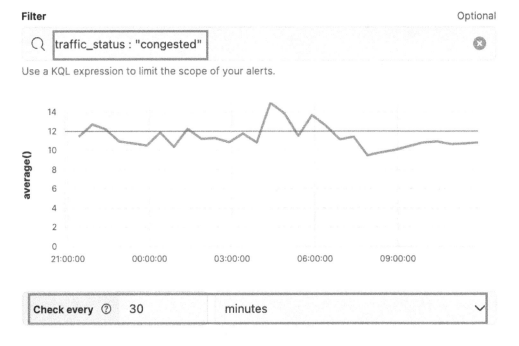

Figure 7.3 – Rule filter and frequency

Now, you are ready to prepare the alert actions. It's very convenient to set up an email action as Elastic Cloud deployments come with a preconfigured **Elastic-Cloud-SMTP** connector. Under **Actions**, select **Email** as the connector type and configure the email action with the following details (refer to *Figure 7.4*).

6. Make sure to select **Elastic-Cloud-SMTP email connector**.

7. Set **Action frequency** to **On status changes** for each alert, and set **Run When** to **Threshold met**.

8. Enter the email address that you used to sign up for on Elastic Cloud.

9. Set **Subject** to **Rennes traffic congestion alert**.

10. Customize the message as follows, then click **Save** to create the rule (the code snippet can be found at `https://github.com/PacktPublishing/Elastic-Stack-8.x-Cookbook/blob/main/Chapter7/snippets.md#alerting-message-snippet`):

```
The measured average value is {{context.value}} km/h
The average vehicle speed for congested locations is less
than 12km h over {{rule.params.timeWindowSize}}{{rule.params.
timeWindowUnit}}
Timestamp: {{context.date}}
```

Actions

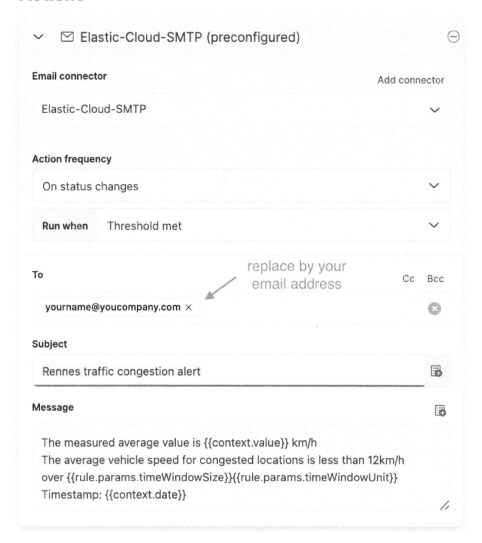

Figure 7.4 – Setting the email rule action

Now, wait and check your inbox for any new alerts, indicating that the threshold has been met and the average vehicle speed for congested locations has fallen below 12 km/h within the last 35 minutes, as shown in *Figure 7.5*:

No Reply - Elastic Alerts <noreply@alerts.elastic.co>   Thu, May 23, 12:50 AM (2 days ago) ☆ ↰ ⋮
to me ▾

The measured average value is 10.314814814814815 km/h

The average vehicle speed for congested locations is less than 12km h over 35{rule.params.timeWindowUnit}}

Timestamp: 2024-05-22T22:50:47.047Z

This message was sent by Elastic. View rule in Kibana.

Figure 7.5 – Example of an email alert

You can also add multiple actions to a single rule. Try exploring action types such as **webhook**, **Slack**, and so on. Click on **Action**, select the connector type of your choice, and then create the connector on the fly.

In the next recipe, we will learn how to monitor our alert rule.

How it works...

As demonstrated in this recipe, Kibana alerting provides a user-friendly interface for creating, managing, and visualizing alerts. Alerts in Kibana are seamlessly integrated with the platform's visualization capabilities, making it easy to set up alerts based on data visualizations. Kibana alerting allows you to define conditions, set triggers, and establish actions through an intuitive graphical interface, thus simplifying the alerting process.

In essence, Kibana alerting works by performing scheduled checks to monitor predefined conditions. When a condition is met, it triggers an alert, leading to the initiation of one or more predefined actions. These actions typically involve communication with Kibana services or third-party integrations. Connectors play a crucial role in facilitating these interactions, enabling seamless integration with external services and applications.

Additionally, in Elastic 8.x, several new enhancements have been introduced to automate the previously manual alert creation processes, offering more flexibility for custom conditions and reducing alert noise. Here is a selected list of new alerting features that were released in Elastic 8.x:

- **Rule as code (introduced in 8.9)**: the possibility to use APIs or Terraform provided by Elastic Stack to manage alerting. (`https://www.elastic.co/guide/en/kibana/current/alerting-apis.html`)

- **Conditional actions (introduced in 8.10)**: Allowing users to build more complex scenarios based on schedules and alert queries. (`https://www.elastic.co/guide/en/kibana/current/create-and-manage-rules.html#defining-rules-actions-details`)

- **Alert actions summarization (introduced in 8.7)**: Alert Action Summarization (introduced in 8.7): Aggregating several actions into one to reduce notification volume. (`https://www.elastic.co/guide/en/kibana/current/release-notes-8.7.0.html#features-8.7.0`)

There's more...

In the following sections, we'll explore best practices for creating effective alerts in Kibana and Elasticsearch. We'll also compare the **Watcher** feature with **Kibana alerting** to understand their differences and how to use them effectively.

Alerting best practices

While Kibana and Elasticsearch provide a comprehensive alerting system, having clear strategies, policies, and processes for alerting is crucial. Additionally, it is essential to have the necessary data upon which to base alerts. Care must be taken to ensure an optimal alerting configuration is used to avoid common issues such as the following:

- Excessive false alerts
- Incorrect or excessive notifications
- Unclear action responsibilities
- Lack of notification acknowledgment
- Ineffective escalation procedures
- Absence of purpose-specific channels
- Insufficient resolution details in notifications

Here are some alerting best practices:

- Ensure that alerts are actionable and contain clear, contextual details for an easy response
- Add tags and identifiers to enhance context
- Alert based on symptoms, not causes, to prioritize immediate issues
- Segregate critical alerts from non-urgent information by using different channels
- Reserve operational channels for notifications that demand urgent action

Alerting versus Watcher

Watcher (`https://www.elastic.co/guide/en/kibana/current/watcher-ui.html`) is a historically built-in feature of Elasticsearch that enables you to automate actions based on specific conditions within your data. It continuously monitors the data and executes predefined actions when certain conditions are met. Watcher utilizes a query language to filter data and can trigger actions such as sending emails, Slack notifications, or invoking webhooks.

Key differences between Watcher and Kibana alerting	Watcher	Alerting
1 alert to multiple instances	✗	✓
Space aware (multi-tenancy)	✗	✓
Scheduled checks run on…	Elasticsearch	Kibana
Actionable alerts (mute and throttle individual alerts)	✗	✓
Integrated into Elastic Observability and Elastic Security	✗	✓

Table 7.1 – Differences between Kibana alerting and Watcher

As shown in *Table 7.1*, Kibana alerting is now the preferred default alerting framework, offering several advantages over Watcher, which has become more of a legacy feature. However, Watcher may still be valuable in situations such as the following:

- **Legacy systems compatibility**: If you have already built and fine-tuned your Watcher alerts, you might prefer to continue using them rather than migrating to the newer Kibana alerting systems

- **Advanced customizability**: If you require complex scripted transformations or programmable and conditional actions within your alerting rules

See also

There are many new features related to alerting introduced in Elastic Stack 8.x:

- Kibana provides a full set of action and connector APIs; the full API documentation can be found at the following address: `https://www.elastic.co/guide/en/kibana/current/actions-and-connectors-api.html`

- To understand more about how to create and configure connectors, please check out the official documentation at this address: `https://www.elastic.co/guide/en/kibana/current/action-types.html`

- Apart from alerting, Kibana also provides a complete system of case management at `https://www.elastic.co/guide/en/kibana/current/cases.html`

- Starting from version 8.11, it's possible to create alerting rules based on ES|QL queries – this blog post gives very detailed examples: `https://www.elastic.co/blog/getting-started-elasticsearch-query-language`

Monitoring alert rules

Setting up effective alerts is essential for prompt detection of issues and maintaining proactive oversight of your data. Monitoring these alerts to identify any malfunctions and ensure no potential problems are overlooked is crucial. In this recipe, we will explore the various tools available within the Kibana alerting framework that enable us to effectively observe and manage rules.

Getting ready

As usual, for this recipe, you'll need an up-and-running Elastic deployment with Elasticsearch, Kibana, and an ML node. To spin up a cluster on **Elastic Cloud** refer to the *Installing the Elastic Stack on Elastic Cloud* recipe in *Chapter 1*.

Also, make sure to have completed the previous recipe — *Creating alerts in Kibana*.

How to do it...

We're going to monitor the activity of the `Rennes Traffic congestion average speed` alerting rule.

1. Go to **Kibana** | **Stack Management** | **Rules** to display the active rules in our environment:

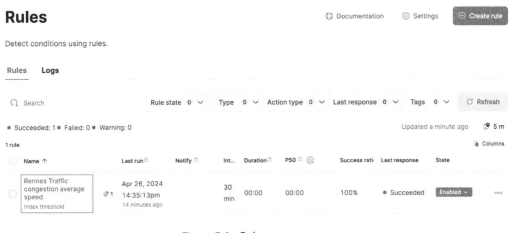

Figure 7.6 – Rules management

2. Click on the **Logs** tab. Since there is only one rule in our environment, all the logs displayed are related to that specific rule. If there were multiple rules active, the logs for each one would be displayed accordingly, allowing you to view and analyze the latest output from each rule.

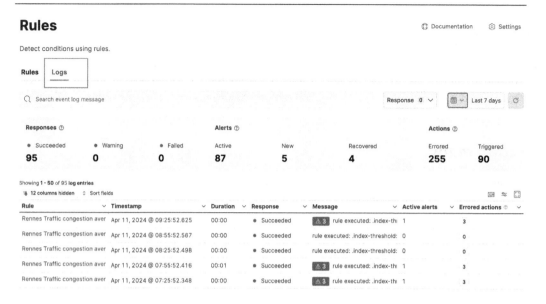

Figure 7.7 – Logs of rules

This tab displays all the logs related to the activity of the rule. At the top, you have some filtering capabilities to search a specific log message, filter on the rule execution response status, and filter on a specific time range. Click on the first log message to see the details of the rule:

Figure 7.8 – Rule execution details

On this page, you are presented with a comprehensive overview of all the crucial information regarding the rule's definition and execution:

- You can easily determine whether the rule is currently enabled or disabled, along with analytical data such as the number of executions in the last 24 hours

- The last execution response is readily available for review, providing immediate insight into the most recent outcome of the rule's execution

- The definition of the rule is also accessible, allowing you to understand the rule's parameters, triggers, and actions at a glance

- This centralized view ensures that you have all the necessary details at your fingertips for effective management and optimization of the rule

3. You can then click on the **History** tab to get more details on the rule execution statistics as shown in *Figure 7.9*:

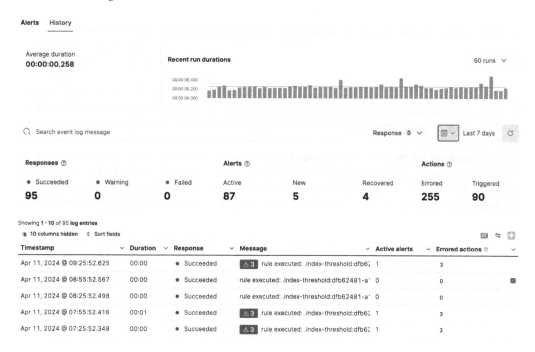

Figure 7.9 – Rule execution history

Here, you'll find many useful pieces of information to monitor rules: the average duration of the rule is a crucial **key performance indicator** (**KPI**) that provides insight into the rule execution performance. It helps in understanding how efficiently the rules are running. Additionally, the **Recent run durations** histogram displays the execution duration of the last 60 rule runs. This range can be adjusted to cover up to the last 120 runs for a more extensive history or reduced to the last 30 or 15 runs for a more recent overview. Below this, there's a list that presents some key KPIs, organized into three sections, offering a comprehensive breakdown to assess and optimize rule performance effectively:

- The outcomes of the most recent rules are categorized into three statuses: **Succeeded**, **Warning**, and **Failed**. It is advisable to investigate any rules that have failed, as this may indicate an issue with the rule's definition or its settings.

- Alerts are categorized into **Active**, **New**, and **Recovered** statuses, which help in understanding the current state of alerts associated with a rule. This breakdown makes it easier to see how many alerts are still active, providing guidance on necessary actions to manage these alerts effectively and maintain system integrity.

- Actions are categorized into **Error** and **Triggered** statuses, which reflect the outcome of actions defined for a rule. This classification allows you to identify any rule actions that failed to execute successfully. Actions that end up in the **Error** category likely indicate that your teams are not being properly notified, which is an issue that requires immediate attention. To gain a better understanding of what went wrong, you can expand the error listings in the table to access more detailed information, ensuring that you can address and rectify the issue promptly.

Now let's introduce an error in our actions to see how we can quickly find it.

4. Head to **Stack Management | Connectors** and click on **Create connector**. Then, select **Email connector**. Enter the following configurations:

- **Connector name**: Non-working Email connector
- **Sender**: cooking-bad@elastic.co
- **Service**: Gmail

Uncheck the **Authentication** toggle so that users are not required to authenticate for this server. Your configuration should look like what is shown in *Figure 7.10*:

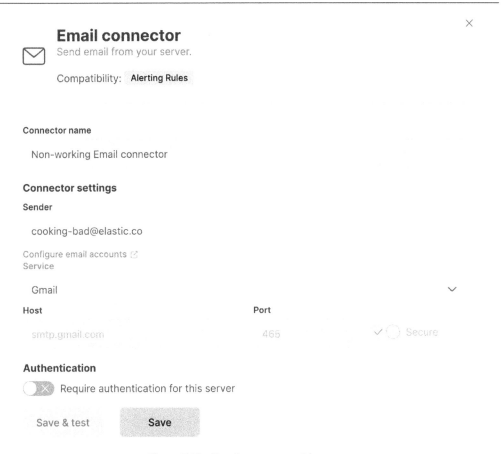

Figure 7.10 – Email connector with error

5. Save the configuration. There's no need to test, as it is obviously going to fail since we're using an email address (cooking-bad@elastic.co) that doesn't exist.

Let's go back to our rule and replace the original SMTP connector with the one we've just created.

6. Navigate to **Stack Management | Rules** and click on the **Rennes Traffic Congestion Average Speed** rule. Then, click on **Edit** in the upper-right corner to open the **Edit Rule** flyout. In the **Actions** section, click on **Email Connector** and choose the **Non-working Email connector** you've just created.

7. Before saving, it's important to adjust the action frequency setting from **On Status Changes** to **On Check Intervals**. This modification ensures that, upon the next execution of the rule, it proceeds through the action step and triggers our designated error connector. This update in the rule configuration is important for ensuring that notifications are consistently sent according to the new frequency setting, enhancing the responsiveness and reliability of the alerting system. The updated rule configuration will look like the one shown in *Figure 7.11*:

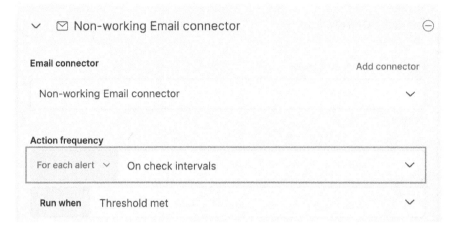

Figure 7.11 – Send email actions configuration

8. Click on **Save** to apply your changes; you will be redirected to the **Rule** page. On the upper right, click on the meatball menu (**...**) to access the menu, and select **Run Rule** to execute the rule immediately as shown in the following screenshot:

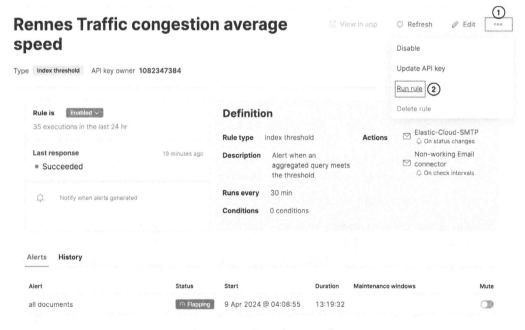

Figure 7.12 – Run rule manually

9. Afterward, navigate to the **History** tab. Here, you should see errors within the log execution table, showing the outcomes of the recent rule execution based on the modifications you've made, including the change in action frequency and the new error connector setup.

Figure 7.13 – Rule execution error

10. Click on the **Warning** sign or the expand icon next to the action marked as **Error** to open it. Then, select **View action errors** to access the detailed page that shows the full error message, including the root cause of why the action ended in error. This step is helpful for diagnosing and understanding the specific issues that led to the failure, allowing for targeted troubleshooting and resolution.

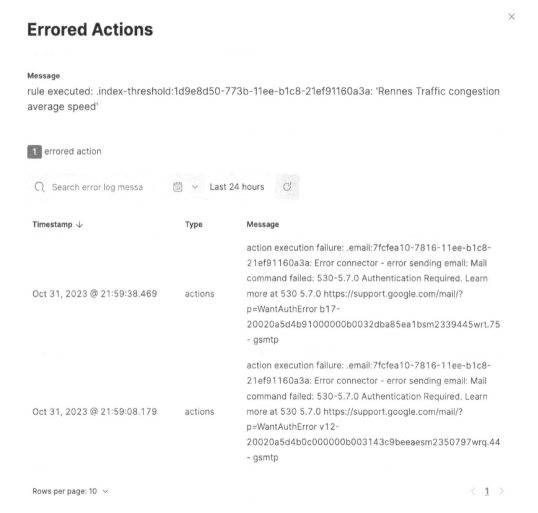

Figure 7.14 – Errored Actions

It's clear that the failure of our action stems from the incorrect connector configuration we've implemented.

This shows the value of the logs feature in monitoring and troubleshooting alerting rules. It enables the creation of an observable alerting system, even in high-load production environments, easing the investigation of performance issues and failures.

How it works...

Behind the scenes, rules and connectors log to the Kibana logger with specific tags. Error messages and events are then captured by the Kibana framework and displayed on the **Rules** page in the Kibana UI.

There's more...

While the Rules logs are especially useful, there are other means and best practices that you can leverage to improve the observability of your rules:

- **Use the test connector UI** to validate the configuration of your connectors to make sure that your actions are going to work properly.

- **Leverage the REST APIs**: A comprehensive collection of HTTP endpoints is available for examining and controlling rules and connectors. Among these, the run connector API is a notable HTTP endpoint for actions. This API can be used to "*test*" an action.

- **Diagnostics for Task Manager**: The alerting capabilities operate using a plugin known as **Task Manager**. This plugin handles the scheduling, execution, and error management of tasks. Consequently, issues within the alerting functionalities may often be identified through the Task Manager system instead of the Rules mechanism.

See also

- Learn how to take your alerting visibility and insights to the next level with this great article: `https://www.elastic.co/blog/apm-kibana-elasticsearch-alerting-insights`

- To learn how to resolve common problems you can encounter with Task Manager, check this page from the official documentation: `https://www.elastic.co/guide/en/kibana/current/task-manager-troubleshooting.html`

- For the full reference on Task Manager diagnostic endpoints, check the following link: `https://www.elastic.co/guide/en/kibana/current/task-manager-troubleshooting.html#task-manager-diagnosing-root-cause`

- Task Manager also offers an internal monitoring mechanism – learn more here: `https://www.elastic.co/guide/en/kibana/current/task-manager-health-monitoring.html`

Investigating data with log rate analysis

This recipe is the first of three where we will cover the **artificial intelligence for IT operations (AIOps)** capabilities of the Elastic Stack. AIOps can be summed up as the application of various machine learning techniques to IT operations with the goal of ultimately enabling teams to respond more quickly and proactively to issues and outages.

Within the stack, AIOps is part of the broader ML capabilities and provides features that leverage advanced statistical methods to understand data trends and behavior.

In this recipe, we'll focus on the log rate analysis feature and learn how we can apply it to detect the reasons behind decreases or increases in the log rates.

Getting ready

To follow along with this recipe, you'll need an up-and-running Elastic deployment with Elasticsearch, Kibana, and an ML node. To spin up a cluster on **Elastic Cloud** refer to the *Installing the Elastic Stack on Elastic Cloud* recipe in *Chapter 1*.

We'll keep using our Rennes traffic dataset, so make sure to have some data in your cluster. If you don't, follow the *Exploring your data in Discover* recipe in *Chapter 6*.

How to do it...

1. Head to **Kibana | Analytics | Machine Learning | AIOps Labs** and click on **Log Rate Analysis**.

2. Now select the **metrics-rennes_traffic-raw** data view. In the time picker, select **last 24 hours** and enter the following query in the search bar to filter the data:

    ```
    traffic_status : "congested" or traffic_status : "heavy"
    ```

    Press *Enter* to run the query.

3. Click on a dip section in the data histogram to activate the **Baseline** and **Deviation** window controls. Try to position the **Deviation** window on the section of the histogram where a dip in log rates is observable, and place the **Baseline** window before that drop, as shown in *Figure 7.15*:

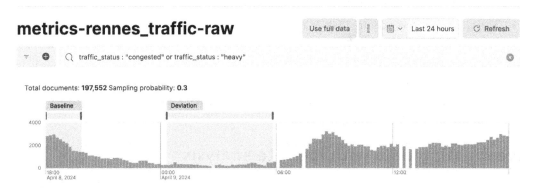

Figure 7.15 – Baseline and Deviation tools for log rate analysis

Execute the analysis by clicking on **Run analysis**. Your results should resemble what is depicted in the following screenshot:

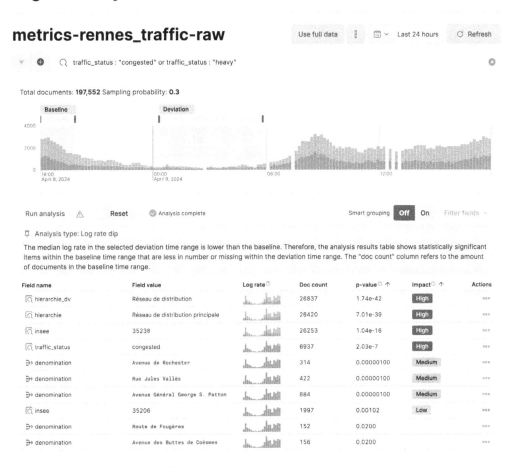

Figure 7.16 – Log rate analysis results

The analysis highlights statistically significant combinations of field values that are linked to the observed decrease, presenting the findings in a table. In our case, we sought to understand the reason behind the drop in events reporting a congested or heavy traffic situation between midnight and 6 a.m. Based on the analysis, the reason is simply that traffic is mostly fluid on main roads ("Reseau de distribution" and "Réscau dc distribution principale") during this period of the day, hence the high impact of those values in the analysis.

The field's table also includes an indicator reflecting the impact level and a sparkline that illustrates the impact's trend in the chart. Moving the cursor over a row provides a more detailed view of its effect on the histogram chart. Options under the **Actions** column allow for further exploration of a field in Discover, deeper analysis in log pattern analysis, or the ability to copy the information of a table row as a query filter directly to the clipboard.

How it works...

The log rate analysis feature is designed to help you quickly identify and diagnose issues related to log rates. By analyzing log data in real time, it can help you find patterns and trends that might otherwise go unnoticed.

The analysis looks at all the events to find patterns in all field's value pairs. It tries to find a correlation between the values and a potential deviation in the logs.

One of the key benefits of the log rate analysis feature is that it provides end-to-end visibility into your log data. By using advanced statistical methods to analyze log rates, it can help you quickly find the root cause of issues and take appropriate action.

The following Elasticsearch features are used to power log rate analysis:

- `significant_terms`: The p_value scoring heuristic is used to pinpoint statistically significant field/value pairs within a foreground and background dataset, serving as a baseline and deviation in log rate analysis. For instance, in the context of web logs, this heuristic can detect specific source IPs or URLs that are driving a surge in log entries.

- `frequent_items_set`: This data mining method is employed to discover common and pertinent patterns, widely used in recommender systems, behavioral analytics, fraud detection, and more. It's applied to uncover clusters of statistically significant field/value pairs that correlate with each other. An example application is in identifying which users are accessing specific URLs.

- `random_sampler`: This aggregation method enables the random sampling of documents in a way that is both statistically solid and quick, facilitating the analysis of terabytes of data. It strikes a balance between speed and accuracy during query execution.

There's more...

Log rate analysis is used as part of a broader analysis and investigation workflow. You can use this capability to quickly understand what is affecting the trend of your data and react accordingly, without having to understand machine learning and go through some configuration steps.

Its applicability spans several areas. For instance, operational monitoring involves scrutinizing log output rates from various IT components including servers and network devices to gauge their health and performance. This analysis is crucial for spotting trends and anomalies in activity levels. In the domain of security, it plays a key role in detecting potential breaches or malicious activities by identifying unusual log generation patterns, such as a spike in failed login attempts. Compliance monitoring also benefits from log rate analysis, particularly in sectors where regulatory standards mandate monitoring log generation as part of auditing procedures, with fraud detection being a prime example. Additionally, it proves invaluable in troubleshooting and root cause analysis, where correlating changes in log rates with specific system changes can quickly pinpoint the start of problems, greatly aiding in their resolution.

See also

- To learn more about using log rate analysis for root cause analysis with logs, check out this blog: `https://www.elastic.co/blog/observability-logs-machine-learning-aiops`

- Log rate analysis has been GA since the release of Elastic Stack 8.12. Here is the blog announcing the feature and how it fits within the broader observability solution: `https://www.elastic.co/blog/elastic-stack-aiops-labs-8-12-ga-of-log-rate-analysis`

Investigating data with log pattern analysis

In this recipe, we will focus on another AIOps feature of Elastic, called log pattern analysis, which is a powerful tool that helps you to find patterns in unstructured log messages and makes it easier to examine your data.

Getting ready

To follow along with this recipe, you'll need an up-and-running Elastic deployment with Elasticsearch, Kibana, and an ML node. To spin up a cluster on **Elastic Cloud** refer to the *Installing the Elastic Stack on Elastic Cloud* recipe in Chapter 1.

We'll keep using our Rennes traffic dataset, so make sure to have some data in your cluster. If you don't, follow the *Exploring your data in Discover* recipe in *Chapter 6*.

We will try to identify patterns in street names (denominations) associated with specific traffic conditions, such as heavy or congested.

How to do it...

1. Head to **Kibana | Analytics|Machine Learning | AIOps Labs** and click on **Log Pattern Analysis**.

2. Select the **metrics-rennes_traffic-raw** data view. You can click on **Use full data** if you want the analysis to run on the whole dataset instead of a limited time range.

3. Enter the following query filter in the query bar: `traffic_status : "congested" or traffic_status : "heavy"`.

4. In the **Category field** section, select **denomination**. Your log pattern analysis configuration will look like the one shown in *Figure 7.17*:

Figure 7.17 – Log pattern analysis configuration

5. Click on **Run pattern analysis**. Wait for the analysis to complete. The results screen should look like the one shown in *Figure 7.18*:

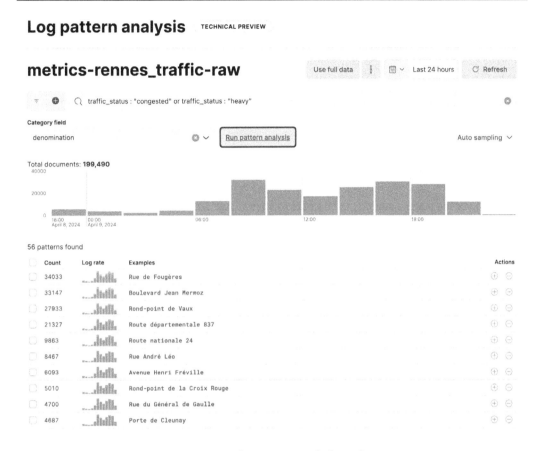

Figure 7.18 – Log pattern analysis results

We observe that street names correlate with a high frequency of documents indicating a heavy traffic status.

How it works...

Under the hood, the **Log pattern analysis** feature uses the Categorize text aggregation feature. This is a multi-bucket aggregation that organizes semi-structured text into distinct groups. Each text field undergoes re-analysis with a custom analyzer, and the tokens produced from this process are categorized into buckets containing text values in similar formats. This type of aggregation is particularly effective with machine-generated text, such as system logs, and it uses only the first 100 analyzed tokens for categorizing the text.

The **Log pattern analysis** feature is designed to help you quickly identify patterns and trends in your log data. By analyzing log data in real-time, it can help you identify the root cause of issues and take appropriate action. This can be especially useful for identifying issues related to application performance, server health, and other critical IT infrastructure components.

To utilize the **Log pattern analysis** feature, you should choose a field to categorize and, if desired, apply any specific filters. Once set up, initiate the analysis. Upon completion, the results will be presented in a flyout panel.

There's more...

You can also use log pattern analysis in Discover as an available action for any text field.

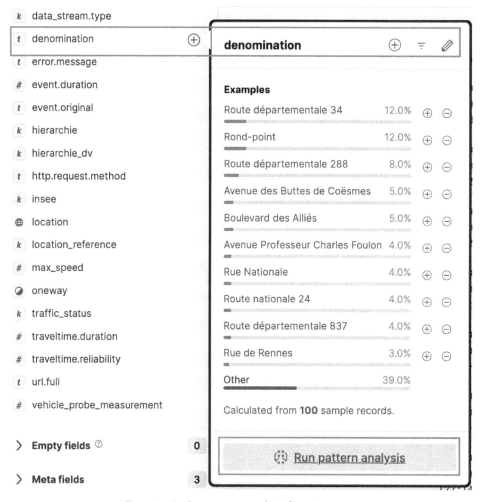

Figure 7.19 – Run pattern analysis from Discover

Log pattern analysis can be used with the **Log rate** feature. This is one of the proposed actions once the log rate analysis has been completed, and can help you correlate any increases or decreases with specific patterns or trends in your data.

Log rate analysis

Figure 7.20 – Log pattern analysis action

Log pattern analysis serves as a powerful tool for managing and interpreting extensive log data by identifying and grouping similar log entries. Its applications are varied and impactful.

For example, in error and exception monitoring, it groups similar error messages or exception logs, facilitating quicker identification of common issues in applications or infrastructure, and thus speeding up root cause analysis and resolution. In addressing performance issues, it categorizes logs related to application or system performance, such as slow database queries or high latency requests. This helps in finding and resolving performance bottlenecks.

In the domain of security, log pattern analysis is instrumental in threat hunting. It identifies patterns indicating security threats, such as repeated access denied errors or unusual application behavior, by clustering similar security-related log messages for in-depth investigation. Additionally, it plays a significant role in user behavior analysis. By grouping logs with similar patterns, it provides insights into common user interactions with a web application, frequent errors encountered, and usage trends.

These examples merely scratch the surface of log pattern analysis's potential. Depending on the specific use case and dataset, numerous other applications can be explored. In essence, log pattern analysis simplifies the complex task of log data management, making it an essential tool in a variety of contexts and analysis.

Investigating data with change point detection

As part of AIOps Labs, **change point detection** helps identify critical moments of change in time series data, which can indicate shifts in patterns or behaviors. It enables early anomaly detection, allowing organizations to respond swiftly to unexpected events or irregularities. Furthermore, it supports data-driven decision-making by providing insights into when and where changes are occurring. It can be applied in various domains such as finance, cybersecurity, industry, and transport, to enable proactive actions and enhance overall data analysis.

In this recipe, we will use change point detection to further explore our Rennes traffic data.

Getting ready

We'll keep using our Rennes traffic dataset, so make sure to have some data in your cluster. If you don't, follow the *Exploring your data in Discover* recipe in *Chapter 6*.

How to do it...

1. Head to **Kibana | Analytics| Machine Learning | AIOps Labs** and click on **Change Point Detection**.

2. From the **Select data view** screen, choose **metrics-rennes_traffic-raw**.

3. First, we aim to detect the change point (dip or spike) for the average speed across different locations. Set the time range to **Last 24 hours** and apply the following settings, after which you will see that the change points have been detected as shown in *Figure 7.21*:

    A. Select **min** for the aggregation.

    B. Choose **average_vehicle_speed** for **Metric field**.

    C. Choose **location_reference** for **Split field**.

D. Set **Change point type** to **dip** and **spike**.

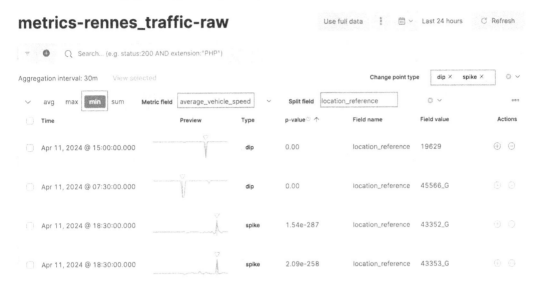

Figure 7.21 – Metrics aggregation and change point type selection

Now, imagine we want to see how the change points correlate between average speed and travel time duration. Let us add a second metric aggregation by clicking the **add** button at the bottom left.

4. We can now set up our second metric aggregation using the following settings, and you will see that the change points have been detected as shown in *Figure 7.22*:

A. Set **max** for the aggregation.

B. Choose **traveltime.duration** for **Metric field**.

C.   Choose **location_reference** for **Split field**.

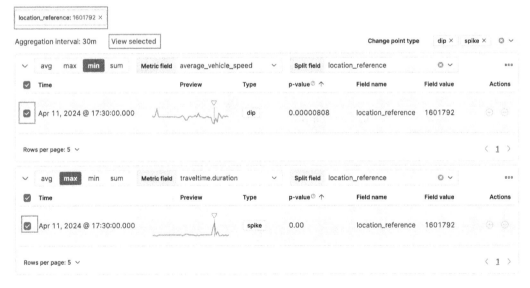

Figure 7.22 – Change point view filtering

As you can see, we now have two metric aggregations running in parallel, giving us a more contextual understanding of change point detection. We can delve even deeper by clicking on the + sign (filter for value) in the first row of our second metric aggregation:

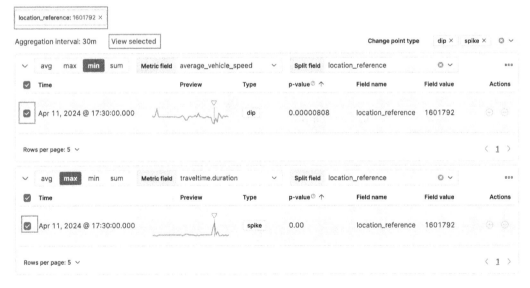

Figure 7.23 – Selecting multiple change point views

5. You can see in *Figure 7.23* that the change point detection metrics have been filtered for our specific location. We can now check the boxes for these two metric aggregations, then click **View selected**. This will bring up a flyout as shown in *Figure 7.24*:

Selected change points   ✕

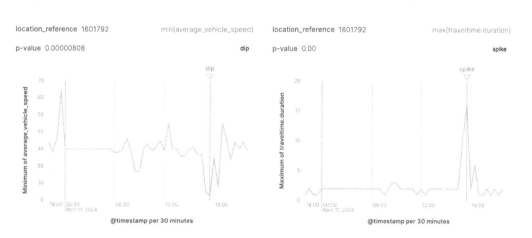

Figure 7.24 – Change point view comparison

You can now hover your mouse over these two charts to understand how this side-by-side view aids in interpreting the influences between change points on average speed and travel time duration for this location. The **p-value** above each chart signifies the magnitude of the change; lower values indicate more significant changes.

How it works...

As you can see, the change point detection UI provides us with a very convenient way to analyze time-series data. It allows us to experiment with different metrics and aggregations and instantly display the results. This feature is powered by **Change Detection Aggregation**, which supports these five types of change point detection:

- `dip`: A significant dip occurs at a point

- `spike`: A significant spike occurs at a point

- `trend_change`: An overall trend change occurs at a point

- `step_change`: A statistically significant step up or down in value distribution that occurs at a point

- `distribution_change`: Overall distribution of the values changes significantly

There's more...

In this recipe, we have learned how to use the AIOps Lab's change point detection to explore our dataset, and understand the detected change points and the possible correlations between them. Once we find a useful metric aggregation chart, we can also embed it into an existing dashboard, or add them to cases by clicking on the meatball menu (**...**) from the upper right of each chart, as shown in *Figure 7.25*:

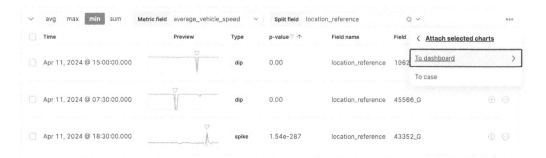

Figure 7.25 – Adding change point chart to dashboard

Real-life example – monitoring network traffic for cybersecurity

In our recipe, we used Rennes traffic data to demonstrate how change point detection can identify traffic congestion by analyzing spike changes in average speed across different locations. Let's explore another type of traffic as an example: monitoring network traffic for cybersecurity.

Suppose you are in charge of monitoring network traffic data for an organization to identify potential security breaches or anomalies that could indicate a cyber-attack. Without change point detection, you might rely on setting static thresholds (such as an unusually high number of login attempts) to identify potential threats. However, this might not detect more subtle or slowly developing threats.

With change point detection, you can identify moments when network behavior changes significantly, which could indicate potential threats such as the following:

- **Increase in traffic volume**: A sudden spike in network traffic could indicate a **Distributed Denial of Service (DDoS)** attack. Change point detection could pinpoint the start of this attack, allowing you to take swift mitigating action.

- **New types of requests**: A change in the pattern of request types (e.g., a sudden increase in POST requests where there are usually GET requests) might suggest that an attacker is trying to exploit a vulnerability in the system.

- **Change in traffic source**: A change point may occur when there's a shift in the geographic source of the traffic, possibly indicating a coordinated access attempt from a new region.

- **Timing and seasonality**: Understanding the normal flow of traffic throughout the day or week is crucial for establishing a baseline. A change point that indicates a deviation from this pattern could signal a security event. In the next recipe, we will explore concrete examples of how to build machine learning jobs that establish baselines for seasonality and prepare the aggregated metrics for monitoring, enabling us to detect anomalies.

By leveraging change point detection in this way, organizations can be more proactive and responsive to emerging threats, ultimately leading to better protection of their digital infrastructure.

See also

In some situations, such as counting occurrences/averages and building data summaries based on change point detection, it can be useful to perform change point aggregation within a Query DSL. Please check out the following documentation: `https://www.elastic.co/guide/en/elasticsearch/reference/current/search-aggregations-change-point-aggregation.html`

Detecting anomalies in your data with unsupervised machine learning jobs

In this recipe, we'll introduce you to the concept of **anomaly detection** and guide you through creating an unsupervised ML job to uncover unusual patterns in your dataset.

But first, what exactly is anomaly detection? Elasticsearch's machine learning anomaly detection feature is a dynamic tool capable of automatically learning the typical behavior of your time series data and pinpointing anomalies as they occur. This feature is equipped to perform sophisticated analysis, enhance root cause investigation, and minimize the occurrence of false positives, ultimately providing automated, real-time anomaly detection for time series data. These techniques are part of the unsupervised machine learning category.

In this recipe, we'll create a machine learning configuration known as a **job** to detect abnormal patterns in our traffic dataset by focusing on data points such as travel time, average speed, and traffic status. Our goal will be to employ machine learning to spot unusually high travel times within our dataset.

We will go through some concepts including **bucket span**, **detectors**, and **influencers**. At the end of this recipe, you'll have a fully working anomaly detection configuration.

Getting ready

To follow along with this recipe, you'll need an up-and-running Elastic deployment with Elasticsearch and Kibana. To spin up a cluster on **Elastic Cloud**, refer to the *Installing the Elastic Stack on Elastic Cloud* recipe in Chapter 1.

> **Important note**
>
> Make sure to have an ML node running in your cloud deployment if you're using Elastic Cloud or a node with an explicit `node.roles=ML` if you're using a self-managed cluster.

We'll keep using our Rennes traffic dataset, so make sure to have some data in your cluster. If you don't, follow the *Exploring your data in Discover* recipe in *Chapter 6*.

Ideally, to achieve the best results with ML, having several days' worth of data is recommended.

How to do it...

1. Head to **Kibana | Analytics | Machine Learning | Anomaly Detection** and click on **Jobs**.

2. Click on **Create anomaly detection job**. You will be presented with the options of **Data View** or **Saved search**. Select the **metrics-rennes_traffic-raw** data view.

3. You are now in the machine learning job creation wizard; select **Single metric** to proceed. We'll set up our anomaly detection job by following these steps:

    I. **Time range**: The first step is to select the time range; click on **Use full data** to run the analysis on the full dataset. Then click **Next**.

    II. **Choose fields**: We will now choose the type of analysis we want our model to perform and the field that is going to be analyzed.

    III. In the drop-down list under the **Choose fields** option, select **High mean(traveltime. duration)**. We're using the `high_mean` function to find unusually high travel time duration.

    IV. Under the **Bucket span** field, click on **Convert to multi-metric job**. This will allow more fine tuning of our configuration.

    V. In the **Split field** option, select the **hierarchie** field. This setting will partition our analysis by the different road hierarchy values present in our dataset.

    VI. In the **Influencers** configuration, add the **traffic_status** field. The role of influencers will be discussed in more detail later on. Your configuration should look like the one shown in *Figure 7.26*. Then click **Next**.

Create job: Multi-metric

Using data view metrics-rennes_traffic-raw

Figure 7.26 – Multi-metric job creation – Choose fields screen

VII. **Job details**: Give your anomaly detection job a name, such as `high_traveltime_duration_by_road_type`.

VIII. Under the **Job description** field, type `Detect unusually high travel time duration` for each road type. Click **Next**.

During the **Validation** phase, the wizard automatically validates the machine learning job configuration. If everything checks out, you should see all the steps validated, as per *Figure 7.27*:

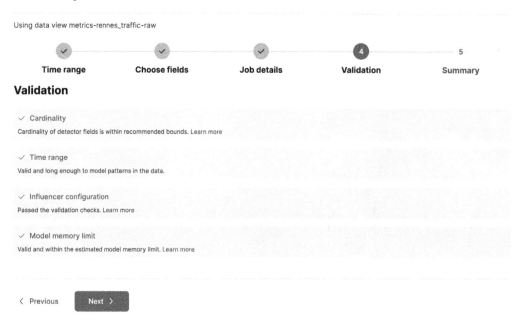

Create job: Multi-metric

Using data view metrics-rennes_traffic-raw

| Time range | Choose fields | Job details | Validation | Summary |

Validation

✓ Cardinality
Cardinality of detector fields is within recommended bounds. Learn more

✓ Time range
Valid and long enough to model patterns in the data.

✓ Influencer configuration
Passed the validation checks. Learn more

✓ Model memory limit
Valid and within the estimated model memory limit. Learn more

‹ Previous   Next ›

Figure 7.27 – Multi-metric job creation validation

IX. Click **Next** to proceed to the last step of the setup.

X. Click on **Summary** to review the summary and click on **Create job** to launch the analysis.

XI. Once the job is created, click **View results** to see the outcome of your anomaly detection analysis.

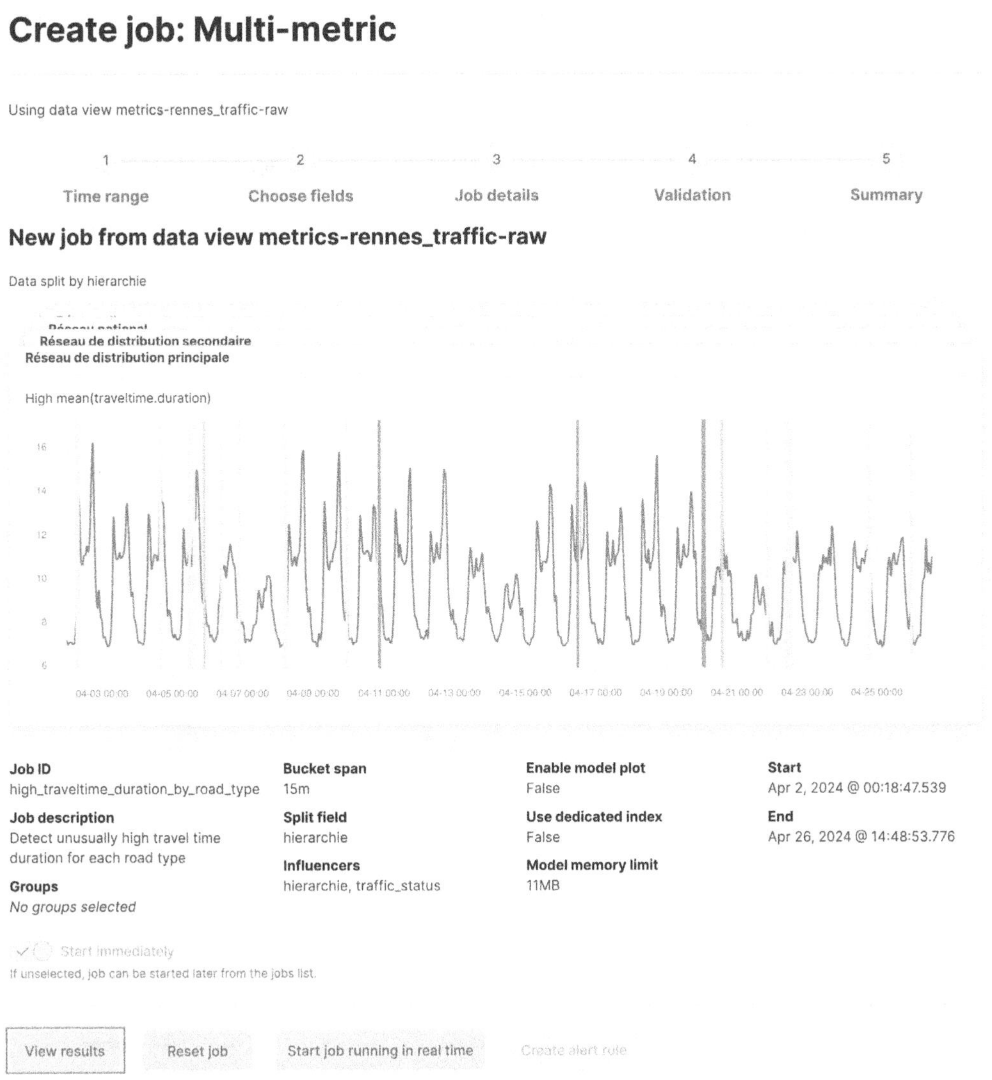

Create job: Multi-metric

Using data view metrics-rennes_traffic-raw

1	2	3	4	5
Time range	Choose fields	Job details	Validation	Summary

New job from data view metrics-rennes_traffic-raw

Data split by hierarchie

Réseau national
Réseau de distribution secondaire
Réseau de distribution principale

High mean(traveltime.duration)

Job ID	**Bucket span**	**Enable model plot**	**Start**
high_traveltime_duration_by_road_type	15m	False	Apr 2, 2024 @ 00:18:47.539
Job description	**Split field**	**Use dedicated index**	**End**
Detect unusually high travel time duration for each road type	hierarchie	False	Apr 26, 2024 @ 14:48:53.776
Groups	**Influencers**	**Model memory limit**	
No groups selected	hierarchie, traffic_status	11MB	

✓ ⚙ Start immediately
If unselected, job can be started later from the jobs list.

View results	Reset job	Start job running in real time	Create alert rule

Figure 7.28 – Multi metric job creation completion

4. Clicking on **View results** will take you to the **Anomaly Explorer** view. It is an excellent tool for visualizing and examining the results of your anomaly detection job.

5.   Upon entering the **Anomaly Explorer**, an anomaly timeline is visible, organized by **Hierarchie**. After running the job we just set up, the **Anomaly Explorer** appears as shown in *Figure 7.29*:

> **Note on anomaly detection results**
>
> The results of anomaly detection can vary based on the dataset you're working with. It's worth noting that typically, having several days' worth of data is preferable for observing meaningful results and identifying anomalies.

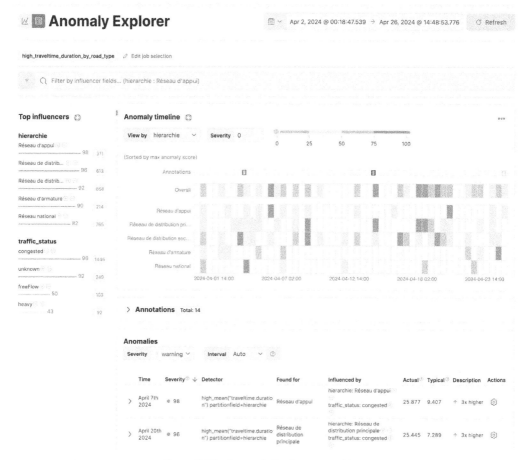

Figure 7.29 – Anomaly Explorer overview

Figure 7.29 indicates that our anomaly detection job has pinpointed some highly unusual events. Now we will look into how to utilize the **Explorer** view to gain further insights into the results.

6. Firstly, there's the **Anomaly timeline** window:

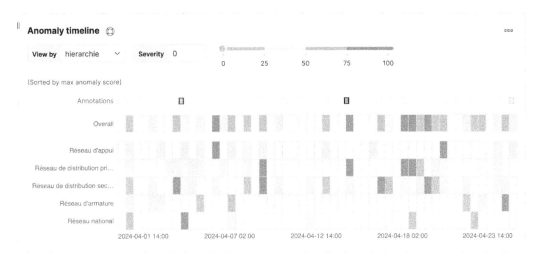

Figure 7.30 – Anomaly timeline

Given that our job is divided based on road hierarchy, separate baselines are established for each unique value of the **hierarchie** field, which is instrumental in detecting anomalies that may have otherwise remained hidden.

In this instance, it is evident that our analysis has uncovered significant anomalies across various road types and dates. Clicking on an anomaly within the timeline provides us with additional information:

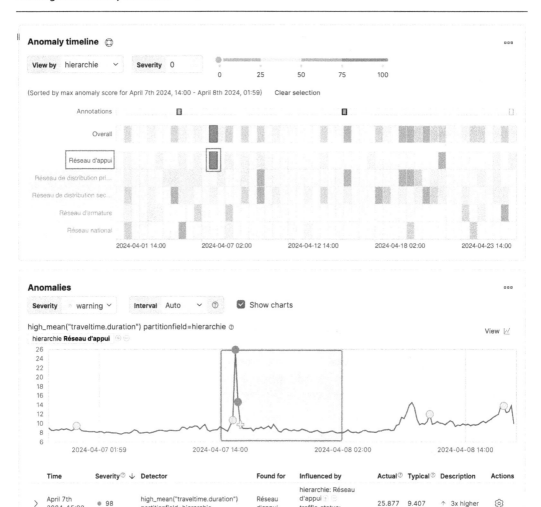

Figure 7.31 – Expand anomaly from the timeline

The preceding screenshot illustrates the detailed view presented when selecting a specific anomaly in the swimlane. It provides comprehensive information associated with the particular anomaly in the table situated beneath the swimlane. Let's take a closer look at some of these details:

I.   **Time**: The date and hour when the anomaly occurred.

II.   **Severity**: Calculated on a scale from **0** to **100**, severity measures how anomalous the data point is. In our case, a score of **97** suggests that the occurrence of the observed travel time duration at this specific time was highly improbable.

III.   **Influenced by**: This section is where the anomaly detection algorithm identifies elements that may have contributed to the anomaly. Recall that when configuring the job, we set two influencers to monitor: **hierarchie** and **traffic_status**. It's interesting to see the value **congested** for **traffic_status** as it could provide a clue as to why the travel time was abnormally high at that moment.

IV.   The additional columns in the table offer further metrics and explanations. Let's zoom in on one particular column: **Description**. In our case it is indicated as 11x higher, which means that the travel time was 11 times higher than the typical value (we can also see both the **Actual** and the **Typical** value fields in the table).

V.   Clicking on the blue arrow expands the view to offer a more detailed description and further details of this anomaly:

Figure 7.32 – Anomaly details from Anomaly Explorer

The details pane is especially useful for understanding the data behind the anomaly. While the computed probability may not be very intuitive, we translate it into an easy-to-understand anomaly score ranging from **0** to **100**, which is also practical for setting up alerts.

7.   Another element to explore is **Top influencers** to the left of the anomaly timeline:

Figure 7.33 – Influencers scores in Anomaly Explorer

We can see that the **congested** value for the **traffic_status** field has the highest score (**97**), which means it strongly contributes to the unusually high travel time that our model has detected.

The analysis shows that the abnormally prolonged travel time duration is influenced by the traffic's condition. Anomaly detection enables the effortless identification of such events and their swift detection as they occur. It would have been challenging to establish a static threshold for detecting such behavior, which is why machine learning-based anomaly detection proves to be exceptionally valuable.

How it works...

The fundamental principles of anomaly detection in the Elastic Stack can be summarized as follows:

- Anomaly detection is performed on time series data through what we call datafeeds. Datafeeds are components that retrieve time series data from Elasticsearch indices and provide it to an anomaly detection job for analysis.

- The analysis splits the time series into buckets using the configured bucket span (basically the granularity of the analysis window). The data within each bucket is then aggregated using a selected detector function—in our case, `'high_mean'` was utilized.

- The values resulting from the detector are used to construct a probability model to modelize our data.

- Values that have a low probability of appearing according to the model are deemed as potential anomalies in the system.

- Calculated probabilities are then mapped to an anomaly score displayed to the user.

- The probability distribution model evolves as more data is seen and computed.

- Depending on the complexity of the data, several models can be constructed independently by using the **Partition fields** option. In our recipe, we split the analysis by the **hierarchie** field to create separate models for different road hierarchies.

This is pretty much how it works on a high level, but there are also some additional parameters incorporated into the model, such as calendar events (planned maintenance, updates, etc.) or step change/shock detection.

Another key aspect worth mentioning is **influencers**.

Influencers play a vital role in anomaly detection. They are the fields within your dataset that are suspected of impacting or contributing to anomalies, and their identification often relies on domain expertise to pinpoint the causes of irregularities. Influencers can be any field within your data, but they must be included in the datafeed query or aggregation to be subject to analysis. While they are not obligatory components of your anomaly detection job's detectors, influencers are frequently utilized to provide context.

Influencers may come from different indices than the fields your detectors operate on, but it's essential that both indices share a common date field, as indicated in the datafeed's `data_description.time_field`.

Machine learning analytics in Elasticsearch evaluate an influencer's effect by temporarily excluding data points associated with that influencer's value to determine whether the anomaly would still be detected without that potential cause. This approach ensures that only statistically significant influencers are reported. Elasticsearch computes an influencer score for each time unit and each influencer field used in the job. These individual scores are then aggregated to determine the overall anomaly score for a given time bucket.

Picking an influencer is strongly recommended for the following reasons:

- It facilitates the task of attributing anomalies to specific causes or sources, which can be essential for root cause analysis.

- It simplifies the interpretation of results by grouping anomalies around common influencers.

There's more...

Elastic ML-based anomaly detection also supports several types of analysis. In the recipe, we went through a metric configuration, but you can also run the following types:

- **Population analysis**: Leveraged when comparing entities against their peers and not against one entity's history. This is especially useful when your dataset has a high cardinality of entities with a mostly homogeneous behavior.

- **Categorization analysis**: Useful for unstructured logs as it groups log messages into categories to detect anomalies within them. It can help find unusual or rare events via log categorization, for example.

- **Geographic analysis**: Used to detect anomalies in the geographic locations and coordinates in the data. For example, it can be useful to detect an activity that occurs in an unusual location, such as a user login or a web request coming from a rarely-seen source location.

One useful capability of anomaly detection is **forecasting**. Once your anomaly detection job has established the normal behavior baselines for your data, you can use this information to forecast future behavior, such as estimating travel time duration for a specific future date. The forecast feature is available from the **Single Metric Viewer** on the **Anomaly detection** page of the **Machine Learning** application.

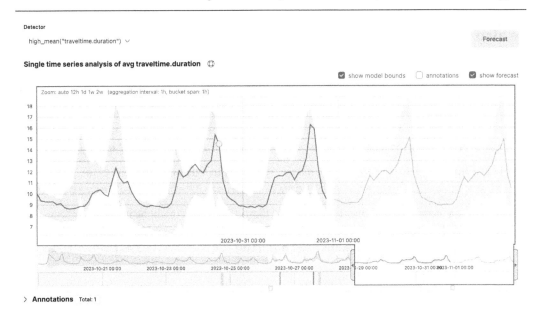

Figure 7.34 – Forecasting travel time duration

You can also run forecasts using the corresponding API: `https://www.elastic.co/guide/en/elasticsearch/reference/current/ml-forecast.html`

Another particularly important aspect of using anomaly detection is **alerting**. You can create anomaly detection alerts based on job results to streamline incident investigation and response. More information on the alerting framework can be found in the *Creating alerts in Kibana* recipe in this chapter.

There are some architectural aspects to consider when using ML anomaly detection:

- ML jobs can be executed on any ML-eligible Elasticsearch node. In Elastic Cloud, these are dedicated instances. For self-managed setups, you must explicitly assign the `"ml"` role to nodes via the `node.roles` parameter.

- The best and recommended practice is to have dedicated ML nodes with enough CPU and RAM. ML jobs run outside of the JVM because they're written in C++.

In terms of indices, both anomaly detection models and results are stored in system indices. Models are stored in the relevant `.ml-state` system index and the results in the `.ml-anomalies` indices. You can configure a dedicated result index especially for large jobs.

See also

Anomaly detection is a vast and rich topic in the context of the Elastic Stack. If this is of particular interest to you, the following is a shortlist of articles worth reading:

- To get a better understanding of how scoring works and how to interpret the various scores presented in the dashboards, check out this article: `https://www.elastic.co/blog/explaining-anomalies-detected-by-elastic-machine-learning`

- Learn how to perform root cause analysis with logs in this blog post: `https://www.elastic.co/blog/reduce-mttd-ml-machine-learning-observability`

- If you are a security practitioner leveraging the Elastic Stack, you might be interested in learning how to apply ML rare analysis from the following post: `https://www.elastic.co/blog/using-elastic-machine-learning-rare-analysis-to-hunt-for-the-unusual`

Creating anomaly detection jobs from a Lens visualization

In the previous recipe, we explored the manual creation of anomaly detection jobs. You may often begin your analysis with insights gathered from dashboards and visualizations to comprehend your data's seasonality. A natural progression from this analysis might be to create an anomaly detection job to monitor the aggregations you've identified in your Lens visualization. In this recipe, we'll demonstrate how to initiate an anomaly detection job directly from within a Lens visualization.

Getting ready

To follow along with this recipe, make sure to have completed the following:

- The *Exploring your data in Discover* recipe in *Chapter 6*

- The previous *Detecting anomalies in your data with unsupervised machine learning jobs* recipe in this chapter

How to do it...

1. Go to **Kibana** | **Analytics** | **Dashboard** and open the dashboard we created in *Chapter 6*, titled **[Rennes Traffic] Overview Dashboard**.

2. Depending on the duration for which you have ingested the **Rennes Traffic** data, adjust the time range to the last X days, ideally more than three days, to observe the seasonality of your data. Locate the **[Rennes Traffic] Average speed & Traffic Status** Lens visualization and examine the data seasonality.

3. From the Lens visualization, click on the **Options** (...) menu, select **More**, and then choose **Create anomaly detection job** as shown in *Figure 7.35*:

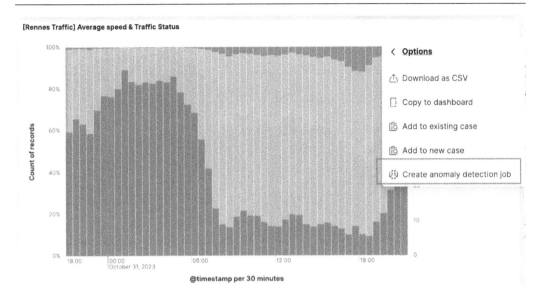

Figure 7.35 – Creating an anomaly detection job from Kibana Lens

4. The Lens visualization has two layers: one is a bar vertical percentage chart showing the count of records broken down by traffic status, and the other is a line chart representing the global average speed. We can create two anomaly detection jobs directly from this visualization:

A. For a multi-metric job, set the job ID to **rennes-traffic-status**. In the **Additional settings** field, set the **Bucket span** to **1h** and click on **Create Job** as shown in *Figure 7.36*:

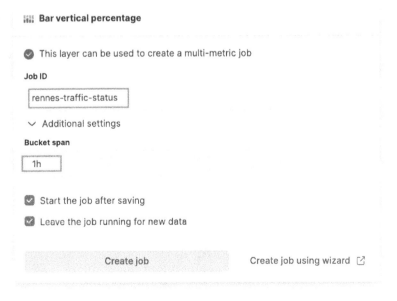

Figure 7.36 – Configuring the multi-metric job

B. For single-metric job, set the job ID to **rennes-traffic-average-speed**. In the **Additional settings** field, set the **Bucket span** to **1h** and click on **Create job** as shown in *Figure 7.37*:

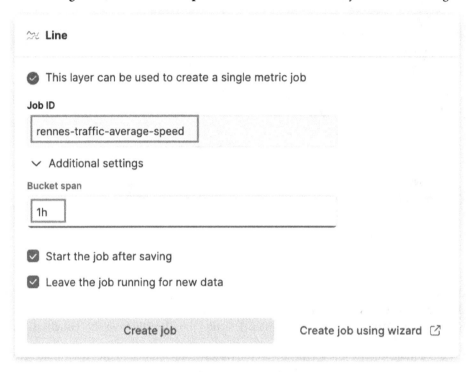

Figure 7.37 – Configuring the single metric ML job

5. Now let's check the result of the first multi-metric job by right-clicking on **View results in Anomaly Explorer** to open it in a new browser tab. This action will take you to the Anomaly Explorer view for your new multi-metric anomaly detection job. Depending on recent traffic data, you might notice some detected anomalies in the swimlane view and **traffic_status** influencers, as depicted in *Figure 7.38*:

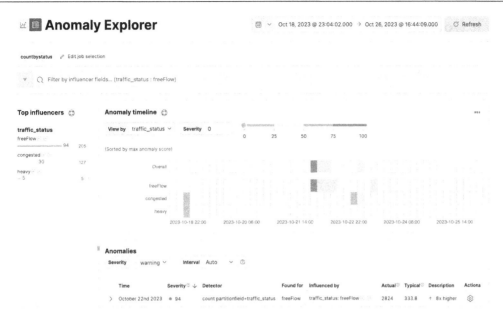

Figure 7.38 – ML Anomaly Explorer for the metric-metric job

6. Return to the original tab of your browser tab and click on **View results in single metric view**. This will bring you to the single-metric view of your new single-metric anomaly detection job. We observe that the mean of the **average_vehicle_speed** field has been correctly chosen as the detector for this ML job. Depending on recent traffic data, some detected anomalies may be visible, as shown in *Figure 7.39*:

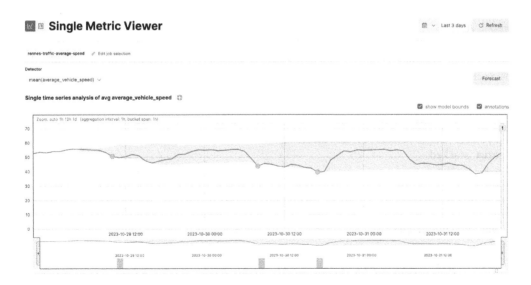

Figure 7.39 – Single metrics view for the created job

How it works...

In this recipe, we've discovered that Kibana can automatically detect aggregations and breakdown values to create single and multi-metric machine learning anomaly detection jobs. The 'count' and 'average' aggregations, in our case, have been accurately recognized and utilized to configure the corresponding anomaly detection functions.

By default, the data feed range used corresponds to the time range set in our dashboards—the last three days, for our purposes.

Once the jobs are created, you can navigate to **Kibana | Analytics | Machine Learning | Anomaly Detection | Jobs** to modify the job. This allows changes to the data feed range, the creation of alert rules, and more, as displayed in *Figure 7.40*:

Figure 7.40 – Editing the existing job

There's more...

Please note that only charts containing a date field on one axis are supported. Moreover, only the following chart types can be used:

- **Area**
- **area_percentage_stacked**
- **area_stacked**
- **Bar**
- **bar_horizontal**
- **bar_horizontal_stacked**
- **bar_percentage_stacked**

- **bar_stacked**

- **Line**

The supported anomaly detection functions include **average**, **count**, **max**, **median**, **min**, **sum**, and **unique_count**.

In this recipe, we created anomaly detection jobs from a Lens visualization on a Kibana dashboard. It's also possible to create them directly from Lens visualizations placed on a Canvas workpad.

8
Advanced Data Analysis and Processing

In the previous chapter, we explored how you can perform **anomaly detection** using an unsupervised learning method for timestamped data within the Elastic Stack. In this chapter, we will shift our focus to additional aspects of the Elastic Stack's **Machine Learning** (**ML**) capabilities, such as **data frame analytics**, as displayed in *Figure 8.1*. Data frame analytics includes unsupervised learning for outlier detection, along with supervised learning methods that employ trained models for both classification and regression predictions:

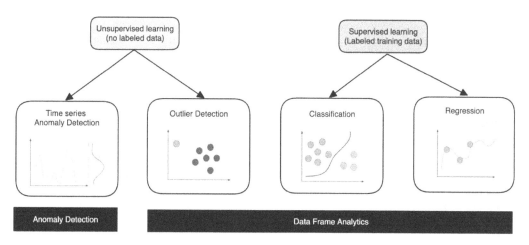

Figure 8.1 – ML in the Elastic Stack

Elasticsearch's supervised learning capabilities provide a robust framework, enabling you to train ML models with labeled training data. Once these models are trained, they can be deployed to predict outcomes or infer patterns in new datasets. This proves particularly useful when dealing with a significant amount of data and when seeking to identify non-obvious patterns or relationships. With Elastic's supervised learning, you stand to uncover data insights that would be challenging or impossible to derive through manual analysis.

In this chapter, we're going to cover the following main topics:

- Finding deviations in your data with outlier detection

- Building a model to perform regression analysis

- Building a model for classification

- Using a trained model for inference

- Deploying third-party NLP models and testing via the UI

- Running advanced data processing with trained models

Technical requirements

To follow the recipes in this chapter, you'll need an Elastic Stack deployment that includes the following:

- Elasticsearch for searching and storing the data

- Kibana for data visualization and stack management (you must be a Kibana user with **All privileges** on Fleet and Integrations)

- An ML node

> **Important note**
>
> Most of the recipes in this chapter rely heavily on the ML capabilities of the Elastic Stack. Ensure you allocate sufficient RAM to handle the various analyses and tasks effectively. If using a trial deployment on Elastic Cloud, consider increasing the RAM capacity of your ML node from the default 1 GB to at least 4 GB. Refer to the *Creating and setting up additional Elasticsearch nodes* recipe in *Chapter 1*. The same recommendation applies if you are operating a self-managed Elastic Stack.

Finding deviations in your data with outlier detection

In data analysis, uncovering meaningful insights often includes identifying patterns, trends, deviations, and anomalies. One useful technique is **outlier detection**—it involves detecting data points that significantly deviate from the majority of the dataset. This helps us identify elements that differentiate normal data points from anomalous ones. Unlike anomaly detection for time series data, as we learned in the previous chapter, we are not concerned with the temporal evolution of the dataset. Instead, we focus on data clusters, evaluating their density and distance using multivariate analysis.

In the first four recipes of this chapter, we will continue to use the Rennes traffic data, which has become familiar to us from previous chapters.

Getting ready

Before delving into the supervised learning intricacies in the context of the Elastic Stack, let's review the ML methodology from end to end, as shown in *Figure 8.2*:

Figure 8.2 – ML end-to-end methodology

A key phase of this pipeline is *Transform your data*. To proceed with data frame analysis, you will often need to pivot your raw continuous timestamped data into new entity-centric indices (features); we will follow this logic for the first three recipes of this chapter. We touched on this topic in the *Setting up pivot data transform recipe* in *Chapter 5*.

Before proceeding, make sure that you've completed the *Exploring your data in Discover* recipe in *Chapter 6*.

We provide a data transform you can use to pivot your Rennes traffic dataset. The Dev Tools commands for the transform can be found in the GitHub repository here: `https://github.com/PacktPublishing/Elastic-Stack-8.x-Cookbook/blob/main/Chapter8/snippets.md#data-transform`

This transform pivots the Rennes traffic dataset on `location`, `hour_of_day`, and `day_of_week` with average aggregations on `average_vehicle_speed` and `traveltime.duration`, a max aggregation on `max_speed`, and a top metrics aggregation on `traffic_status`. The following screenshot shows a preview table of the transform:

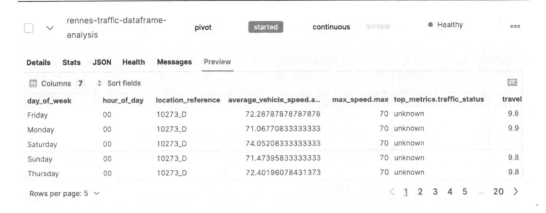

Figure 8.3 – Rennes traffic data frame results example

To create the transform, go to **Kibana** | **Management** | **Dev Tools** and execute the two commands to create and start the data transform job: `rennes-traffic-dataframe-analysis`.

To validate that the data transform has been successfully created, follow these steps:

1. In Kibana, navigate to **Stack management** | **Kibana** | **Data views**, then click on **Create data view**. Create a new data view using the `rennes-traffic-dataframe-analysis` index pattern, as shown in *Figure 8.4*:

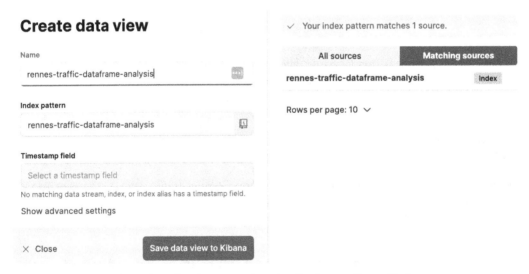

Figure 8.4 – Creating a new data view for the transformed index

2. You can now go to **Analytics | Discover** and select the newly created data view. Verify that it is populated with the results from the data transform, as shown in *Figure 8.5*:

Figure 8.5 – Rennes traffic data frame data view in Discover

Once you have successfully set up the data transform and created the data view for the destination index, you can proceed to the main part of the recipe.

How to do it...

The following steps will show you how to set up a data frame analytics job in Kibana to identify unusual data within our dataset. We'll use the transformed data view from the Rennes traffic data and apply ML techniques to detect outliers:

1. In Kibana, go to **Machine Learning | Data Frame Analytics | Jobs**.

2. Click on **Create data frame analytics job** and on the following page, select the transformed data view, **rennes-traffic-dataframe-analysis**, as the source data view.

3. Set **Job type** to **outlier detection** and proceed with the following steps:

    A. In the **Configuration** section, retain the default values and click **Continue**.

    B. In the **Additional Options** section, increase the model memory limit to **50mb** and retain the other default values. Then, click **Continue**.

    C. Scroll to the **Job details** section, set Job_ID as rennes-traffic-dataframe-outlier, then click **Continue**.

    D. In the **Validation** section, click **Continue**.

E. Click on the **Create** button to create the job. At this stage, you should see the job creation and start acknowledgments, as shown in *Figure 8.6*:

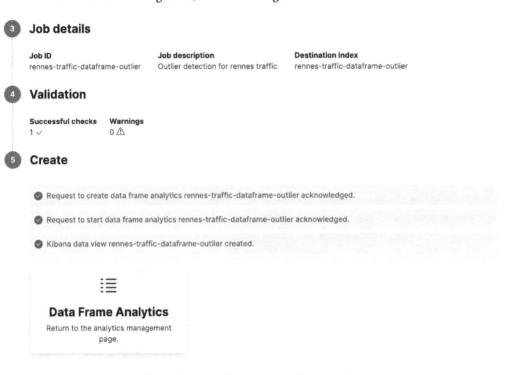

Figure 8.6 – Creating an outlier detection job

F. Click on **Data Frame Analytics** and you should see the overview of the ML job configuration and the progress of the ML job (we have provided the JSON object of the reference job configuration in the GitHub repository: `https://github.com/PacktPublishing/Elastic-Stack-8.x-Cookbook/blob/main/Chapter8/snippets.md#sample-outlier-job`).

4. When the status of the job changes to **stopped** (it should take only a few seconds), click on the view icon to view the results, as shown in *Figure 8.7*:

Figure 8.7 – Outlier detection job status

5. Now, you can begin exploring the results of your outlier detection job. In the **Scatterplot matrix** pane, under the **Fields** option, select **max_speed.max** and **traveltime.duration.avg** and set the sample size to **10000**, as illustrated in *Figure 8.8*. This will reveal the two-dimensional relationship between these two aggregations, highlighting outliers with larger and distinct dots:

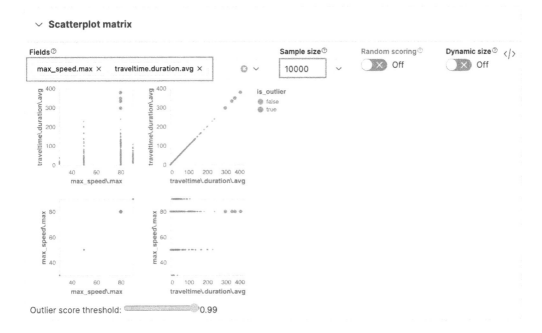

Figure 8.8 – Outlier scatterplot matrix

6. Scroll down and you will see a data table sorted by outlier score, displaying location, hour, and day. Click on **2 columns hidden** and ensure that you select the **traveltime.duration.avg** column (you can also click on the **Show all** link). The data frame analytics job creates an index containing the original data, along with outlier scores for each document. These scores indicate the level of deviation of each entity from the others:

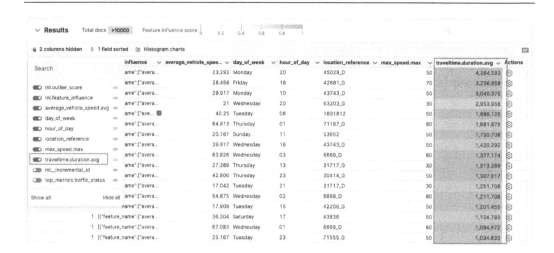

Figure 8.9 – Outlier scatterplot matrix

ml.outlier_score spans from **0** to **1**. The higher the value, the more likely it is that the entity is an outlier.

Furthermore, each document receives annotations in the form of feature influence values assigned to individual fields. These values sum up to **1**, signaling the relative importance of each field in classifying the entity as either an outlier or an inlier. As an example, dark shading on the field suggests that **traveltime.duration.avg** was the most influential factor in identifying the outlier.

How it works...

Outlier detection focuses on identifying atypical data points within a dataset. The unusualness of a data point is determined by its proximity to other points and the density of nearby clusters. Normal points typically have many neighbors, while outliers are situated further from these clusters.

The method to calculate this unusualness is based on distance, using K-nearest neighbor algorithms. The outcome is expressed as an outlier score—the higher the outlier score, the more atypical the point is considered.

While pinpointing outliers, Elasticsearch also assists in interpreting the reasons behind a data point being an outlier by highlighting the most influential field contributing to its unusualness.

See also

For more insights into benchmarking outlier detection, refer to this blog post: `https://www.elastic.co/blog/benchmarking-outlier-detection-in-elastic-machine-learning`

Building a model to perform regression analysis

Regression analysis within the Elastic Stack, facilitated by data frame analytics, provides a powerful means of deriving insights from complex datasets. The Elastic Stack's advanced statistical analysis methods allow users to examine relationships between diverse data points in detail.

In data frame analytics, regression techniques are vital for estimating continuous values, such as sales figures or temperature readings, using patterns derived from historical data. With these methods, businesses and researchers can anticipate trends, discern the primary influences on outcomes, and make well-informed decisions—all in the scalable, real-time environment of the Elastic Stack.

Regression and classification are part of the supervised learning category of ML capabilities provided by the Elastic Stack, as illustrated in *Figure 8.10*:

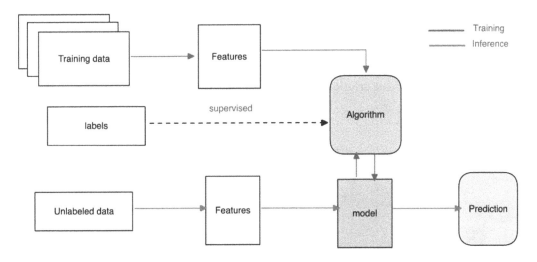

Figure 8.10 – Supervised learning process overview

Once a model is trained and evaluated to be effective, it can be deployed for prediction purposes.

In this recipe, we will construct a regression model to estimate the average travel time duration based on factors such as the location, the day of the week, and the hour of the day. You'll learn how to train a regression model and evaluate its performance.

Getting ready

Make sure you have worked through the following recipes:

- *Creating a Logstash pipeline* in *Chapter 5*
- *Creating visualizations from runtime fields* in *Chapter 6*

Also, ensure you have applied the transform on the Rennes traffic dataset as described in the *Getting ready* section of the *Finding deviations in your data with outlier detection* recipe.

How to do it...

Let's begin in Kibana, where we'll perform all the steps for this recipe:

1. Go to **Machine Learning | Data Frame Analytics | Jobs**.

2. Click on **Create job**, and on the following page, select **rennes-traffic-dataframe-analysis** as the source data view.

> **Important**
>
> `rennes-traffic-dataframe-analysis` is the resulting index from the `rennes-traffic-dataframe-analysis` transform job. Make sure you have created and run the data transform as described in the *Technical requirements* section of this chapter.

3. Now, in the **Data Frame creation** wizard, select **Regression** under **Configuration**, as shown in *Figure 8.11*:

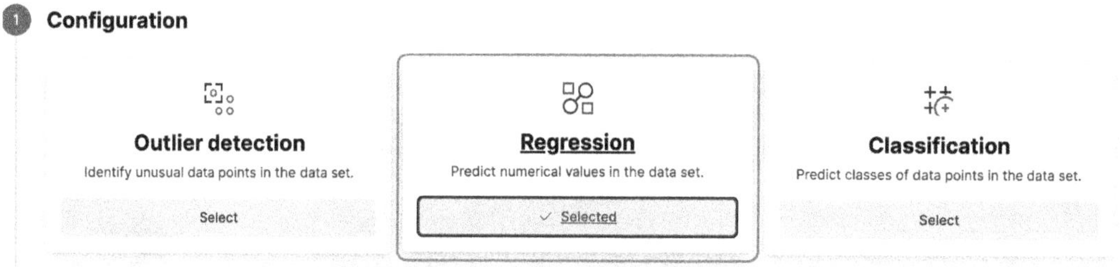

① Configuration

Outlier detection	**Regression**	**Classification**
Identify unusual data points in the data set.	Predict numerical values in the data set.	Predict classes of data points in the data set.
Select	✓ Selected	Select

Figure 8.11 – Regression analysis selection

In the **Dependent variable** field, select **traveltime.duration.avg**. This is the variable we're going to predict with our regression model.

4. Next, in the **Included fields** panel, select the following fields, as shown in *Figure 8.12*:

    *   **max_speed.max**
    *   **day_of_week**
    *   **hour_of_day**

- **location_reference**

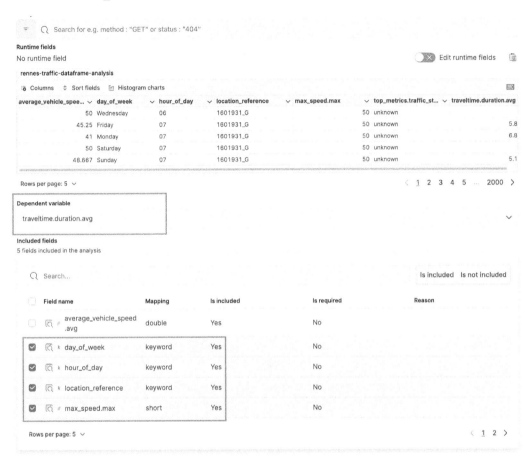

Figure 8.12 – Regression configuration section 1

We are only selecting certain fields because these are the ones we'll use for actual prediction. It's best practice to train the model with the same set of features that will be used for prediction to maintain consistency and reliability. If it's necessary to train with more features, careful consideration should be given to manage any feature discrepancies during prediction.

5. In the **Training percent** panel, you can choose the percentage of the data you want to use for training the regression model. Let's set it to 50 and click on the **Continue** button.

Important note on the training percentage

To determine the right training percentage, start by evaluating the size of your dataset. If you have over 100,000 documents, a starting percentage of around 10 or 15% is recommended. You can refine this percentage based on the quality of your analysis results.

6. In the **Additional options** section, set **Feature importance values** to 4. Then, click on the **Continue** button:

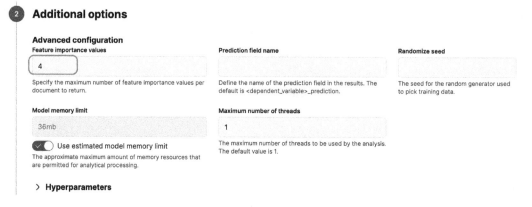

Figure 8.13 – Regression configuration section 2

Feature importance values help in understanding the influence of each input feature on the predictive model's output. Setting it to 4 means that for each prediction made by the model, the top four features (variables) that had the most significant impact on that specific prediction will be identified and reported. It's important not to set too many fields for feature importance, as it can drastically impact the performance of your model. The goal here is to focus on the significant values that affect a particular prediction.

This section also contains hyperparameters. We will not change anything here as we will rely on the ML job to evaluate the best combination of values through a process of hyperparameter optimization. You can see those values in the analysis stats once the job is completed.

7. In the **Job details** section, provide the following job ID: `rennes-traffic-dataframe-regression`. Enter a brief description and then click on **Continue** to move on to the **Validation** section.

    In this section, you can also change the name of the index where the results of the analysis will be stored.

8. In the **Validation** section, the wizard performs some checks on the configuration before launching the model.

> **Important note**
>
> You will certainly receive a warning message regarding feature importance during the validation steps, as shown in *Figure 8.14*. Feature importance is calculated for each document in the analysis, hence the notification when using a large training dataset.

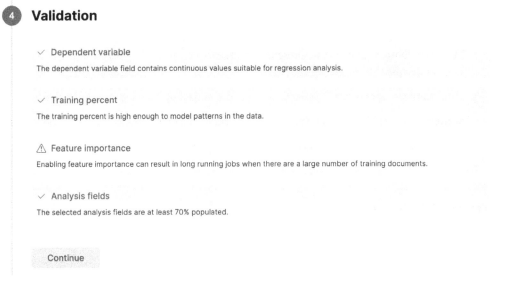

If you get a warning during that phase regarding the training percentage, try reducing it using the rule of thumb we previously mentioned. Once the validation has successfully completed, click on **Continue** to proceed to the next and last section.

9. Lastly, in the **Create** section, we're good to launch our model. Leave the default **Start immediately** option, then click on **Create** and return to the **Data Frame Analytics Jobs** page:

Figure 8.15 – Regression configuration section 5

10. On the **Data Frame Analytics Jobs** page, you will see the newly created job running (we have provided the JSON object of the reference job configuration in the GitHub repository: `https://github.com/PacktPublishing/Elastic-Stack-8.x-Cookbook/blob/main/Chapter8/snippets.md#sample-regression-job`). Based on the number of documents in the dataset and the percentage of training you've configured, the job can take some time to complete (during our tests, it took approximately 10 minutes). Once it has completed, you can click on the view icon to explore the results. At this stage, we will explore the results of our analysis. The results exploration page is divided into four sections:

- **Analysis**: This section provides job details, the corresponding JSON configuration, and log messages, which are useful for monitoring the job's status and progress.

- **Model evaluation**: This section presents key metrics on the model's performance and fitting. We will discuss this further in the recipe.

- **Total feature importance**: This section shows the impact of the selected field on the analysis.

- **Scatterplot matrix**: This visual representation illustrates relationships between the various fields.

- **Results**: The results documents are displayed in a table.

Figure 8.16 shows the overall structure of the results page:

> **Important note on results**
> Keep in mind that the actual results you see in your environment may differ from the screenshots, as they depend on how long you have been ingesting data and on real-time values.

Figure 8.16 – Regression results exploration page

We'll now examine specific sections of this page, namely **Model evaluation**, **Total feature importance**, and the **Results** table.

- The **Model evaluation** section on the regression analysis results page delivers a concise assessment of your regression model's performance:

Figure 8.17 – Regression results model evaluation

This section typically encompasses key metrics such as **Mean Squared Error (MSE)**, **Root Mean Squared Error (RMSE)**, and **Mean Absolute Error (MAE)**. These metrics are essential for quantifying the discrepancy between the predicted values generated by your model and the actual values observed in your dataset. They are vital for gauging the accuracy and efficacy of your model. Generally, lower values for these metrics signify a model that more accurately fits the data.

For our scenario, we have an R-squared value of **0.582** for the generalization error, indicating our model explains about 58.2% of the variance in the dependent variable—a reasonable outcome. R-squared is valuable for understanding the amount of variance in the dependent variable predictable from the independent variables.

- Moving on to the **Total feature importance** section, we previously touched upon the critical aspect of feature importance as it helps in understanding the influence of each input feature on the predictive model's output. Let's see how it pans out for our model:

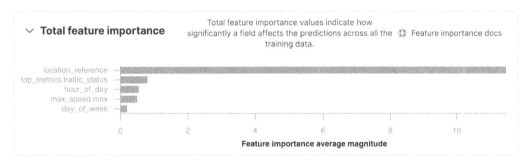

Figure 8.18 – Regression results – Total feature importance

Here, we note that among the four fields, **location_reference** significantly impacts the prediction with an average magnitude of **8.1**.

Now we have a good understanding of the key metrics behind the evaluation of our model and the total feature importance. Let's have a look at the actual results displayed in a table, as illustrated in *Figure 8.19*:

∨ **Results** Total docs >10000								
Showing documents for which predictions exist								
⌾ 3 columns hidden ↕ Sort fields 📊 Histogram charts								
ml.is_traini... ∨	ml.traveltime.dur... ∨	traveltime.duration.avg ∨	ml.feature_importance ∨	average_vehicle_spee... ∨	day_of_week ∨	hour_of_... ∨	location_referen... ∨	Actions
true	4.331	3.364	[{"feature_name":["day_of...	53.909	Tuesday	06	1601794_D	⚙
true	4.688	5	[{"feature_name":["day_of...	35.833	Friday	07	1601794_D	⚙
true	4.799	3	[{"feature_name":["day_of...	58.333	Monday	07	1601794_D	⚙
true	4.312	4.333	[{"feature_name":["day_of...	29.667	Sunday	07	1601794_D	⚙
true	4.752	4.167	[{"feature_name":["hour_...	37.833	Thursday	07	1601794_D	⚙
true	5.112	5.5	[{"feature_name":["day_of...	33	Friday	08	1601794_D	⚙

Figure 8.19 – Regression results table

You can see in the table the predicted values and how they compare with the actual ones. You also have the details of feature importance for each document in the `ml.feature_importance` columns. In the `Actions` column, you can click on the gear wheel icon and then click on **View in Discover** to open one of the documents in **Discover**.

- In **Discover,** you can expand the documents and have a look at `ml.feature_importance`, for example. You can build a quick visualization to compare the predicted and actual values:

    i.   To remove the selected document, clear any selected document from the search field and click **Update** to refresh and display all documents in the index.

    ii.   On the left panel, click on the **ml.traveltime.duration.avg_prediction** field, then click on **Visualize**; you'll be redirected to **Lens**.

    iii. Once in Kibana Lens, on the right panel, leave the median of **ml.traveltime.duration. avg_prediction** in the vertical axis section.

    iv.   In the vertical axis section, also add the **traveltime.duration.avg** field with the median function below the **ml.traveltime.duration.avg_prediction** field.

    v.   Add the **hour_of_day** field to the horizontal axis section and select **Top values**. Change **number of values** from **5** to **24**, then rank by **Alphabetical**.

    vi.   Change the visualization type from **Bar Vertical** to **Line**. You should now see a visualization with two lines: one for the predicted values and the other one for the actual values, as shown in *Figure 8.20*:

Figure 8.20 – Lens visualization for regression results

In our testing, predicted and actual values showed similar trends over a 24-hour period. We've trained and built a regression model—our next logical step is using the model for inference to make real-time predictions, which will be covered in the *Using a trained model for inference* recipe.

How it works...

Performing regression analysis using Elastic ML is a process that can be broken down into the steps outlined in *Figure 8.21*:

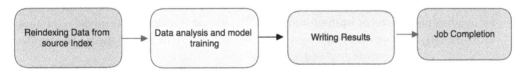

Figure 8.21 – Regression analysis phases

A regression analysis, like any data frame job, is a persistent Elasticsearch task; the task works by going through the following phases:

1. **Reindexing data from source index**: The job starts by loading the source data from the specified index or indices into memory. This data is used for training the model.

2. **Data analysis and model training**: Depending on the type of job (regression, classification, etc.), the job performs the appropriate analysis on the loaded data. In this phase, the job may

include steps such as analyzing the data for patterns, training an ML model, and tuning this model to optimize performance.

3. **Writing results**: Once the analysis is complete, the results are written to the specified destination. This can include information such as the model's predictions, evaluation metrics, and feature importance scores.

4. **Job completion**: After the results are written, the job is completed. You can then review the results, use them for inference, or integrate them into larger workflows within the Elastic Stack.

Regression analysis within the data frame analytics framework is a sophisticated process designed to establish relationships between a dependent variable and one or more independent variables. We have broken down the applied methodology in a concise manner here:

- **Data preparation and ingestion**: This step involves getting data ready and putting it into Elasticsearch. The data, including features and the target for analysis, must be of good quality and relevant because it affects how well the model works.

- **Creating the data frame analytics job**: Through Kibana's ML tool, you can set up a job for data frame analytics. This job outlines the analysis type, what data to use, and where to save the results. It's important to specify what you're trying to predict and what data you'll use for the prediction.

- **Feature selection and model training**: The Elastic Stack automatically picks the most relevant features for predicting the outcome and trains a regression model with them. The process involves creating a formula that best predicts the target from the features while handling potential issues such as multicollinearity and overfitting.

- **Model evaluation and results interpretation**: After training, the Elastic Stack uses metrics such as R-squared and MSE to gauge the model's performance. The analysis outcomes, such as predictions and which features mattered most, are saved and can be looked at in Kibana.

- **Operationalization and continuous learning**: Once the model is evaluated, it can be used for ongoing predictions or be improved upon. The Elastic Stack supports model retraining with new data, allowing the model to get better and stay up to date with changing data trends.

In summary, regression analysis in the context of the Elastic Stack's data frame analytics is a comprehensive process that involves data preparation, job configuration, automatic feature selection, model training, and evaluation. This process is streamlined and integrated within the Elastic Stack, providing an efficient and user-friendly approach to predictive analytics.

There's more...

In the *How to do it...* section of this recipe, we provided an overview of how you can use the **Results Explorer** to evaluate your model. Another way to do it is to leverage the *Evaluate data frame analytics API* especially if you want to analyze those metrics outside of Kibana in a Jupyter notebook, for example.

The data frame analytics wizard in Kibana features a tool called **Analytics Maps**. It presents, in a visual map fashion, the various steps and phases applied to a specific dataset in the context of supervised learning and outlier detection.

Regression employs a collective learning approach akin to XGBoost, an advanced version of gradient boosting. This technique merges decision trees with gradient-boosting principles. In XGBoost, a series of decision trees are sequentially trained, with each tree enhancing its accuracy by learning from the errors made by the preceding trees in the ensemble. With every new iteration, the added trees refine the overall decision-making capability of the collective group of trees. Through standard settings, this regression method focuses on minimizing a specific loss function, known as the MSE loss.

Regarding the types of feature variables applicable in these algorithms, there are three main categories: **numerical**, **categorical**, and **Boolean**. It's important to note that array types are not compatible with these algorithms.

Training and evaluating an ML job is very important but we do not stop there. The next step is to deploy the trained model and use it for prediction through a process called **inference**. We will cover this in the *Using a trained model for inference* recipe of this chapter.

See also

- If you want to dive deeper into how to interpret and validate ML models using feature importance in the Elastic Stack, check out this blog: `https://www.elastic.co/blog/feature-importance-for-data-frame-analytics-with-elastic-machine-learning`

- To better understand the steps required to build a supervised learning pipeline in the Elastic Stack, read this excellent article: `https://www.elastic.co/blog/machine-learning-models-supervised-learning-pipeline`

- For a very extensive dive into ML within the Elastic Stack, check out this book: `https://www.packtpub.com/product/machine-learning-with-the-elastic-stack-second-edition/9781801070034`

Building a model for classification

In this recipe, we will perform another type of analysis: **classification**. Classification analysis within the context of data frame analytics in the Elastic Stack is a powerful ML technique used to categorize data into predefined classes or groups.

This process involves training a model on a dataset with known class labels, thereby enabling the model to learn how to categorize new, unseen data. In the Elastic Stack, classification is commonly applied to tasks such as spam detection, customer segmentation, and sentiment analysis.

In this recipe, we will train a model to classify traffic according to the **free-flow**, **heavy**, **congested**, and **unknown** categories using the Rennes traffic dataset, based on features such as *location*, *hour of the day*, *day of the week*, and *maximum authorized speed*.

Getting ready

Make sure you have worked through the following recipes:

- *Exploring your data in Discover* in *Chapter 6*
- *Building a model to perform regression analysis* in this chapter (while this is not a technical prerequisite, the configuration steps in the *Building a model to perform regression analysis* recipe closely mirror those needed for classification, so it would be beneficial to familiarize yourself with them beforehand)

Additionally, make sure you have applied the transform on the Rennes traffic dataset as described in the *Getting ready* section of the *Finding deviations in your data with outlier detection* recipe in this chapter.

How to do it...

Let's head to Kibana, where we'll perform all the actions for this recipe:

1. In Kibana, go to **Analytics | Machine Learning | Data Frame Analytics | Jobs**.

2. Click on **Create job** and on the following page, select `rennes-traffic-dataframe-analysis` as the source data view.

> **Important note**
>
> `rennes-traffic-dataframe-analysis` is the resulting index from the `rennes-traffic-dataframe-analysis` transform job. Make sure to have created and run the data transform as described in the *Technical requirements* section of this chapter

3. We are now in the data frame creation wizard. Select **Classification** under **Configuration**:

Figure 8.22 – Classification job type selection

4. In the **Dependent variable** field panel, select **top_metrics.traffic_status**. This variable is crucial as it will be used for the classification prediction, serving as the target outcome the model aims to predict based on the input features. To streamline the analysis and focus on the most relevant data, include only specific fields that are pertinent to your model's predictive capabilities (make sure to uncheck **traveltime.duration** on the second page):

- day_of_week

- hour_of_day

- location_reference

- max_speed.max

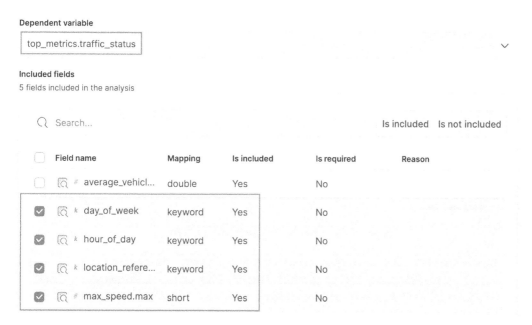

Figure 8.23 – Classification configuration section 1

5. Below the scatterplot matrix, in the **Training percent** slide bar, select the percentage of the data you want to use for training the regression model. You can set it to **20** and then click on **Continue**.

> **Important note on the training percentage for classification**
>
> We've already explained the rationale behind allocating the right training percentage (in the *Building a model to perform regression analysis* recipe). For the classification, we're setting a percentage of 20, which is far below the 50% we used previously in the regression jobs. This is simply because classification analysis tends to take more time to complete. Hence, we're using a smaller percentage to speed up the analysis. Of course, this is a process of trial and error, so feel free to adjust the value based on your use case and the number of documents.

6.   In the **Additional options** section, set **Feature importance values** to 4. Refer to the previous recipe for a detailed explanation of feature importance.

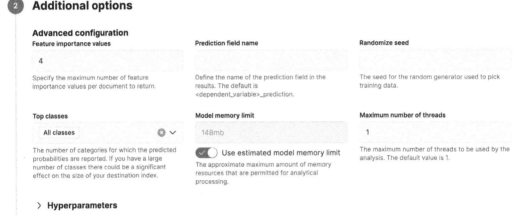

Figure 8.24 – Classification additional options

This section also contains hyperparameters; we will not change anything here, as we will rely on the ML job to evaluate the best combination of values through hyperparameter optimization. You can see those values in the analysis stats once the job is completed. Click on **Continue**.

7.   In the **Job details** section, set the following ID: `rennes-traffic-dataframe-classification`. Enter a short description in the **Job description** field, and then click on **Continue** to move to the **Validation** section.

8.   In this section, you can also change the name of the index by turning off the **Use job ID as destination index name** and **Use results field default value: "ml"** toggles where the results of the analysis will be stored.

Figure 8.25 – Classification job details

9. In the **Validation** section, the wizard performs some checks on the configuration before creating the job:

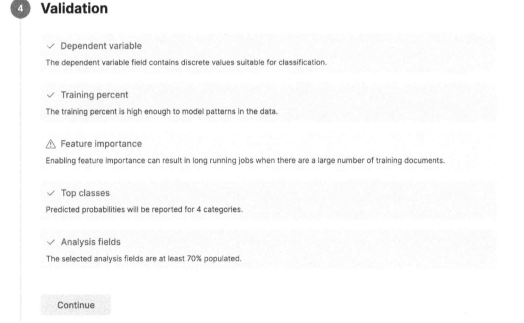

Figure 8.26 – Classification configuration section 4

> **Important note**
> You will certainly receive a warning message regarding feature importance during the validation steps. Feature importance is calculated for each document in the analysis, hence the notification when using a large training dataset.

If you see a warning during that phase regarding the training percentage, try reducing the percentage using the rule of thumb we previously mentioned. Once the validation has successfully completed, click on **Continue** to proceed to the next and final section.

10. Lastly, in the **Create** section, we're now good to launch our model. Leave the default **Start immediately** option, then click on **Create** and return to the **Data Frame Analytics Jobs** page:

 **Create**

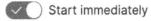 Start immediately
If unselected, job can be started later by returning to the jobs list.

Figure 8.27 – Classification configuration section 5

11. On the **Data Frame Analytics Jobs** page, you'll observe the job you've just created is now running (we have provided the JSON object of the reference job configuration in the GitHub repository: `https://github.com/PacktPublishing/Elastic-Stack-8.x-Cookbook/blob/main/Chapter8/snippets.md#sample-outlier-job`). It will take some time; during our test, it took approximately 10 minutes. The duration of this job depends on the dataset's size and the training data percentage you've configured. Once the job is complete, you can click on the view icon to delve into and analyze the results of your analytics job.

At this stage, we can explore the results of our analysis. The **Results Explorer** page is divided into four sections:

- **Analysis**: This essentially provides details on our job, corresponding JSON config, and log messages. It is useful for monitoring the status and progress of the job.

- **Model evaluation**: This provides important information on the performance and fitting of our model. More on this later in the recipe.

- **Scatterplot matrix**: This shows in a visual way the relationship between various fields.

- **Results**: The documents are displayed in a table.

Figure 8.28 shows the overall structure of the results page:

Figure 8.28 – Classification Results Explorer page

Next, we will dive into specific sections of this page, namely **Model evaluation**, **Total feature importance**, and the **Results** table.

Section 1 – Model evaluation

The **Model evaluation** section in the classification results explorer provides critical insights into the performance of your classification model. This section typically includes key metrics such as accuracy, precision, recall, and F1 score, which are essential for assessing the model's ability to correctly classify data:

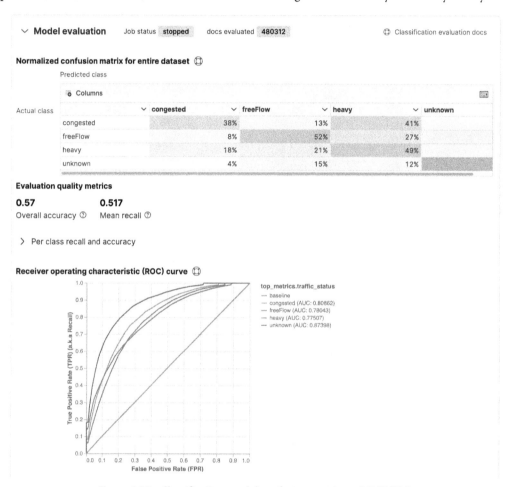

Figure 8.29 – Classification model confusion matrix and AUC ROC

Let us focus on the **Normalized confusion matrix for entire dataset** and **Receiver operating characteristic (ROC) curve** sections, as seen in *Figure 8.29*:

- **Multiclass confusion matrix**: Compares the actual versus predicted classifications and is split into four parts: true positives, true negatives, false positives, and false negatives. A normalized confusion matrix presents these values as proportions or percentages rather than raw counts. This normalization allows for easier comparison across datasets of different sizes and is particularly useful in evaluating class imbalances.

- **ROC curve**: The ROC curve is a graph that shows how well a system that classifies things into two groups works as we change its threshold for making decisions. It draws the **True Positive Rate (TPR)** versus the **False Positive Rate (FPR)** for different thresholds. The **Area Under the Curve (AUC)** measures how good the model is at telling the difference between the two groups. An AUC of 1 represents a perfect classifier, while an AUC of 0.5 suggests no discriminative power, equivalent to random guessing

Section 2 – Total feature importance

We previously touched upon the critical aspect of feature importance as it helps in understanding the influence of each input feature on the predictive model's output. Let's see how it pans out for our model:

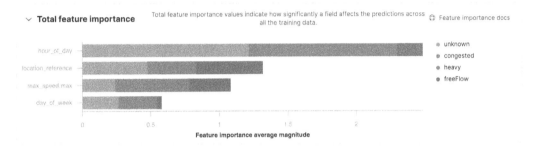

Figure 8.30 – Regression results – Total feature importance

We can observe that among the four fields, `hour_of_day` is the one that had the most significant impact on the prediction with an average magnitude above 2.5; `location_reference` has also quite a significant impact on the prediction.

Section 3 – Results

Now that we have a good understanding of the key metrics behind the evaluation of our model and the total feature importance, let's have a look at the actual results displayed in a table, illustrated in *Figure 8.31*:

ml.is_training	ml.top_metrics.traffic...	top_metrics.traffic_st...	ml.prediction_probabi...	ml.feature_importance	ml.prediction_score	ml.top_classes	average_vehicle_spee...	Actions
true	congested	unknown	0.134	[{"feature_name":["day_of...	0.134	[{"class_score":[0.13385...	34.176	⚙
true	freeFlow	heavy	0.429	[{"feature_name":["day_of...	0.049	[{"class_score":[0.04900...	18.882	⚙
true	freeFlow	unknown	0.515	[{"feature_name":["day_of...	0.059	[{"class_score":[0.05884...	21.412	⚙
true	heavy	unknown	0.336	[{"feature_name":["day_of...	0.072	[{"class_score":[0.07202...	70.412	⚙
true	heavy	unknown	0.387	[{"feature_name":["day_of...	0.083	[{"class_score":[0.08293...	70.824	⚙
true	congested	heavy	0.147	[{"feature_name":["day_of...	0.147	[{"class_score":[0.14738...	30	⚙
true	congested	freeFlow	0.083	[{"feature_name":["day_of...	0.083	[{"class_score":[0.08260...	106.059	⚙
true	freeFlow	freeFlow	0.883	[{"feature_name":["day_of...	0.101	[{"class_score":[0.10077...	94.235	⚙

Figure 8.31 – Classification results table

You can see in the table the predicted traffic status classes and how they compare with the actual ones. You also have the details of feature importance for each document in the `ml.feature_importance` columns. In the `Actions` column, you can click on the gear wheel of any document and then click on **View** in **Discover** to open the document in **Discover**.

In **Discover**, you have the option to expand a document to examine details such as `ml.feature. importance`, which gives insight into the features that most significantly impact the model's predictions.

Following a similar approach to the previous recipe, you can swiftly create a visualization to compare the predicted values against the actual values:

1. To remove the selected document, clear any selected document from the search field and click **Update** to refresh and display all documents in the index.

2. Find and click on the **top_metrics.traffic_status** field on the left navigation menu, then click on **Visualize**. This action will take you to **Lens**, where you can create and customize your visualization.

3. In **Lens**, the displayed visualization will be a **Bar vertical** chart stacked with the top five values of **top_metrics.traffic_status** on the horizontal axis and **Count of records** values on the vertical axis.

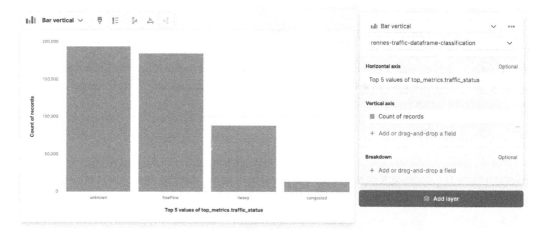

Figure 8.32 – Top five values of top_metrics.traffic_status

4. Click on **Add layer** on the right panel to add a new layer of the visualization type; in the horizontal axis, add the **ml.top_metrics.traffic_status_prediction** field. On the vertical axis, add **Count of records**. Change the visualization type from **Bar vertical stacked** to **Bar vertical**. The resulting chart will look like *Figure 8.33*:

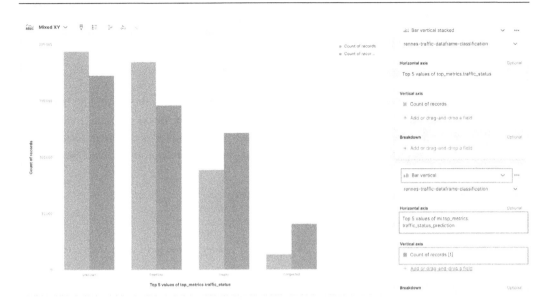

Figure 8.33 – Lens visualization for classification results

We can see based on the number of documents how our model fared in predicting the traffic status classes.

> **Note**
>
> Leveraging Kibana Lens to build this type of visualization provides a quick and easy way to see how predicted classes by the trained model compare to the existing labels present in your data, but the best and most accurate way to evaluate the results of the model is to rely on the multiclass confusion matrix and the AUC ROC.

We have successfully trained and developed a classification model to predict traffic status based on average travel time duration. In the forthcoming recipe titled *Using a trained model for inference*, we will explore how to effectively utilize this model to make actual predictions

How it works...

Classification analysis within the Elastic Stack, particularly in Elasticsearch's data frame analytics, involves the same structured process as the one described earlier for regression analysis.

The **data frame** process involves initially preparing and importing data into Elasticsearch for proper indexing. Users then configure a classification task in Kibana by selecting the source and destination indexes and identifying the variables to predict. This step is followed by preparing the data, including converting categorical variables and normalizing the dataset. The system proceeds to train a classification model with the defined variables and target. Upon completion, the model's performance is assessed, and the outcomes are stored in Elasticsearch for further analysis and visualization in Kibana, facilitating an efficient workflow from data preparation to evaluation.

The following are the algorithms and techniques used under the hood:

- **Ensemble learning**: Elasticsearch often employs ensemble methods, combining multiple models (such as decision trees) to improve prediction accuracy, and robustness. Decision trees work by splitting the data based on feature values, creating a tree-like model of decisions. *Figure 8.34* shows how a basic decision tree can be used to illustrate its application to our use case:

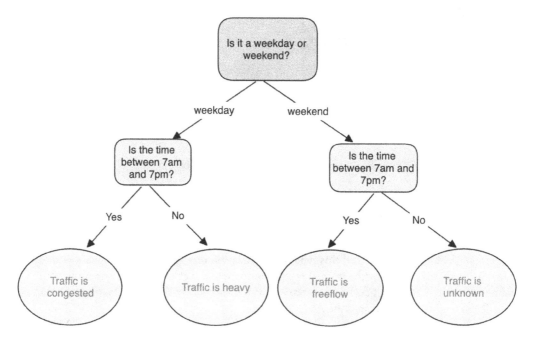

Figure 8.34 – Example of a simplified decision tree for classification

- **Handling class imbalance**: Techniques to handle imbalanced datasets are used, ensuring that minority classes are appropriately represented in the model.

- **Regularization**: To prevent overfitting, regularization techniques are applied, which keeps the model general enough to perform well on unseen data.

In essence, classification analysis in the Elastic Stack involves a comprehensive workflow from data ingestion to model training and evaluation, utilizing advanced algorithms and techniques to ensure accurate and reliable predictions.

There's more...

When creating a classification job, you must specify the field with the classes you aim to predict, called the dependent variable, which should have no more than 30 classes. All other relevant fields are automatically included as feature variables, which serve as input for the model. Keep in mind that the job's runtime and resource usage increase with the number of feature variables involved.

To manage resources effectively, consider using model memory limits and data frame analytics memory estimation. Optimize the performance by adjusting the model training data size and utilizing advanced features such as hyperparameters and early stopping

You can add custom URLs to data frame analytics jobs to enhance the interpretability and practicality of the analysis.

These custom URLs can be configured to point to Kibana dashboards or other web pages, providing direct links to additional insights or detailed visualizations relevant to the analysis results. For instance, in our job analyzing traffic data, a custom URL could link to a dashboard displaying the traffic status. This feature significantly improves the user experience, enabling seamless navigation between analytical results and more in-depth, contextual information, thereby facilitating a more comprehensive and efficient data analysis workflow.

See also

- To get more information on how classification results are benchmarked in Elastic ML, read this article: `https://www.elastic.co/blog/benchmarking-binary-classification-results-in-elastic-machine-learning`

- For an example of classification analysis using Jupyter Notebook, head to this GitHub repository: `https://github.com/elastic/examples/tree/master/Machine%20 Learning/Analytics%20Jupyter%20Notebooks`

- To learn how we've combined supervised and unsupervised ML for enhanced detection, check out this great article: `https://www.elastic.co/blog/supervised-and-unsupervised-machine-learning-for-dga-detection`

Using a trained model for inference

Now that you have trained some models using Elastic, we will look at how you can use them for prediction through a process called **inference**. It is a process of using trained ML models against incoming data in a continuous way. In the Elastic Stack, this process happens essentially through an inference processor in ingest pipelines or pipeline aggregation.

In this recipe, we'll build upon the classification model we trained in the previous recipe, configure it into an ingest pipeline processor, and use it for prediction.

Getting ready

Make sure you have worked through the following recipes:

- *Creating a Logstash pipeline* in *Chapter 5*
- *Creating visualizations from runtime fields* in *Chapter 6*
- *Building a model to perform regression analysis* in this chapter

The snippets for this recipe can be found at `https://github.com/PacktPublishing/Elastic-Stack-8.x-Cookbook/blob/main/Chapter8/snippets.md#using-trained-model-for-inference`.

How to do it...

To use a trained model on incoming data through inference, you need to *deploy* the model first, which essentially boils down to making the model usable as a function:

1. Navigate to **Machine Learning | Model Management | Trained models** in Kibana. Look for the `rennes-traffic-dataframe-classification` model and click **Deploy model** in the actions menu on the right:

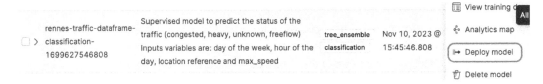

Figure 8.35 – Deploy trained models

2. This will open the **Deploy analytics model** configuration page:

Deploy analytics model

Figure 8.36 – Deploy analytics model configuration

3.   Begin by inputting a name and description for your job. If desired, you can specify a target field. If the pre-filled information aligns with your requirements, click **Continue**.

4.   On the **Configure processor** page, you can set up the inference configuration. You can usually leave this at its default settings:

Deploy analytics model

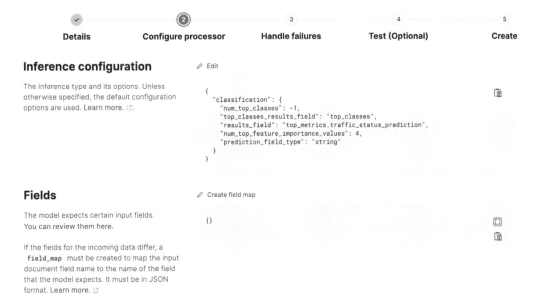

Figure 8.37 – Deploy model step 2

5. Below the inference configuration, you'll notice a **Fields** section. This is important to configure if the incoming documents don't have the input fields the model anticipates. Click **Continue**.

6. On the **Handle failures** page, define how the pipeline should react to errors in problematic documents through the on_failure configurations. By default, the document is stored with the failure context. You can add custom on_failure instructions to execute a particular set of processors, for instance.

7. The **Test (optional)** page gives you a good opportunity to test your pipeline at this point. It's pre-populated with a sample document. Click **Simulate pipeline** to check the result, then click **Continue**:

Deploy analytics model

| Details | Configure processor | Handle failures | Test (Optional) | Create |

Test the pipeline results

This is an optional step. Run a simulation of the pipeline to confirm it produces the anticipated results. Learn more. ⌁
Check for the target field ml.inference.top_metrics.traffic_status_prediction for the prediction in the Result tab.
The provided sample document is taken from the source index used to train the model.

Figure 8.38 – Deploy model step 4 – test

8. On the **Create** page, you can review your pipeline. If desired, copy and save the pipeline configuration for reuse, possibly via an API call in Dev Tools. If all is as it should be, finalize the creation by clicking **Create pipeline**.

Your inference pipeline should have been created successfully:

Deploy analytics model

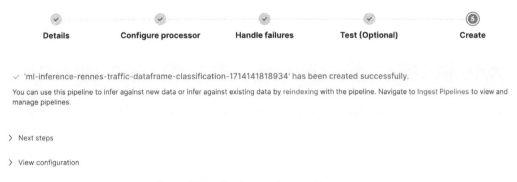

'ml-inference-rennes-traffic-dataframe-classification-1714141818934' has been created successfully.

You can use this pipeline to infer against new data or infer against existing data by reindexing with the pipeline. Navigate to Ingest Pipelines to view and manage pipelines.

› Next steps

› View configuration

Figure 8.39 – Deploy model step 5 – create

Our classification model is now successfully deployed and is ready to be used within the inference pipeline. This allows you to begin applying it to new data and making predictions based on what the model has learned.

To facilitate predictions, we've provided a Streamlit-based Python application. Find it in our GitHub repository: `https://github.com/PacktPublishing/Elastic-Stack-8.x-Cookbook/tree/main/Chapter8/traffic-prediction-app`. The app allows user-friendly interaction with the deployed model, making it possible to input data and receive predictions through a graphical interface.

After downloading the app and installing the required dependencies, do the following:

1. Fill the `.env` file with the values corresponding to your environment (Elasticsearch connection information and inference model ID).

2. After that, execute the following command to install the dependencies:

   ```
   $ pip install -r requirements.txt
   ```

3. To launch the app, run the following command:

   ```
   $ streamlit run rennes_traffic_predict.py
   ```

4. This will open your default web browser to the page at `http://localhost:8501`. Here, you can interact with the application. Select a location, an hour of the day, a day of the week, and a maximum speed. Then, click **Predict** to summon the classification pipeline. The traffic status prediction based on your criteria will be presented to you:

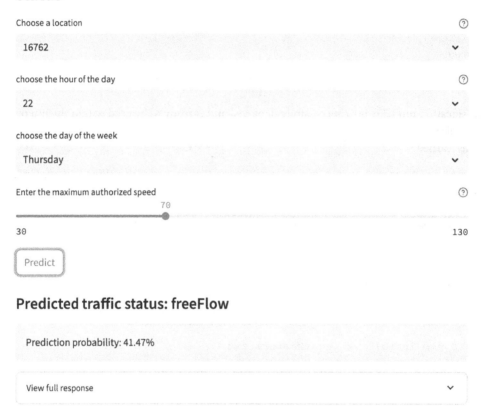

Figure 8.40 – Prediction Streamlit application

By clicking on **View full response**, you can access the actual JSON response from the Elasticsearch API. This is particularly insightful, as it offers detailed information on how the predicted values are distributed across the various traffic status classes.

How it works...

Inference in Elastic Stack 8 is technically executed by deploying a trained model to an ML node. During the deployment phase, you define an ingest pipeline that will be used for inference. The model, trained on historical data, is stored in Elasticsearch. When new data comes in, it's sent to the inference processor or API, which applies the model to this data to make predictions or classifications.

For example, you can use it in the processor of an ingest pipeline or in a pipeline aggregation within a search query. The process is deeply integrated with Elasticsearch's data-handling and storage capabilities, allowing for efficient and scalable inference operations.

There's more...

While we've been focusing on a deployed model trained inside Elasticsearch so far, it's worth noting that you can also import third-party models into the Elastic Stack. This can be done using the **Eland Python client**, which enables models trained using libraries such as `scikit-learn`, `XGBoost`, and `LightGBM` to be serialized and then used for inference in Elasticsearch.

You can also deploy third-party NLP models. This will be covered as part of the next recipe: *Deploying third-party NLP models and testing via the UI.*

See also

For more information on third-party models you can deploy in the stack with the Eland client, check out this great article on Search Labs: `https://www.elastic.co/search-labs/may-2023-launch-machine-learning-models`

Deploying third-party NLP models and testing via the UI

In the previous recipe, we learned how to use a trained ML model for inference. With the Elastic Stack, we can extend the application of trained models to **Natural Language Processing** (**NLP**) use cases. In this recipe, we will learn how to deploy a third-party NLP model and test it.

Getting ready

Ensure you have Python 3.11 or later installed on your system and an Elastic 8.x deployment running. We recommend installing `pip` for easy package management, which you can do by following the instructions at `https://pip.pypa.io`.

For this recipe, you will need the Cloud ID as well as the username and password for basic authentication to your Elastic Cloud deployment. Recall that we saved the password for our default user, **elastic**, in the *Deploying the Elastic Stack on Elastic Cloud* recipe in *Chapter 1*. You can find instructions for locating your Cloud ID in the management console in that same recipe.

The snippets of this recipe can be found at `https://github.com/PacktPublishing/Elastic-Stack-8.x-Cookbook/blob/main/Chapter8/snippets.md#deploying-third-party-nlp-models-and-testing-via-ui`.

How to do it...

We aim to import some compelling NLP-trained models from `https://huggingface.com/`, compatible with the Wikipedia movie plot dataset introduced in *Chapter 2*. We'll then test these models using Kibana's dedicated **Trained Models UI**:

1. From the terminal, install Eland (the Python Elasticsearch client for data exploration and analysis) with the following command:

    ```
    $ pip install 'eland[pytorch]'
    ```

2. After installing Eland, verify the installation with this command:

    ```
    $ eland_import_hub_model --help
    ```

    Make sure that you can see the usage guide as shown in *Figure 8.41*:

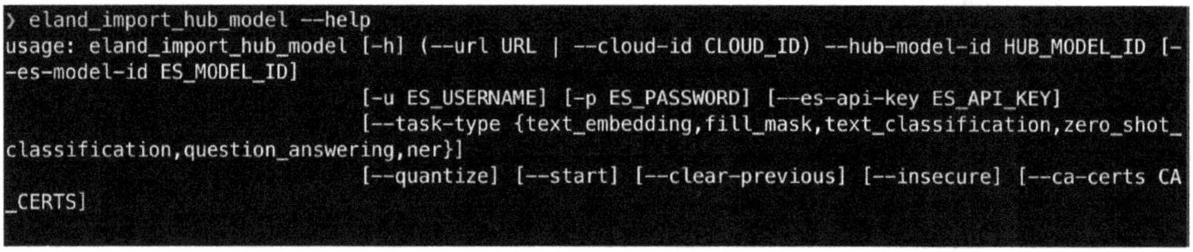

```
) eland_import_hub_model --help
usage: eland_import_hub_model [-h] (--url URL | --cloud-id CLOUD_ID) --hub-model-id HUB_MODEL_ID [-
-es-model-id ES_MODEL_ID]
                              [-u ES_USERNAME] [-p ES_PASSWORD] [--es-api-key ES_API_KEY]
                              [--task-type {text_embedding,fill_mask,text_classification,zero_shot_
classification,question_answering,ner}]
                              [--quantize] [--start] [--clear-previous] [--insecure] [--ca-certs CA
_CERTS]
```

Figure 8.41 – Checking the Eland installation

After confirming the installation, we can now import the first NLP models from Hugging Face. Let's import our first model for **Named Entity Recognition** (**NER**). The model we will use is one of the most popular NER models offered by Hugging Face (`https://huggingface.co/dslim/bert-base-NER`).

3. To import our first NER model, use the following terminal command (replace `<ES_CID>`, `<ES_USER>`, and `<ES_PWD>` with your Elastic Cloud deployment's Cloud ID, username, and password, respectively):

    ```
    $ eland_import_hub_model \
    --cloud-id <ES_CID> \
    -u <ES_USER> -p <ES_PWD> \
    --hub-model-id dslim/bert-base-NER \
    --task-type ner --start
    ```

4. Next, let's import our second model for text classification. This model is specifically designed for movie genre prediction and is available at `https://huggingface.co/nickmuchi/distilroberta-base-movie-genre-prediction`. To import the text classification model, use the following terminal command (again, replace `<ES_CID>`, `<ES_USER>`, and `<ES_PWD>` with the credentials of your Elastic Cloud deployment):

```
$ eland_import_hub_model \
--cloud-id <ES_CID> \
-u <ES_USER> -p <ES_PWD> \
--hub-model-id nickmuchi/distilroberta-base-movie-
genreprediction
\
--task-type text_classification --start
```

5. Once the models are imported, navigate to **Kibana | Analytics | Machine Learning | Model management | Trained models**. To find the NER model, type `ner` into the search bar. You should see that the `dslim_bert-base-ner` model has been successfully imported and is now deployed. To test the model, click on the **...** button to bring up the actions menu, as shown in *Figure 8.42*:

> **Note**
>
> If you can't see the recently imported models, there might be a warning at the top of the screen indicating **ML job and trained model synchronization required**. You need to click on the **Synchronize your jobs and trained models** link and then **Synchronize** to import the missing models.

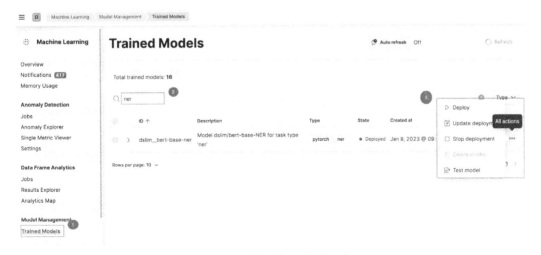

Figure 8.42 – Trained models list view

6. We can now test our model using a sample movie plot text provided at the following address: `https://github.com/PacktPublishing/Elastic-Stack-8.x-Cookbook/blob/main/Chapter8/snippets.md#sample-movie-plot`. Click on **Test model** to open a flyout for testing the trained model. In **Input text**, copy and paste the sample movie plot text and then click on the **Test** button. Observe that named entities have been correctly recognized by the model, as depicted in *Figure 8.43*:

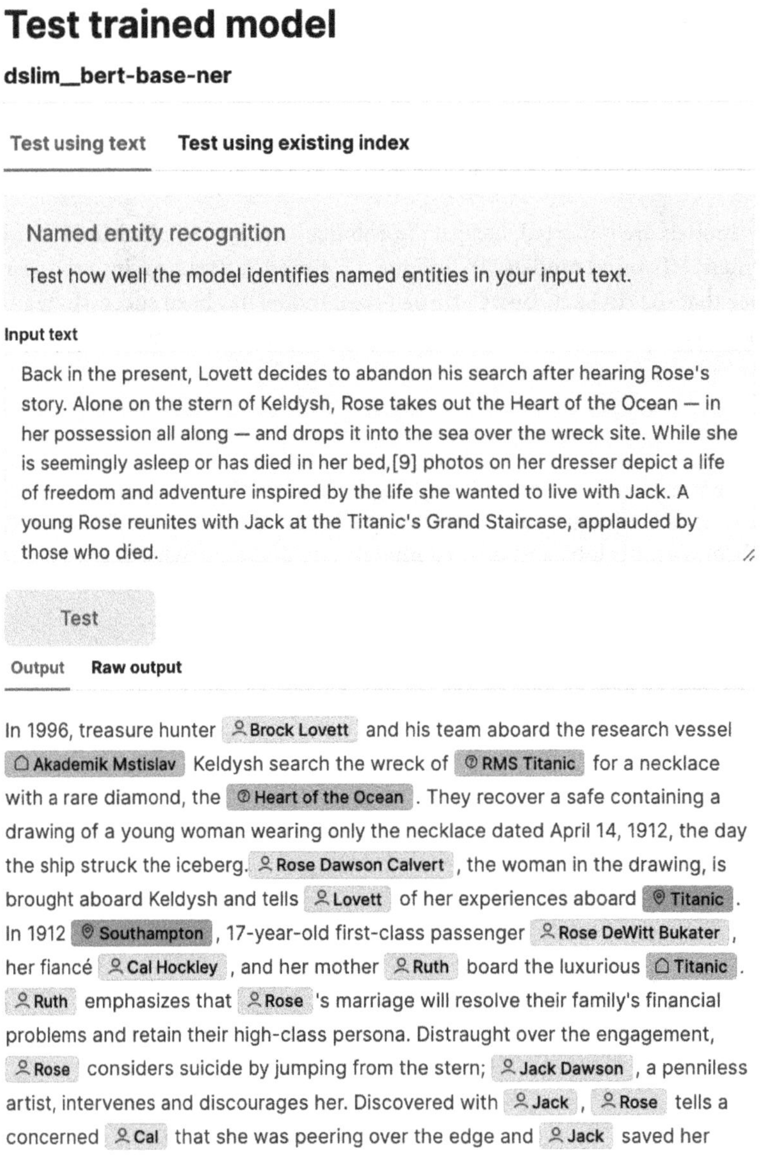

Test trained model

dslim__bert-base-ner

Test using text   **Test using existing index**

Named entity recognition

Test how well the model identifies named entities in your input text.

Input text

Back in the present, Lovett decides to abandon his search after hearing Rose's story. Alone on the stern of Keldysh, Rose takes out the Heart of the Ocean — in her possession all along — and drops it into the sea over the wreck site. While she is seemingly asleep or has died in her bed,[9] photos on her dresser depict a life of freedom and adventure inspired by the life she wanted to live with Jack. A young Rose reunites with Jack at the Titanic's Grand Staircase, applauded by those who died.

Test

Output   **Raw output**

In 1996, treasure hunter ⚲ **Brock Lovett** and his team aboard the research vessel ⌂ **Akademik Mstislav** Keldysh search the wreck of ⌾ **RMS Titanic** for a necklace with a rare diamond, the ⌾ **Heart of the Ocean** . They recover a safe containing a drawing of a young woman wearing only the necklace dated April 14, 1912, the day the ship struck the iceberg. ⚲ **Rose Dawson Calvert** , the woman in the drawing, is brought aboard Keldysh and tells ⚲ **Lovett** of her experiences aboard ⌾ **Titanic** . In 1912 ⌾ **Southampton** , 17-year-old first-class passenger ⚲ **Rose DeWitt Bukater** , her fiancé ⚲ **Cal Hockley** , and her mother ⚲ **Ruth** board the luxurious ⌂ **Titanic** . ⚲ **Ruth** emphasizes that ⚲ **Rose** 's marriage will resolve their family's financial problems and retain their high-class persona. Distraught over the engagement, ⚲ **Rose** considers suicide by jumping from the stern; ⚲ **Jack Dawson** , a penniless artist, intervenes and discourages her. Discovered with ⚲ **Jack** , ⚲ **Rose** tells a concerned ⚲ **Cal** that she was peering over the edge and ⚲ **Jack** saved her

Figure 8.43 – NER model test

Let's continue with our second text classification model that was imported earlier.

7. In the search bar, type `movie` to locate the movie genre prediction model. Using the actions menu, as we did with the NER model, click on **Test model**. Within the flyout's **Input text** field, copy and paste the same sample movie plot text and click on the test button. The text classification model predicts the movie genres based on the plot text and ranks them by their prediction probability, as shown in *Figure 8.44*:

Test trained model   ✕

nickmuchi__distilroberta-base-movie-genre-prediction

boat deck. The lifeboats have departed and passengers are falling to their deaths as the stern rises out of the water. The ship breaks in half, lifting the stern into the air. Jack and Rose ride it into the ocean and he helps her onto a wooden panel buoyant enough for only one person. He assures her that she will die an old woman, warm in her bed. Jack dies of hypothermia[8] but Rose is saved. With Rose hiding from Cal en route, the RMS Carpathia takes the survivors to New York City where Rose gives her name as Rose Dawson. Rose says she later read that Cal committed suicide after losing all his money in the Wall Street Crash of 1929. Back in the present, Lovett decides to abandon his search after hearing Rose's story. Alone on the stern of Keldysh, Rose takes out the Heart of the Ocean — in her possession all along — and drops it into the sea over the wreck site. While she is seemingly asleep or has died in her bed,[9] photos on her dresser depict a life of freedom and adventure inspired by the life she wanted to live with Jack. A young Rose reunites with Jack at the Titanic's Grand Staircase, applauded by those who died.

romance	0.419
adventure	0.258
mystery	0.098
crime	0.0732
family	0.0429

Figure 8.44 – Text classification model test

How it works...

In this recipe, we imported two third-party NLP models from Hugging Face using the Python Elasticsearch client Eland. Once the models were successfully imported and deployed, we navigated to the dedicated ML UI in Kibana to test the models' inference capabilities. The summary of the flow is explained in *Figure 8.45*:

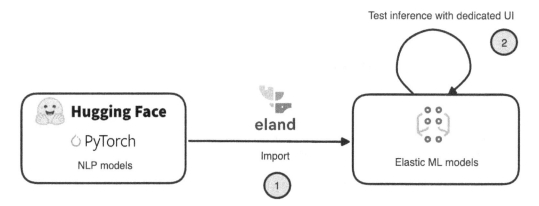

Figure 8.45 – NLP model import and test flow

There's more...

In the final recipe of this chapter, we'll explore how to use NLP models for advanced data processing. Beyond NER and text classification, Elasticsearch supports additional NLP model types, including fill-mask, question-answering, text similarity, zero-shot text classification, and text embedding.

If you're interested in experimenting with different Hugging Face NLP models in Elasticsearch, here is a list of sample models you can try:

- Text embedding: `https://huggingface.co/sentence-transformers/msmarco-MiniLM-L-12-v3`

- Zero-shot classification: `https://huggingface.co/typeform/distilbert-base-uncased-mnli`

- Question answering: `https://huggingface.co/deepset/tinyroberta-squad2`

See also

- The full documentation of Eland can be found at the following link: `https://eland.readthedocs.io`

- The full documentation of supported third-party models can be found at the following link: `https://www.elastic.co/guide/en/machine-learning/current/ml-nlp-model-ref.html`

- In the next chapter, we will be learning how to use text embedding models for semantic search scenarios

Running advanced data processing with trained models

In the previous recipe, we learned how to import third-party trained NLP models and test them using the dedicated UI. This recipe will show us how to use these models during the data ingestion phase to execute advanced data processing.

Getting ready

Make sure that you have completed the previous recipe, *Deploying third-party NLP models and testing via the UI*.

We will use a shortened version of the movie dataset used in previous chapters. Make sure to download and save the file from the following URL: `https://github.com/PacktPublishing/ Elastic-Stack-8.x-Cookbook/blob/main/Chapter8/dataset/wiki_movie_ plots_short.csv`

The snippets for this recipe can be found at `https://github.com/PacktPublishing/ Elastic-Stack-8.x-Cookbook/blob/main/Chapter8/snippets.md#running- advanced-data-processing-with-trained-models`.

How to do it...

Our goal is to incorporate advanced data processing while ingesting our movie plot dataset. We'll use the inference processor based on the trained NER model that we imported in the previous recipe, and we'll combine it with other processors to anonymize named entities in the plot field. Let's get started:

1. We will use the file import process: go to **Kibana** | **Analytics** | **Machine Learning** | **Data visualizer** | **File** and drag and drop `wiki_movie_plots_short.csv` to the drop zone, then on the next screen, click on the **Import** button.

2. On the data import screen, select the **Advanced** tab, then enter `movie-anonymized` for **Index name** and keep the **Create data view** option selected.

3. Replace the content in the mapping text area with the mapping provided in the following URL: `https://github.com/PacktPublishing/Elastic-Stack-8.x-Cookbook/ blob/main/Chapter8/snippets.md#movie-anonymized-mapping`. Replace the content in the ingest pipeline text area with the pipeline definition provided in the following URL: `https://github.com/PacktPublishing/Elastic-Stack-8.x-Cookbook/ blob/main/Chapter8/snippets.md#movie-anonymized-ingest-pipeline`. Here we utilize the inference processor among other processors in the ingest pipeline. We will explain more in the *How it works...* section. At this stage, your screen should look like *Figure 8.42*. Click on the **Import** button:

Data Visualizer

wiki_movie_plots_short.csv

Figure 8.46 – Importing movie data with a customized pipeline

4. Once the import is complete, verify that the movie data has been successfully imported and click on **View index in Discover**, as shown in *Figure 8.47*:

Figure 8.47 – Verifying the successful import

5.   The **movie-anonymized** data view will be selected automatically. You can choose the movie with the title `Terrible Teddy, the Grizzly King` by using the search bar and expand the document with the *toggle to expand* button, as shown in *Figure 8.48*. You should see that the named entities in the `anonymized_plot` field have been recognized and anonymized by the ingest pipeline, as illustrated in *Figure 8.48*:

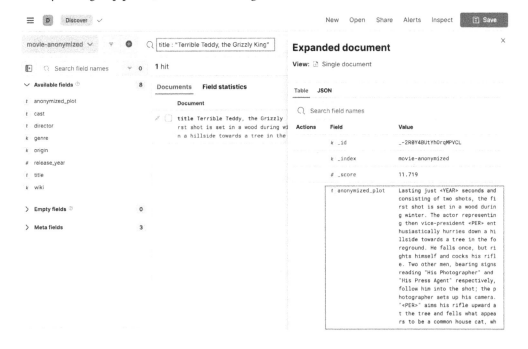

Figure 8.48 – Checking the plot anonymization

How it works...

To anonymize our movie data, we employed a dedicated ingest pipeline featuring multiple processors, specifically the following:

- An inference processor to identify named entities with the ML model imported in the previous recipe. It enriches them with entity tags in the `plot` field and saves them to a new field called `anonymized_plot`.

- A script processor to remove the original named entities and retain only the entity tags.

- A redact processor to recognize regular patterns, such as email addresses or phone numbers, then replace them with tags. We used it to identify year patterns in the `plot` field and replaced them with a common `<YEAR>` tag.

- Other processors we used (`csv`, `set`, `convert`, and `remove`) extract data from CSV, set the `anonymized_plot` field, convert `release_year` into a numeric value, and remove unnecessary fields after processing. The whole ingestion flow is illustrated in *Figure 8.49*:

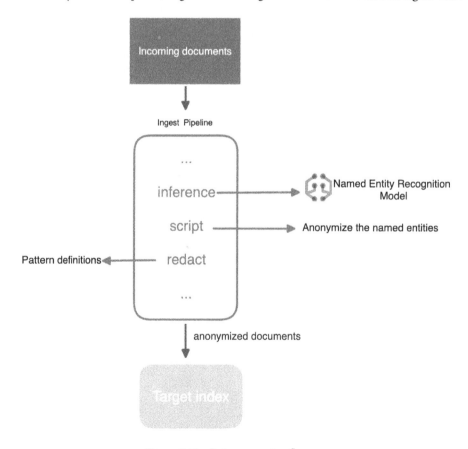

Figure 8.49 – Data processing flow

There's more...

In this recipe, we crafted the ingest pipeline using the File Uploader and Data Visualizer. You can also verify the ingest pipeline through stack management: go to **Stack Management | Ingest Pipeline**, search for `movie-anonymized` to locate the ingest pipeline, and click on the **Edit** button in the action menu, as demonstrated in *Figure 8.50*:

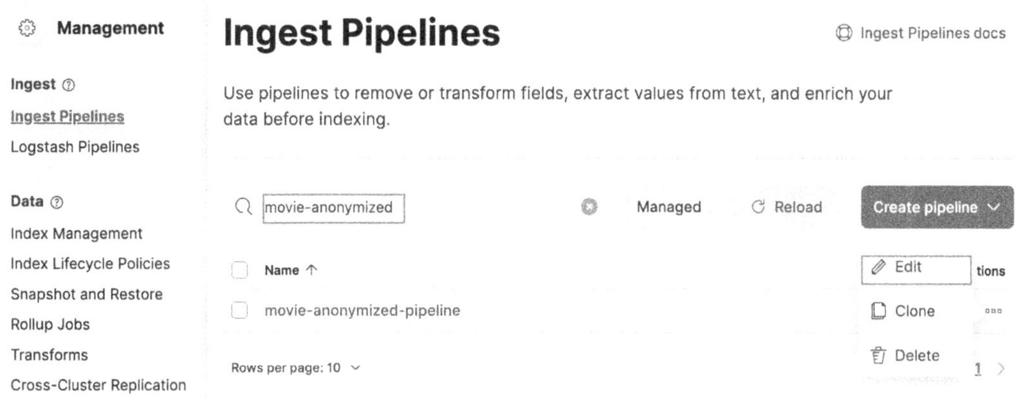

Figure 8.50 – Editing the ingest pipeline

This action directs you to the ingest pipeline's detailed UI, where you can visually inspect all the processors in action. We can pinpoint the processors that were uniquely introduced for this recipe: inference, script, and redact.

As shown in *Figure 8.51*, it is also possible to directly test the ingest pipeline from the UI for debugging purposes.

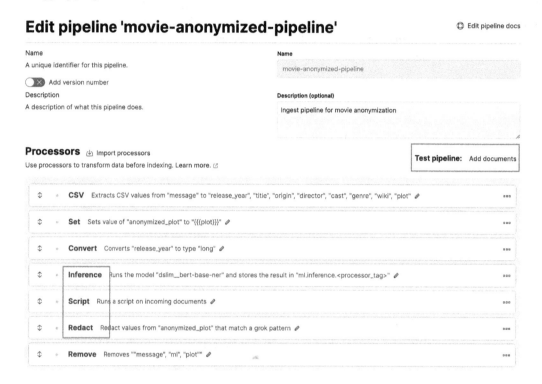

Figure 8.51 – Ingest pipeline detail

> **Performance consideration**
>
> Implementing an inference processor demands more computational resources and may result in latency during data ingestion, which is of particular concern for real-time data flows. Thus, it's advisable to tune ML nodes and models for optimal performance.

See also

- The comprehensive documentation of the inference processor is accessible at `https://www.elastic.co/guide/en/elasticsearch/reference/current/inference-processor.html`

- Rather than assessing the ingest pipeline via the UI, you can leverage the Simulate pipeline API. For complete instructions, visit `https://www.elastic.co/guide/en/elasticsearch/reference/current/simulate-pipeline-api.html`

9

Vector Search and Generative AI Integration

In previous chapters, we learned how to build traditional search applications using Query DSL and how to import third-party **natural language processing** (**NLP**) models for advanced data processing. In this chapter, we will expand upon these techniques with the latest features of Elastic Stack 8, which include vector search, hybrid search, and integrations with **large language models** (**LLMs**), to develop advanced search applications and sophisticated chatbot applications powered by Generative AI. *Table 9.1* shows different search strategies supported in Elastic Stack 8 and their use cases:

Search strategies	Use cases
Lexical search (also known as BM25 search; BM stands for best matching)	• Traditional search applications based on keyword matching
Vector search	• Semantic search applications that retrieve semantically relevant results rather than performing simple keyword matching • Recommendation systems applications suggest products or content based on user's behavior or preferences • Similarity detection applications identify duplicate or plagiarized content • Identifying duplicate or plagiarized content

Search strategies	Use cases
Hybrid search	• Search applications enhanced by more nuanced ranking, where traditional keyword search is augmented with semantic context for better relevance • A multi-modal search that retrieves information from various data types (text, image, and sound) that have been encoded into vectors
Retrieval-augmented generation (RAG)	• Question-answering applications: Customer support automation, educational tools, and interactive information retrieval systems • Chatbots: Automate customer service interactions and provide real-time assistance • Document summarization: Generate concise summaries of lengthy documents to aid quick comprehension

Table 9.1 – Search strategies and use cases

In this chapter, we will cover the following main topics:

- Implementing semantic search with dense vectors
- Implementing semantic search with sparse vectors
- Using hybrid search to build advanced search applications
- Developing question-answering applications with Generative AI
- Using advanced techniques for RAG applications

Technical requirements

To apply the recipes in this chapter, you'll need an Elastic Stack deployment (version 8.12 or later) that includes the following:

- Elasticsearch to search and store data
- Kibana for data exploration and stack management
- A machine learning node
- An Enterprise Search node

You will also need a local environment capable of running Python scripts and React applications. Make sure that you meet the following requirements:

- Python version 3.x

- Node.js version 16 or higher

- Ollama version 0.1.xx (required for the *Developing question-answering applications with generative AI* and *Using advanced techniques for RAG applications* recipes)

The code snippets for this chapter are available at `https://github.com/PacktPublishing/ Elastic-Stack-8.x-Cookbook/blob/main/Chapter9/snippets.md`.

Implementing semantic search with dense vectors

In this recipe, we will explore the fundamentals of vector search. We'll demonstrate how to implement semantic search for the movie Search Application we built in *Chapter 3*. Elasticsearch supports both dense and sparse vectors – these are the two primary types of data representations commonly used to convert text or other types of data into a numeric form known as vectors. For this recipe, our focus will be on dense vectors. In the next recipe, we will delve into sparse vectors and examine the key characteristics of each type.

Getting ready

We will reuse some concepts that we learned in *Chapter 3* and *Chapter 8*:

- Ensure that you are familiar with search templates and search applications

- Ensure that you understand the concept of importing third-party machine learning models

- The snippets for this recipe are available at `https://github.com/PacktPublishing/ Elastic-Stack-8.x-Cookbook/blob/main/Chapter9/snippets. md#implementing-semantic-search-with-dense-vectors`

How to do it...

In this recipe, we will revisit the movie dataset, focusing specifically on films released in the 1990s (`https://github.com/PacktPublishing/Elastic-Stack-8.x-Cookbook/blob/ main/Chapter9/search/dataset/wiki_movie_plots_90s.csv`). This time, we will first deploy a machine learning model that enables us to generate dense vector representations of our data. Then, we will ingest the movie data through a new pipeline, using this model to create vector representations, which are also known as **embeddings** or **vector embeddings**. After the ingestion, we will use a single Search Application to test various search templates based on both lexical and vector search, allowing us to observe the differences between them.

Let's begin by deploying the trained machine learning model for dense vectors:

1. In Kibana, navigate to **Analytics | Machine Learning**. Once you're on the **Machine Learning overview** page, go to **Model Management | Trained Models** using the left navigation bar. Click on the **Add a trained model** button, and a flyout will appear. Scroll down to **E5** and select **Intel and Linux optimized**, as shown in *Figure 9.1*, and then click on **Download**:

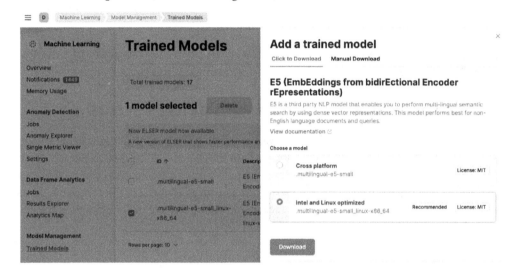

Figure 9.1 – Download the trained dense vector model

> **Note**
>
> The **Intel and Linux optimized** option is the default choice if you deployed your Elastic Stack on Elastic Cloud; otherwise, choose **Cross platform** for environments that are not based on Intel and Linux.

2. Observe how the state of the model changes to **Downloading** once you initiate the download. After the download is complete, the state will switch to **Ready to Deploy**. Then, under the **Actions** column, click on the meatballs menu (the three dots) to reveal the available options, and then select **Deploy** to proceed with the deployment of the model, as shown in *Figure 9.2*:

Figure 9.2 – Starting the trained model

3. On the next screen, you will see the model deployment configuration page. Since we will work with a small dataset, we can retain the default values and simply click on the **Start** button to initiate the model's deployment, as shown in *Figure 9.3*:

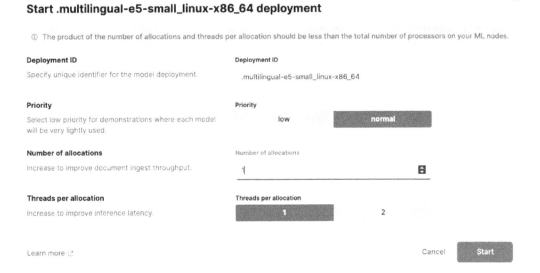

Figure 9.3 – Configuring the trained model

4. At this stage, our model has been successfully downloaded and deployed. Now, we can prepare to ingest our movie data. We will use the Python script located at `https://github.com/PacktPublishing/Elastic-Stack-8.x-Cookbook/blob/main/Chapter9/search/scripts/densevector_ingest.py`.

5. Before running the script, let's inspect the code to understand the most important parts that we have added for the vector search use case. First, locate the definition of the index mapping. As you can see in the following snippet, we define the mapping for the new `plot_vector` field by specifying its type, the number of dimensions, and the similarity algorithm:

```
mappings = {
    "properties": {
        ...
        "plot_vector": {
            "type": "dense_vector",
            "dims": 384,
            "index": "true",
            "similarity": "dot_product"
        }
    }
}
```

6. Next, let's examine the ingest pipeline snippet. As you can see, we define an ingest pipeline that processes incoming documents by running them through an inference process, using the machine learning model we deployed in the previous steps. This process generates dense vector representations in the field `plot_vector` from the `plot` text field. After the generated vector is copied to the `plot_vector` field, the pipeline cleans up by removing the model's raw inference data from the document:

```
SENTENCE_TRANSFORMERS_MODEL_ID = ".multilingual-e5-small_linux-
x86_64"

# Define ingest pipeline
ingest_pipeline_processors = {
    "processors": [
        {
            "inference": {
                "field_map": {
                    "plot": "text_field"
                },
                "model_id": SENTENCE_TRANSFORMERS_MODEL_ID,
                "target_field": "ml.inference.plot_vector",
                "on_failure": [
                    #failure processing part
                ]
            }
        },
        {
            "set": {
                "field": "plot_vector",
                "if": "ctx?.ml?.inference != null && ctx.
ml.inference['plot_vector'] != null",
                "copy_from": "ml.inference.plot_vector.
predicted_value",
                "description": "Copy the predicted_value to
'plot_vector'"
            }
        },
        {
            "remove": {
                "field": "ml.inference.plot_vector",
                "ignore_missing": True
            }
        }
    ]
}
```

7. Let's also inspect the bulk ingest method at the end of the script; note that the chunk_size parameter is set to 100, while the default value is 500. A smaller chunk size for bulk ingest is recommended for an ingestion process that involves compute-intensive ingest pipelines, such as an ingest pipeline which generates vector representations:

```
print("Indexing documents...")
progress = tqdm.tqdm(unit="docs", total=number_of_docs)
successes = 0
for ok, action in streaming_bulk(
        client=es, chunk_size=100, index=INDEX_NAME,
        actions=generate_actions(),
):
    progress.update(1)
    successes += ok
print("Indexed %d/%d documents" % (successes, number_of_docs))
```

8. Now, we can begin the ingestion process. First, adjust the connection parameters in the .env file (if you are not familiar with this process, refer to *Chapter 2*'s recipe, *Adding Data from the Elasticsearch Client*). Also, make sure that you have correctly configured the path to the dataset file. Then, open your terminal, navigate to the Chapter9/search/scripts directory, and run the following commands:

```
$ pip install -r requirements.txt
$ python densevector_ingest.py
```

You should have the result in the console output shown in *Figure 9.4*:

```
Loading dataset...
number of docs:  2254
Creating ingest pipeline...
Pipeline ml-inference-plot-vector created.
Creating index...
Indexing documents...
100%|██████████| 2254/2254 [00:14<00:00, 159.20docs/s]
Indexed 2254/2254 documents
```

Figure 9.4 – The console output of the dense vector ingestion script

Once the documents are correctly ingested, inspect them in Kibana.

9. First, go to **Stack Management | Kibana Data Views** and click on **Create data view**. In the flyout, set the name and the index pattern to `movies-dense-vector` and click **Save data view to Kibana**, as shown in *Figure 9.5*:

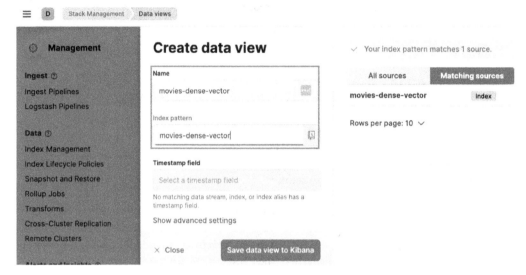

Figure 9.5 – Creating the dense vector data view

10. Now, in **Kibana**, go to **Analytics | Discover** and select the `movies-dense-vector` data view. Click on the expand button of any document and locate the `plot_vector` field in the expanded document, as shown in *Figure 9.6*. As you can see, the ingested document's vector field contains the dense vector representation, as expected:

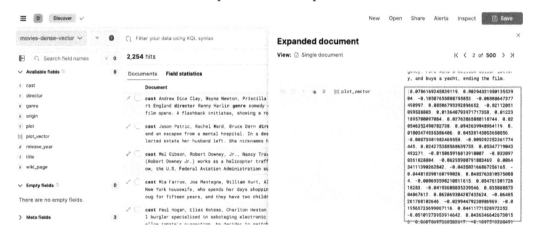

Figure 9.6 – Inspecting the vector field

Before building the Search Application for the ingested documents, let's first test out a basic query to do the semantic search.

11. Go to Dev Tools and execute the following query (the snippet is available at `https://github.com/PacktPublishing/Elastic-Stack-8.x-Cookbook/blob/main/Chapter9/snippets.md#dense-vector-query`). As you can see, Elasticsearch uses the deployed machine learning model to transform the input text "`romantic moment`", into a dense vector. Then it performs a **k-nearest neighbor** (**kNN**) search on the "`plot_vector`" field of the indexed documents to find the five documents whose "`plot_vector`" values are closest to the query vector generated by the model:

```
GET movies-dense-vector/_search
{
  "knn": {
    "field": "plot_vector",
    "k": 5,
    "num_candidates": 50,
    "query_vector_builder": {
      "text_embedding": {
        "model_id": ".multilingual-e5-small_linux-x86_64",
        "model_text": "romantic moment"
      }
    }
  },
  "fields": [ "title", "plot" ]
}
```

In the results console, you will see that five results have been returned, as illustrated in *Figure 9.7*. Note that the "plot" field of the first result aligns with the semantic meaning of our "romantic moment" query, even though there are no tokens that directly match the query keywords:

```
                                                              200 - OK   94 ms
{
  "took": 8,
  "timed_out": false,
  "_shards": {
    "total": 1,
    "successful": 1,
    "skipped": 0,
    "failed": 0
  },
  "hits": {
    "total": {
      "value": 5,
      "relation": "eq"
    },
    "max_score": 0.9105685,
    "hits": [
      {
        "_index": "movies-dense-vector",
        "_id": "QsrpzI0B3e9oy_b1JtYu",
        "_score": 0.9105685,
        "_source": {
          "cast": "Corey Haim, Ami Dolenz",
          "plot": "A college freshman Ramsy, played by Corey Haim experiences love for
            the first time in the 1960s when he asks out Joy, played by Ami Dolenz.",
          "director": "Redge Mahaffey",
          "origin": "American",
          "release_year": "1995",
          "genre": "comedy",
          "plot_vector": [
            0.07327594608068466,
            -0.026357507333159447,
            -0.022570980712771416,
            -0.04700397327542305,
            0.07931853085756302,
            0 01CC01170077777774
```

Figure 9.7 – Testing the vector search

We can now begin to build our Search Application using the search template that we learned about in *Chapter 3*.

12. First, we will create our Search Application with a template based on lexical search, as we did in *Chapter 3*. This will enable us to compare the results between the lexical and semantic searches later on. Go to **Dev Tools**, copy the snippet from https://github.com/PacktPublishing/Elastic-Stack-8.x-Cookbook/blob/main/Chapter9/snippets.md#search-template-with-lexical-search, and then paste it into the left console.

This creates a Search Application that facilitates dynamic search queries based on a lexical search (BM25 search) for the index we just created, as well as other input parameters for filtering, sorting, and aggregation:

```
# search template with lexical search
PUT _application/search_application/movie_vector_search_
application
{
  "indices": ["movies-dense-vector"],
  "template": {
    "script": {
      "lang": "mustache",
      "source": """
        {
          "query": {
            "bool": {
              "must": [
              {{#query}}
              {
                "multi_match" : {
                  "query": "{{query}}",
                  "type": "phrase_prefix",
                  "fields": [ "title^4", "plot", "cast",
                              "director"]
                }
              }
              {{/query}}
              ],
              ...
            }
          """,
      "params": {
        ...
      }
    }
  }
}
```

13. Once the preceding command is executed, the Search Application is created. Navigate to **Search | Search Applications**, and you will be able to locate the `movie_vector_ search_application`, as shown in *Figure 9.8*. After verification of the existence of the Search Application, we can proceed with the steps shown at `https://github.com/ PacktPublishing/Elastic-Stack-8.x-Cookbook/blob/main/Chapter9/ search/vector-search-application/README.md`, similar to what we did in the *Building search experience with the Search Application client* recipe in *Chapter 3*.

These steps allow us to initialize our React application in our cloned GitHub repository (located in `Chapter9/search/vector-search-application`), generate an API key for the Search Application from Kibana, and finally, adjust the `SearchApplicationClient` connection parameters in the file located at `Chapter9/search/vector-search-application/src/App.tsx`:

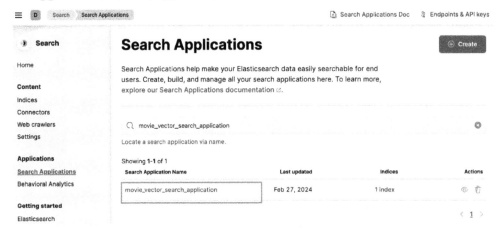

Figure 9.8 – Configuring the Search Application

14. We can then start our React application by navigating to `Chapter9/search/vector-search-application` in the terminal and running the following command:

```
$ yarn start
```

Now, we can test our semantic Search Application in the browser at `http://localhost:3000`. Try entering the query `love story and a jewel onboard a ship while travelling across Atlantic` in the search bar. We can observe the Search Application behavior when using lexical/BM25 search: movies, with titles that contain keywords such as `love` or `story` having a higher relevance, as shown in *Figure 9.9*:

Figure 9.9 – A Search Application with lexical search

15. Let's try dense vector search. You can find the snippet at this address: `https://github.com/PacktPublishing/Elastic-Stack-8.x-Cookbook/blob/main/Chapter9/snippets.md#search-template-with-dense-vector-search`. Copy and paste it into the **Kibana Dev Tools** console, and then execute the command.

This time, we will replace the previous template, based on lexical search, with a new template that uses **k-nearest neighbors (k-NN)** search to find relevant documents, based on dense vector similarity:

```
# search template with dense vector search
PUT _application/search_application/movie_vector_search_
application
{
  "indices": [
    "movies-dense-vector"
  ],
  "template": {
    "script": {
      "lang": "mustache",
      "source": """
      {
          "knn": {
            "field": "{{knn_field}}",
            "k": "{{k}}",
            "num_candidates": {{num_candidates}},
            "filter": {{#toJson}}_es_filters{{/toJson}},
            "query_vector_builder": {
              "text_embedding": {
                "model_id": ".multilingual-e5-small_linux-
x86_64",
                "model_text": "{{query}}"
              }
            }
          },
          "fields": {{#toJson}}fields{{/toJson}},
          "aggs": {
            "genre_facet": {
              "terms": {
                "field": "genre",
                "size": "{{agg_size}}"
              }
            },
            "director_facet": {
              "terms": {
                "field": "director.keyword",
```

```
            "size": "{{agg_size}}"
          }
        }
      },
      "from": {{from}},
      "size": {{size}}
    }
    """,
    "params": {
      "knn_field": "plot_vector",
      "query": "",
      "k": 20,
      "num_candidates": 100,
      "_es_filters": [],
      "fields": ["title", "plot"],
      "agg_size": 5,
      "from": 0,
      "size":9
    }
  }
}
```

16. Our Search Application is now updated with the new template that uses k-NN search, without the need for a restart. Let's refresh our browser page at http://localhost:3000. Try entering the same query into the search bar – love story and a jewel onboard a ship while travelling across the Atlantic. As you can see in *Figure 9.10*, semantic search enables us to find the famous movie **Titanic**, which matches the semantics of our query. Note that the semantic search also supports search filters and aggregations:

Figure 9.10 – A Search Application with vector search

17. Thanks to the multilingual capability of our model, we can process queries in other languages even though our dataset is entirely in English. Try entering the following French query into the search bar – `histoire d'amour sur un bateau de luxe en océan impliquant un bijou` (literally, A love story on a luxury ship in the ocean involving a jewel). Note that the vector search successfully identifies the movie **Titanic**:

Figure 9.11 – Testing the multilingual search

How it works...

The process of this recipe can be divided into three main steps, as shown in *Figure 9.12*:

1. **Model deployment**: Select and deploy the **Sentence Transformer** model (in our recipe, we used the E5 model) to the Elasticsearch machine learning nodes. Once deployed, it can be used to generate vectors during ingestion or build vectors from natural language during the query.

2. **Vector generation and ingestion**: Define the mapping for the vector field, and use the inference processor to generate dense vectors through the ingest pipeline before ingesting data into Elasticsearch.

3. **Semantic search**: Create a search template that constructs query vectors from natural language and leverages the k-NN search for semantic searches. This template is then used in the end user Search Application.

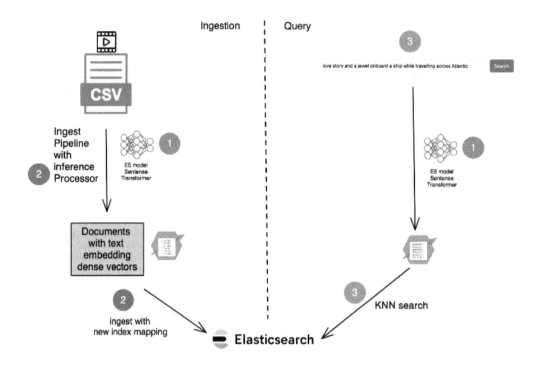

Figure 9.12 – The main steps of vector search

Now that we've progressed from the foundational steps to the advanced capabilities of vector search, let's delve into the role of the NLP model.

The NLP model

As demonstrated in this recipe, Elastic offers a built-in NLP model, named E5, which enables us to generate dense vector representations for our "plot-vector" field. Additionally, as we discovered in *Chapter 8*, it is possible to import third-party NLP models from huggingface.com. The selection of a model for vector search involves weighing up various factors, including accuracy, efficiency, language support, scalability, performance overhead, and input data type. You can find a list of compatible third-party NLP models at the following link: https://www.elastic.co/guide/en/machine-learning/current/ml-nlp-model-ref.html#ml-nlp-model-ref-text-embedding.

Once a model is chosen, we must configure the mapping for the vector field to correspond with the model's defined number of dimensions. In our case, the E5 model has 384 dimensions. As of version 8.12, Elasticsearch has expanded its support to handle up to 4,096 dimensions.

Beyond the number of vector dimensions, models also specify one or more supported similarity functions, such as `dot_product`, `cosine`, or `l2_norm`, which are utilized to compute vector similarity during the query phase. In our recipe, we chose the "`dot_product`" function. When a model supports multiple similarity functions, it is beneficial to experiment with different functions using sample queries to determine the most suitable one for your specific dataset.

k-NN search

Let's take a look at how k-NN search operates, using the initial example from our recipe. Elasticsearch includes k-NN search within its search API options. The k-NN search offers specific parameters such as the following:

- `k`: Indicates the number of results to retrieve.

- `num_candidates`: Specifies the number of approximate nearest neighbor candidates to evaluate on each shard. This is a very important parameter that allows us to find the balance between search accuracy and performance.

- `query_vector_builder`: Allows the on-the-fly generation of the query vector from text; otherwise, we must supply the numeric representation of the query vector in the `query_vector` parameter:

```
GET movies-dense-vector/_search
{
  "knn": {
    "field": "plot_vector",
    "k": 5,
    "num_candidates": 50,
    "query_vector_builder": {
      "text_embedding": {
        "model_id": ".multilingual-e5-small_linux-x86_64",
        "model_text": "romantic moment"
      }
    }
  },
  "fields": [ "title", "plot" ]
}
```

As you can see in the preceding query, when executing k-NN search, we start by defining `query vector` which can either be pre-computed or generated on the fly from a text query. The k-NN search then calculates the distance between the query vector and other vectors, using the chosen similarity function. After determining the distances from a predefined number of candidates, it sorts them based on their proximity to the query vector and selects the top k-closest points, deemed `k-nearest neighbors`.

Prior to version 8, Elasticsearch utilized the brute-force method for k-NN search, computing distances between the query vector and every vector in the dataset – a process that could be quite time-consuming and computationally demanding.

From version 8 onwards, Elasticsearch has adopted the **Hierarchical Navigable Small World** (HNSW) algorithm for **Approximate Nearest Neighbor** (ANN) searches, favoring efficiency and speed. This method uses a layered graph structure, with each layer being a subset of the preceding one and the bottom layer containing all elements. Searches start at the top layer with fewer points and proceed downward, significantly reducing the number of distance computations and speeding up the search process. Further details are available at this blog post: `https://www.elastic.co/blog/ introducing-approximate-nearest-neighbor-search-in-elasticsearch-8-0`.

There's more...

In the following sections, we will see some further techniques and considerations to optimize performance, enhance precision, and expand the applicability of vector search.

Chunking

In this recipe, we generated vectors from the movie plots using the E5 multilingual model, which truncates the plots to only the first 500 tokens. Given the varying lengths of movie plots, with some being quite long, this truncation may lead to a loss of information. To mitigate this, we could opt for models that handle more tokens or apply the chunking method, which will be introduced later in this chapter in the *Using advanced techniques for RAG applications* recipe.

A byte-sized vector

Our recipe shows that vectors are represented by arrays of 32-bit floating-point numbers, ranging from -1 to 1. With the 8.6 release, Elasticsearch introduced a new type of vector known as the byte-sized vector, which reduces storage costs with only a minimal loss in precision. More information on this new vector type can be found at `https://www.elastic.co/blog/save-space-with- byte-sized-vectors`.

Model deployment scalability

In our recipe, for the sake of simplicity, we deployed our model with one allocation and one thread per allocation to our machine learning node. Model allocations serve as the work units for NLP tasks, such as using the inference processor to generate vectors. To enhance scalability for the ingestion, models can be configured with multiple allocations to utilize several machine learning nodes, and the number of threads per allocation can also be adjusted to make better use of each machine learning node's allocated processors. Detailed documentation is available at `https://www.elastic.co/ guide/en/machine-learning/current/ml-nlp-deploy-model.html`.

Other use cases

In our recipe, we used vector search to implement a semantic Search Application. Vector search is suitable for various other use cases as well, including image similarity searches, duplicate detection systems, recommendation systems, and fraud detection. You can find a few practical examples at the Elastic Search Labs at `https://www.elastic.co/search-labs/tutorials/examples`.

See also

- Detailed documentation on k-NN search is available at the following link: `https://www.elastic.co/guide/en/elasticsearch/reference/current/knn-search.html#knn-search`

- Detailed documentation on the k-NN search API can be found at the following link: `https://www.elastic.co/guide/en/elasticsearch/reference/current/search-search.html#search-api-knn`

- For index mapping, there are additional parameters that allow us to tune the index's performance and accuracy; more details are available at the following link: `https://www.elastic.co/guide/en/elasticsearch/reference/current/knn-search.html#knn-indexing-considerations`

- To learn more about vector search with Elasticsearch, also check out a book dedicated to this topic, *Vector Search for Practitioners with Elasticsearch*, by Bahaaldine Azarmi and Jeff Vestal, published by Packt in 2023

Implementing semantic search with sparse vectors

In this recipe, we'll introduce another vector search strategy – the sparse vector. In Elastic, it's implemented through a proprietary model developed by search and machine learning experts at Elastic, called **Elastic Learned Sparse EncodeR (ELSER)**. This model gives you the ability to perform semantic search right out of the box. We will learn how you can apply it to our dataset and what the benefits of leveraging such a strategy are versus dense vector search.

Getting ready

To follow along in this recipe, you'll need to meet the following requirements:

- Make sure that you have completed the *Implementing semantic search with dense vector* recipe, as we will rely on the same dataset and Search Application

- As we will use the ELSER v2 sparse embedding model, which requires Elastic Stack version 8.11, make sure to have an Elastic deployment running in at least this version.

- Also, be mindful to have machine learning nodes with at least 4 GB of RAM in your deployment

The snippets of this recipe are available at `https://github.com/PacktPublishing/Elastic-Stack-8.x-Cookbook/blob/main/Chapter9/snippets.md#implementing-semantic-search-with-sparse-vectors`.

Now that you're all set, let's start this recipe.

How to do it...

The following steps will guide you to set up the ELSER model inside your Elastic deployment; once it is completed, we will explore how you can use it to generate embeddings on the movie dataset and then perform some semantic search.

> **A note on model management**
>
> If you completed the previous recipe on implementing semantic search with dense vectors, the e5 model will already be deployed in your cluster. For those on a cloud trial, you will need to stop the e5 model deployment before proceeding with this recipe. To do this, navigate to **Kibana**, and then to **Machine Learning | Model Management | Trained Models**. In the table that appears, find the `.multilingual-e5-small_linux-x86_64` model. Click on the meatball menu (three dots) to the right of the model entry, and from the actions menu that pops up, select **Stop deployment**, as shown in *Figure 9.13*. Before proceeding further, ensure you wait for the model's state to change to **Ready to deploy**.

Figure 9.13 – Stopping the multilingual E5 model

1. Head to **Kibana | Machine Learning | Model Management | Trained Models**. In the table displayed on the page, you will see two models at the top, both starting with `.elser_model_2`. The presence of two flavors of the same model is intended to offer an optimized version specifically for machine learning nodes used in Elastic Cloud. Locate the model named

`.elser_model_2_linux-x86_64` and click on **Download** under the `Actions` column of the table, as illustrated in the following figure:

Figure 9.14 – The Trained Models table with .elser_model_2

2. After the download is complete, the state will switch to **Ready to Deploy**. Then, under the `Actions` column, click on the meatballs menu and select **Deploy**. Leave the default settings. Upon successful deployment, the status of the model will update to **Deployed**.

Figure 9.15 – The .elser_mode_2 model deployed

As we did in the previous recipe, we will need to generate sparse embeddings for our movie dataset using ELSER. We provide a script for this in the cookbook's GitHub repository: `https://github.com/PacktPublishing/Elastic-Stack-8.x-Cookbook/blob/main/Chapter9/search/scripts/sparsevector_ingest.py`. The main differences with the script used in the previous recipes are:

- Ingest pipelines
- A mapping type for vector fields

Let's have a look at those specific parts in the script. First, the ingest pipeline used for ELSER looks like this:

```
elser_ingest_pipeline_processors = {
    "processors": [
        {
            "inference": {
                "model_id": ELSER_MODEL_ID,
                "input_output": [
                    {
                        "input_field": "plot",
                        "output_field": "plot_sparse_vector"
                    }
                ],
                ...
            }
        }
    ]
}
```

Here, the main difference is that we used the input_output field in the inference processor to list the fields in our dataset that are going to be used for inference (token generation with ELSER).

> **Note**
>
> This option, as mentioned in the documentation at https://www.elastic.co/guide/en/elasticsearch/reference/8.12/inference-processor.html, is incompatible with target_field and field_map, which we used previously to generate the dense vector embeddings.

Now, let's have a look at the mapping for the output field, where the corresponding embeddings will be stored – plot_sparse_vector and title_sparse_vector. The type defined for the vector field is sparse_vector, in contrast to dense_vector, which we used previously. We will later explain in more detail what sets the two approaches apart.

Now that we have reviewed the most important part of the script, we can execute it. Before doing so, make sure you have updated the .env file with your deployment information. After that, run the following command in your terminal:

```
$ python sparsevector_ingest.py
```

The script can take a while to ingest and process the dataset, but once it is complete, you should see the following output:

```
Loading dataset...
number of docs:  2254
Creating ingest pipeline...
Pipeline ml-inference-plot-vector-sparse created.
Creating index...
Deleting existing movies index...
Indexing documents...
100%|███████████| 2254/2254 [03:21<00:00, 11.17docs/s]
Indexed 2254/2254 documents
```

Figure 9.16 – The sparse vectors' ingest script output

3. Once the documents are correctly ingested, we will inspect them in Kibana. Go to **Stack Management | Kibana Data Views** and click on **Create data view**. In the flyout, set the name and the index pattern to `movies-sparse-vector`, and then create the data view.

4. Then, go to **Analytics | Discover** and select the `movies-sparse-vector` data view. Click on one of the documents to expand it and take a look at the `plot_sparse_vector` field, which contains the embeddings generated by ELSER for the `plot` field:

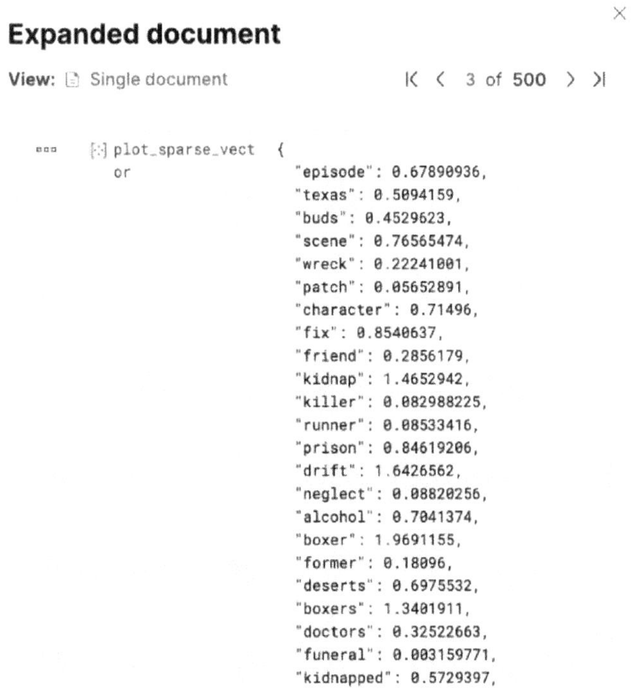

Expanded document   ✕

View: Single document   K 〈   3 of **500**   〉 〉|

```
□□□   [::] plot_sparse_vect   {
            or                      "episode": 0.67890936,
                                    "texas": 0.5094159,
                                    "buds": 0.4529623,
                                    "scene": 0.76565474,
                                    "wreck": 0.22241001,
                                    "patch": 0.05652891,
                                    "character": 0.71496,
                                    "fix": 0.8540637,
                                    "friend": 0.2856179,
                                    "kidnap": 1.4652942,
                                    "killer": 0.082988225,
                                    "runner": 0.08533416,
                                    "prison": 0.84619206,
                                    "drift": 1.6426562,
                                    "neglect": 0.08820256,
                                    "alcohol": 0.7041374,
                                    "boxer": 1.9691155,
                                    "former": 0.18096,
                                    "deserts": 0.6975532,
                                    "boxers": 1.3401911,
                                    "doctors": 0.32522663,
                                    "funeral": 0.003159771,
                                    "kidnapped": 0.5729397,
```

Figure 9.17 – The plot_sparse_vector field embeddings

Note the difference between this embedding, which contains tokens and associated float values, and the dense vector embeddings composed of floating values.

5. Now that we have our target dataset, we can start using it for semantic search with our Search Application; however, before that, let's perform the following query in **Kibana | Dev Tools | Console** (the snippet can be found at `https://github.com/PacktPublishing/Elastic-Stack-8.x-Cookbook/blob/main/Chapter9/snippets.md#semantic-search-by-using-the-text_expansion-query`):

```
GET movies-sparse-vector/_search
{
  "query":{
    "text_expansion":{
      "plot_sparse_vector":{
        "model_id":".elser_model_2_linux-x86_64",
        "model_text":"romantic moment"
      }
    }
  }
}
```

6. To search on generated embeddings, we use a `text_expansion` query. It's a special query designed for sparse vector fields. It takes the target field – in our case, `plot_sparse_vector` – with `model_id` and the query text. Let's run the query and look at the first result, as shown in the following figure:

```
1 ▾ {
2     "took": 69,
3     "timed_out": false,
4 ▾   "_shards": {
5       "total": 1,
6       "successful": 1,
7       "skipped": 0,
8       "failed": 0
9 ▾   },
10 ▾  "hits": {
11 ▾    "total": {
12        "value": 2253,
13        "relation": "eq"
14 ▾    },
15      "max_score": 8.141485,
16 ▾    "hits": [
17 ▾      {
18          "_index": "movies-sparse-vector",
19          "_id": "8CDu9o08dWUTibnuoQdi",
20          "_score": 8.141485,
21 ▾        "_source": {
22            "plot_sparse_vector": {▦},
228           "cast": "Jeanne Moreau, Jude Law, Claire Danes",
229           "plot": "The film is told through the stories of two women: Nana, a grandmother, and Daisy, her
              granddaughter. Daisy tells Nana of her strong and blossoming romance with a young man named Ethan
              and her problems at school because she's Jewish. Nana tells the story of her young life when she was
              sent to a ghetto and then a concentration camp. The romantic love feelings she has for the boy are
              indeed strong and genuine, but the romantic love he has for her is questionable. He lets his friends
              judge her from the outside, not for who she is on the inside, and when she turns out to not be like
              every other girl he breaks up with her. Daisy is sad so she goes and sees Nana and takes her anger
              out on her. She then runs away and tries to kill herself but she does not. At the end, she tries to
              see him again but he looks at her for a long time and walks away with his friends. She stands there;
              heartbroken, sad and crying, realizing that maybe it was not meant to be and she walks away happy.",
```

Figure 9.18 – The "romantic moments" results on sparse vector movies

7. Now, let's update the search template of our `movie_vector_search_application` with a new one, available at `https://github.com/PacktPublishing/Elastic-Stack-8.x-Cookbook/blob/main/Chapter9/snippets.md#using-elser-in-search-application-template`. In the following code block, you can see the main differences between the template used for dense vector search and this one for sparse vector:

```
PUT _application/search_application/movie_vector_search_
application
{
  "indices": [
    "movies-sparse-vector"
  ],
  "template": {
    "script": {
      "lang": "mustache",
      "source": """
      {
          "query":{
              "text_expansion":{
                  "plot_sparse_vector":{
                      "model_id":"{{elser_model_id}}",
                      "model_text": "{{query}}"
                  }
              }
          },
          ...
      """,
      "params": {
        "elser_model_id": ".elser_model_2_linux-x86_64",
        "query": "",
        ...
      }
    }
  }
}
```

8. By applying the new search template, we instruct our Search Application to use the sparse vector movies index. This setup involves a new query approach, leveraging text expansion, and applying the ELSER model to process the actual user search query.

9. We can test the same query we used on dense vector embeddings, `love story and a jewel onboard a ship while travelling across Atlantic`, and observe the results, as illustrated in the following screenshot:

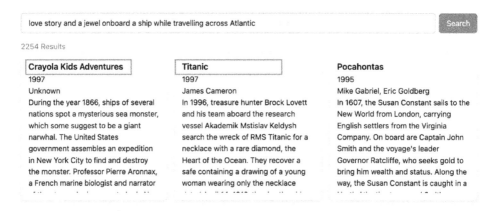

Figure 9.19 – The ELSER results in the Search Application

As you can see, we have the same top two results, albeit in a slightly different order – **Crayola Kids Adventures** and **Titanic**. This outcome is largely attributed to the way the ELSER model operates, which will be elaborated upon in the *How it works* section. Most importantly, we've successfully implemented semantic search on our dataset using ELSER. While the underlying technology differs significantly from dense vector embeddings, we still achieve almost identical results. This highlights the effectiveness of ELSER in facilitating semantic search without needing to adapt the domain of your data.

How it works...

ELSER is a neural network model that relies on a sparse retrieval technique to unlock semantic search capabilities. Most importantly, it's based on a technique called **term expansion**. Simply put, term expansion expands a query with learned related terms, similar to what is performed by BM25 in lexical search. By identifying contextual importance between terms, the model uses that knowledge to return token/weight pairs that are conceptually related to the input. Let's have a look at the following figures to better understand how it works:

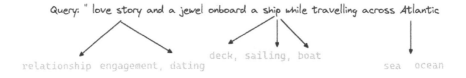

Figure 9.20 – A text expansion diagram

Figure 9.20 illustrates how text expansion operates. Those terms have been learned by ELSER to co-occur frequently within various training datasets:

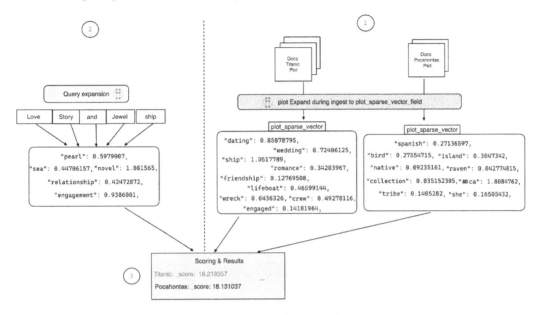

Figure 9.21 – ELSER in operation

Semantic search using ELSER can be broken down into the following phases:

1. During the ingest phase, the model performs expansion on documents on the field configured in the ingest pipelines. The output is a list of tokens with a weight associated that are stored using Elasticsearch rank feature types.

> **A note on the rank_feature type**
>
> The `rank_feature` type in Elasticsearch is a special field used to boost relevance scores of documents based on numerical features, enhancing search result rankings by incorporating document-specific signals, such as popularity or freshness, directly into the relevance-scoring process.

2. At query time, term expansion is also applied to the query to generate the corresponding tokens/weights.

3. A score is calculated for each document by considering the weight of each token. The documents are then ranked according to these scores, from the highest to the lowest, as illustrated in *Figure 9.21*.

Sparse models such as ELSER address one of the primary limitations associated with dense vector retrieval, domain adaptation, which typically necessitates extensive fine-tuning. Dense vector models often require retraining or adjustment to perform optimally within specific domains or on specialized datasets. In contrast, sparse models, using text expansion techniques, can more easily adapt to different domains without the need for significant customization.

The following is a table that highlights the main differences between these two approaches within the context of the Elastic Stack:

ELSER	Dense vector
Implements semantic search out of the box with great relevance	Large choice of supported text embedding models
Out-performs on text only – question-answer pairs, weather records, and medical text	Multimodal (text, images, audio, video, etc.)
Generalizes across domains without training	Requires domain adaptation to perform well
English only	Multilingual with E5
Not indexed using HNSW but traditional Lucene indices	Uses HNSW to store vectors
Does not need to be in memory for real-time search	Requires vectors to fit in RAM for fast search

Table 9.2 – The ELSER versus Dense vector main differences and use cases

We recommend consulting the official documentation for a more comprehensive understanding of how ELSER operates: `https://www.elastic.co/guide/en/machine-learning/current/ml-nlp-elser.html#elser-tokens`.

There's more...

As a Sparse vector model, ELSER has a limitation in the number of tokens it supports. Currently, it only encodes the first 512 tokens of a field, which means that for larger fields, you might need to introduce techniques such as **chunking**, which we will discuss in the *Using advanced techniques for RAG applications* recipe. The latest version of ELSER (v2 at the time of writing) also supports a technique called **quantization**, which allows compressed vectors to save memory.

You can also perform hybrid search and combine ELSER's sparse retrieval techniques with traditional honed approaches, such as BM25. There'll be more on this in the next recipe, *Using hybrid search to build advanced search applications*.

See also

In this recipe, we focused on applying ELSER to your dataset and saw how you can enjoy out-of-the-box semantic search, but if you want to deep dive into the inner technical workings of sparse retrieval, we will recommend looking at the following blog series:

- *Part 1, Steps to improve search relevance*: `https://www.elastic.co/blog/improving-information-retrieval-elastic-stack-search-relevance`

- *Part 2, Benchmarking passage retrieval*: `https://www.elastic.co/blog/improving-information-retrieval-elastic-stack-benchmarking-passage-retrieval`

- *Part 3, Introducing Elastic Learned Sparse Encoder, our new retrieval model*: `https://www.elastic.co/search-labs/blog/articles/may-2023-launch-information-retrieval-elasticsearch-ai-model`

Using hybrid search to build advanced search applications

In the previous recipes, we learned the fundamentals of dense vector search and sparse vector search. In this recipe, we will explore how to build advanced search applications enhanced by nuanced ranking, where traditional keyword search is augmented with semantic context. Hybrid search leverages the benefits of both lexical and vector search, enabling a Search Application to match exact keywords while also understanding the broader context or meaning of a query. It also serves as the foundation for RAG applications, which we will learn about later in this chapter.

Getting ready

Make sure that you have gone through the first two recipes in this chapter:

- *Implementing semantic search with dense vectors*

- *Implementing semantic search with sparse vectors*

Make sure that the React Search Application `vector-search-application` is up and running. We will reuse the concept of the search template to test the behavior of hybrid search.

The snippets of this recipe are available at `https://github.com/PacktPublishing/Elastic-Stack-8.x-Cookbook/blob/main/Chapter9/snippets.md#using-hybrid-search-to-build-advanced-search-applications`.

How to do it...

In this recipe, we will begin by testing lexical search and vector search to understand how the relevance score functions for each search method. Then, we will explore how to execute hybrid search with **score boosting** and **reciprocal rank fusion** (**RRF**).

Let's start by examining the relevance score for lexical search.

In Kibana, navigate to **Dev Tools**, use the snippet provided at `https://github.com/PacktPublishing/Elastic-Stack-8.x-Cookbook/blob/main/Chapter9/snippets.md#bm25-search-relevance-score-test`, copy and paste it into the left-hand console, and then execute the search query. Once the results are displayed, you will see that the document scores are computed according to the BM25 algorithm and that `max_score` is approximately 47, as shown in *Figure 9.22*. Scroll down in the right panel to observe which movies are ranked highest for this query with lexical search:

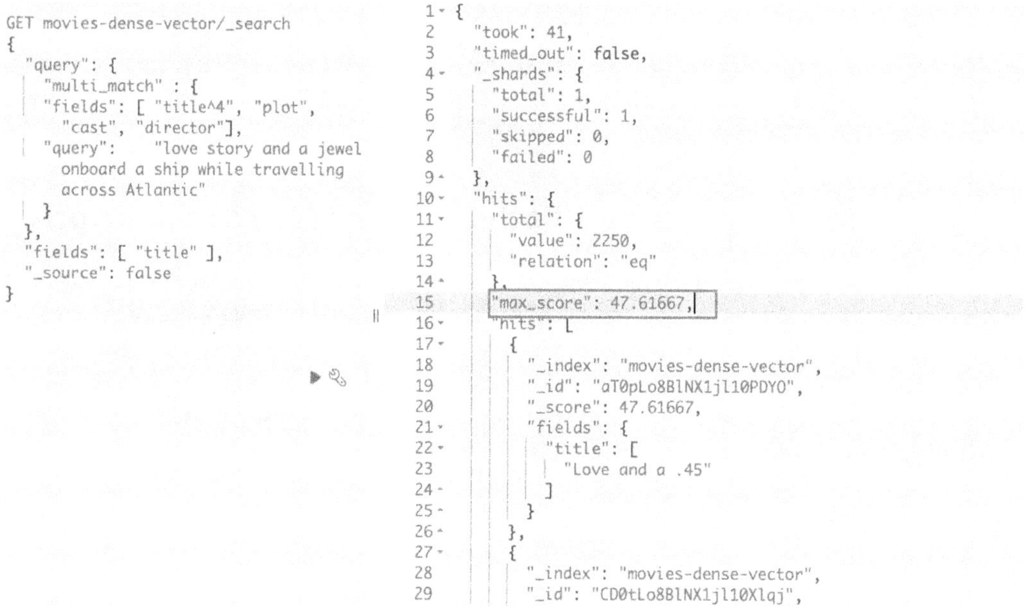

```
GET movies-dense-vector/_search
{
  "query": {
    "multi_match" : {
    "fields": [ "title^4", "plot",
      "cast", "director"],
    "query":    "love story and a jewel
      onboard a ship while travelling
      across Atlantic"
    }
  },
  "fields": [ "title" ],
  "_source": false
}
```

```
1 - {
2     "took": 41,
3     "timed_out": false,
4 -   "_shards": {
5       "total": 1,
6       "successful": 1,
7       "skipped": 0,
8       "failed": 0
9 -   },
10 -  "hits": {
11 -    "total": {
12        "value": 2250,
13        "relation": "eq"
14 -    },
15      "max_score": 47.61667,
16 -    "hits": [
17 -      {
18          "_index": "movies-dense-vector",
19          "_id": "aT0pLo8BlNX1jl10PDYO",
20          "_score": 47.61667,
21 -        "fields": {
22 -          "title": [
23              "Love and a .45"
24 -          ]
25 -        }
26 -      },
27 -      {
28          "_index": "movies-dense-vector",
29          "_id": "CD0tLo8BlNX1jl10Xlqj",
```

Figure 9.22 – Lexical search relevance score test

1. Now, let's do the same with the k-NN query. Use the snippet at `https://github.com/PacktPublishing/Elastic-Stack-8.x-Cookbook/blob/main/Chapter9/snippets.md#vector-search-relevance-score-test`, copy and paste it into the left-hand console, and then execute the search query. Once the results are displayed, you will see that, this time, the document scores are calculated based on the similarity function, and the `_score` value of the closest vector is approximately 0.91, as shown in *Figure 9.23*. Scroll down in the right panel to observe which movies are ranked highest for this query with k-NN search:

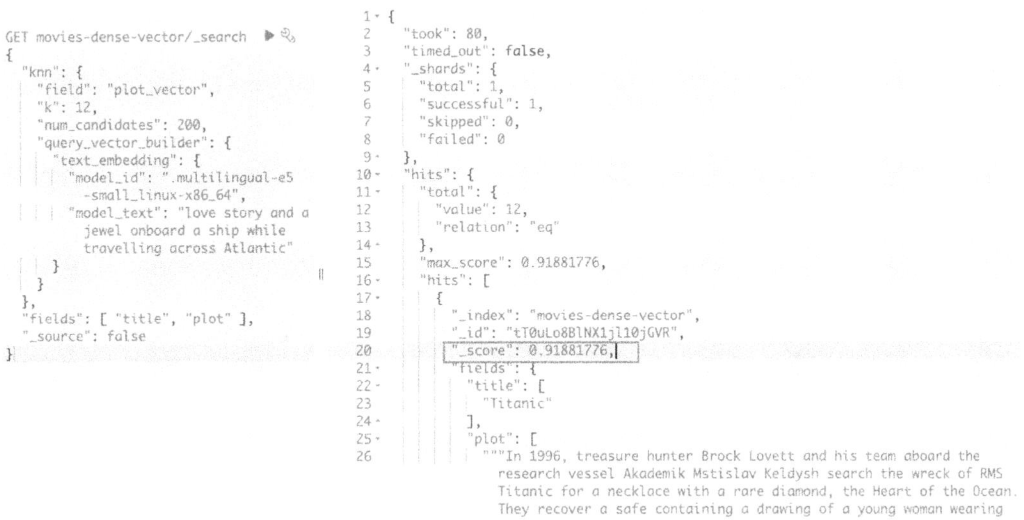

```
GET movies-dense-vector/_search  ▶ ⚐
{
  "knn": {
    "field": "plot_vector",
    "k": 12,
    "num_candidates": 200,
    "query_vector_builder": {
      "text_embedding": {
        "model_id": ".multilingual-e5
          -small_linux-x86_64",
        "model_text": "love story and a
          jewel onboard a ship while
          travelling across Atlantic"
      }
    }
  },
  "fields": [ "title", "plot" ],
  "_source": false
}
```

```
 1 · {
 2     "took": 80,
 3     "timed_out": false,
 4 ·   "_shards": {
 5       "total": 1,
 6       "successful": 1,
 7       "skipped": 0,
 8       "failed": 0
 9 ·   },
10 ·   "hits": {
11 ·     "total": {
12         "value": 12,
13         "relation": "eq"
14 ·     },
15       "max_score": 0.91881776,
16 ·     "hits": [
17 ·       {
18           "_index": "movies-dense-vector",
19           "_id": "tT0uLo8BlNX1jl10jGVR",
20           "_score": 0.91881776,
21 ·         "fields": {
22 ·           "title": [
23               "Titanic"
24 ·           ],
25 ·           "plot": [
26               """In 1996, treasure hunter Brock Lovett and his team aboard the
                 research vessel Akademik Mstislav Keldysh search the wreck of RMS
                 Titanic for a necklace with a rare diamond, the Heart of the Ocean.
                 They recover a safe containing a drawing of a young woman wearing
```

Figure 9.23 – The k-NN search relevance score test

At this stage, we want to update our Search Application with a hybrid search-based search template. As you may have observed, lexical search and vector search each provide a distinct set of results governed by different scoring systems. Hybrid search involves combining these two result sets. To merge the two sets equitably, we can adjust the boost parameter for each search option.

2. Locate the snippet at `https://github.com/PacktPublishing/Elastic-Stack-8.x-Cookbook/blob/main/Chapter9/snippets.md#hybrid-search-using-dense-vector-and-bm25-search-with-boost`, and then copy and paste it into the left-hand console of Dev Tools.

3. As you can see in our search template, we have set `boost_bm25` to 1 and `boost_knn` to 50. This decision is based on the difference in scores between the highest-scoring documents of each result set, as we observed in the previous steps of this recipe.

4. After reviewing the snippet, execute the command in **Dev Tools**:

```
# Hybrid search using dense vector and bm25 search with boost
PUT _application/search_application/movie_vector_search_
application
{
  "indices": [
    "movies-dense-vector"
  ],
  "template": {
    "script": {
      "lang": "mustache",
      "source": """
```

```
{
    "query": {
      "multi_match" : {
...
        "boost": "{{boost_bm25}}"
      }
    },
    "knn": {
      ...
      "boost": "{{boost_knn}}"
    },
    "fields": {{#toJson}}fields{{/toJson}},
    "aggs": {
      ...
    }
  }
""",
  "params": {
    ...
    "boost_bm25" : 1,
    "boost_knn": 50,
    ...
  }
}
}
}
```

Our Search Application is now updated with the hybrid search-based search template.

5. In your browser, navigate to `http://localhost:3000` and enter the same query we used in the previous recipes – `love story and a jewel onboard a ship while travelling across the Atlantic`. As you can see in *Figure 9.24*, **Titanic**, which originally scored approximately **0.91** in the k-NN search result set, now appears among the top results the hybrid search, thanks to the boost:

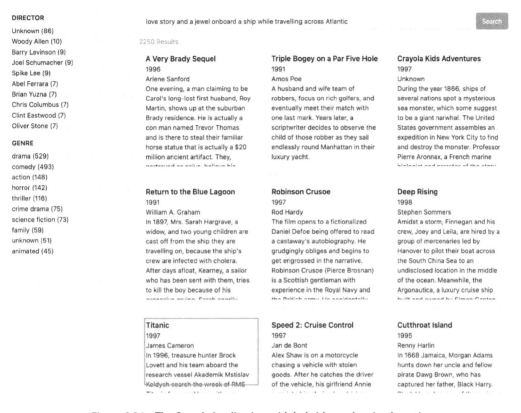

Figure 9.24 – The Search Application with hybrid search using boosting

6. Try to set `boost_bm25` and `boost_knn` to 1 in the snippet, and execute the previous command again. Once done, test the Search Application with the same query; you will see that **Titanic** disappears from the initial results.

   As you may have noticed in the previous step, merging result sets from lexical search and k-NN search using boosting can be quite challenging, especially when it comes to choosing the correct boost values. Let's try another method, this time with RRF.

7. Locate the snippet at `https://github.com/PacktPublishing/Elastic-Stack-8.x-Cookbook/blob/main/Chapter9/snippets.md#hybrid-search-using-dense-vector-and-bm25-search-with-rrf` and copy it to the left-hand console in **Dev Tools**.

8. In the new search template, you'll see that we have removed the parameters related to boosting. Instead, we have introduced the **rrf** option. We will discuss the `window_size` and `rank_constant` parameters later in this recipe.

After reviewing the snippet, execute the following command in **Dev Tools**:

```
# Hybrid search using dense vector and bm25 search with boost
with rrf
PUT _application/search_application/movie_vector_search_
application
{
  "indices": [
    "movies-dense-vector"
  ],
  "template": {
    "script": {
      "lang": "mustache",
      "source": """
      {
          "query": {
            ...
          },
          "knn": {
            ...
            }
          },
          "rank": {
            "rrf": {
              "window_size": {{rrf.window_size}},
              "rank_constant": {{rrf.rank_constant}}
            }
          },
          "fields": {{#toJson}}fields{{/toJson}},
          "aggs": {
            ...
          },
          ...
      }
      """,
        "params": {
          ...
          "rrf": {
            "window_size": 50,
            "rank_constant": 20
          }
        }
    }
  }
}
```

Our Search Application is now updated with the hybrid search-based search template, by using RRF.

9. In your browser, navigate to `http://localhost:3000` and enter the same query we used before – `love story and a jewel onboard a ship while travelling across the Atlantic`. As you can see in *Figure 9.25*, this time, the merged result set provides a more balanced representation of the top-ranked documents from both the lexical search and k-NN search result sets:

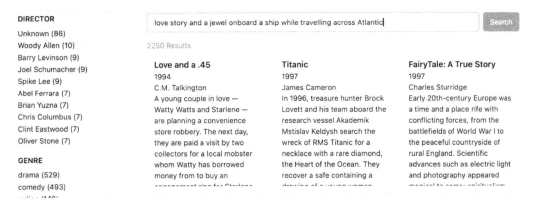

Figure 9.25 – The Search Application with hybrid search using boosting

How it works...

In this recipe, we have seen that Elasticsearch offers a very intuitive method to implement hybrid search – we simply need to add both regular lexical search and k-NN search within the same query, using a single search API endpoint. With hybrid search, the most critical aspect to understand is how different merging strategies work, particularly between linear boosting and RRF. Let's see how they work exactly.

Linear boosting

Linear boosting involves assigning weight (i.e., boost values) to different parts of a search query to prioritize certain results over others. In our example, lexical search, which uses the BM25 algorithm, and vector search, using k-NN, each produce their own sets of scores. By boosting, you adjust the scores before they are merged for each document:

```
final_score = boost_bm25*bm25_score + boost_knn*knn_similarity_score
```

This approach requires initially examining each of the search methods separately with sample queries to understand the differences in terms of the scores, as we did in the first steps of our recipe. We can then attempt to find the right balance for the boost values to ensure that neither search method inappropriately dominates the results.

As we can see in our recipe, finding the right balance for the boost scores is very challenging, especially because the scoring systems and the score granularity differ significantly between lexical search and k-NN search. Nevertheless, boosting can be very useful in scenarios where you want to prioritize results from one search method, such as lexical search for exact-match scenarios.

RRF

RRF is an algorithm used to combine rankings from different systems (in our case, lexical and vector search) by making use of the ranks of documents, rather than their raw scores. This is advantageous, as it normalizes the difference between the scoring scales of the two systems, which can vary significantly.

Here's the general process:

1. Perform both lexical and vector search separately, obtaining ranked lists of documents from each.

2. Use the `window_size` parameter to define the number of individual results per query; higher values may improve relevance but could affect performance.

3. Apply the RRF formula to combine these rankings into a single score for each document. The formula is `1/(rank_constant + rank)`, where `rank` is the position of the document in the list and `rank_constant` is a parameter you can adjust (the default value is set to the value of `60`). Setting `rank_constant` to a lower value will assign more significance to higher-ranked documents.

4. Use these combined scores to produce a merged ranked list of results. Documents that rank highly in both searches will generally receive the highest combined scores.

5. You can adjust the `window_size` and `rank_constant` parameters of the hybrid search-based search template that uses RRF (e.g., setting `window_size` to `100` and `rank_constant` to `60`), as we did at the end of our recipe, and then test the Search Application again. This will help you understand how these changes affect the behavior of the Search Application.

There's more...

In addition to what we have explored in this recipe, there are some further aspects of hybrid search, as follows:

*   **Hybrid search with multiple k-NN searches**: Elasticsearch supports multiple vector fields within a single index, which makes it possible to conduct multiple k-NN searches within a single hybrid search query. You can find an example at the following link: `https://www.elastic.co/guide/en/elasticsearch/reference/current/knn-search.html#_search_multiple_knn_fields`.

    This capability enables the implementation of multi-modal search use cases, allowing for the retrieval of information from various data types (e.g., text, image, sound) that have been encoded into vectors.

- **Hybrid search with sparse vector**: In our recipe, the hybrid search queries utilize dense vector search. We also provided a snippet for sparse vector hybrid search at the end of the snippet file, which you can access here: `https://github.com/PacktPublishing/Elastic-Stack-8.x-Cookbook/blob/main/Chapter9/snippets.md#hybrid-search-using-elser-and-bm25-search-with-rrf`.

See also

- The detailed documentation for hybrid search can be found at the following link: `https://www.elastic.co/search-labs/tutorials/search-tutorial/vector-search/hybrid-search`

- The following blog post offers a comprehensive guide that covers the various aspects (such as ingestion pipeline, inference, mapping, and knn search) we learned in the first three recipes of this chapter. It also includes other considerations, such as sizing, performance testing, and query tuning: `https://www.elastic.co/search-labs/blog/articles/vector-search-implementation-guide-api-edition`

Developing question-answering applications with Generative AI

We have explored various vector retrieval techniques throughout this chapter that the Elastic Stack offers to perform semantic searches. In this recipe, we will begin to leverage these building blocks to design a question-answering application. Our journey will involve the integration of four powerful tools:

- **LangChain**: This is a versatile library designed to chain language models to external knowledge sources, offering a streamlined approach to building complex language applications

- **Elastic Search Relevance Engine (ESRE)**: This is known for its vector and hybrid search capabilities, which will enhance the retrieval aspect of our application

- **Ollama**: This is a deployment solution that enables the integration of powerful local LLMs, allowing us to incorporate a robust language model into our application

- **Streamlit**: This allows for the rapid creation of simple and intuitive user interfaces

Throughout this recipe, we will also introduce the concept of RAG. This approach enables us to apply the powerful generative capabilities of LLMs to confidential data stored in Elasticsearch, resulting in increased accuracy and reduced instances of erroneous or fabricated information ("hallucinations").

Getting ready

Make sure to meet to have completed the *Using hybrid search to build advanced search applications* recipe.

Additionally, you will need to install Ollama on your local machine. Visit `https://ollama.com/` to download and install the version appropriate for your operating system.

Once Ollama is installed, launch your preferred terminal, and enter the following command to verify that the installation was successful and everything operates correctly:

```
$ ollama --help
```

You should get the following result:

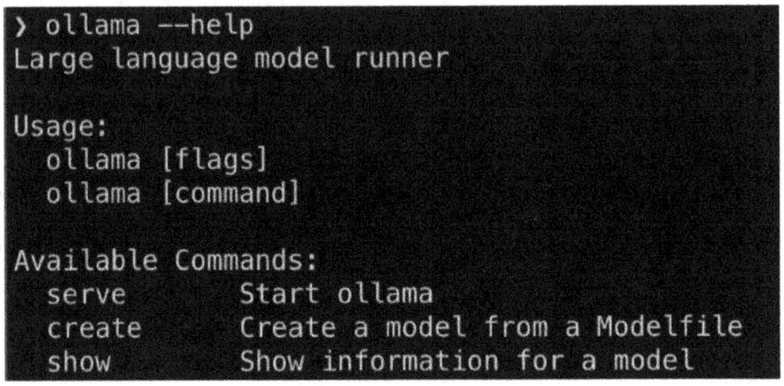

```
> ollama --help
Large language model runner

Usage:
  ollama [flags]
  ollama [command]

Available Commands:
  serve       Start ollama
  create      Create a model from a Modelfile
  show        Show information for a model
```

Figure 9.26 – Ollama's help output

Note that the output shown in the figure is not complete, but it gives you an idea of what to expect.

Next, we will need to deploy an LLM; Ollama allows you to run a wide variety of open LLMs locally on your machine. For our recipe, we will proceed with Mistral 7B (`https://mistral.ai/news/announcing-mistral-7b/`). Run the following command:

```
$ ollama run mistral
```

Once the model is ready, you will be prompted to send a message, as shown in *Figure 9.27*:

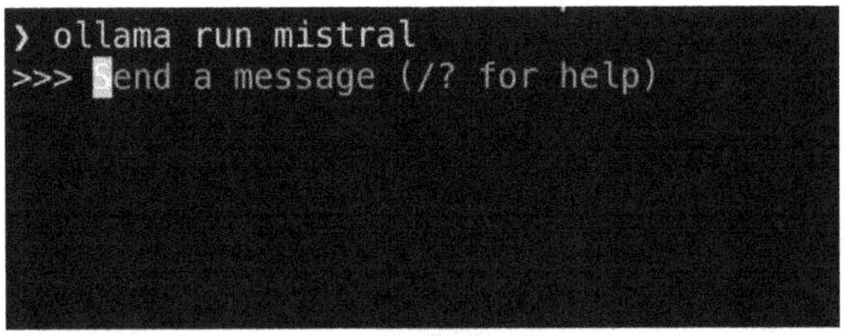

```
> ollama run mistral
>>> Send a message (/? for help)
```

Figure 9.27 – Running the Mistral model with Ollama

Now that you're all set, let's start with the good stuff.

The snippets of this recipe can be found at `https://github.com/PacktPublishing/Elastic-Stack-8.x-Cookbook/blob/main/Chapter9/snippets.md#developing-question-answering-applications-with-generative-ai`.

How to do it...

We want to build a simple question-answering application on top of our movie dataset, with two objectives – first, to achieve a more natural interaction with our data, and second, to demonstrate the RAG pattern with Elasticsearch as the vector store and search engine:

1. In the terminal where Ollama is running, enter the prompt `Which film talks about a love story and a precious jewel on board a large ocean liner while traveling across the Atlantic?` and wait for the LLM to answer. Note that due to the nature of the LLM, the answer may vary, but in our case, we got the following answer:

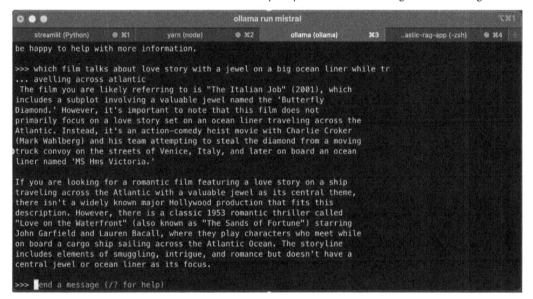

Figure 9.28 – Asking a sample question in Ollama

The LLM does not seem to know which movie we are referring to. Let's see how we can improve things with RAG.

2. We will deploy the **Question Answering** (**QA**) application, available in the GitHub repository at `https://github.com/PacktPublishing/Elastic-Stack-8.x-Cookbook/tree/main/Chapter9/rag/`. In the folder, open the `.env` file, update the environment variables according to your configuration, and then run the following commands to start the application:

```
$ pip install -r requirements.txt
$ streamlit run qa_app.py --server.port 8501
```

Once the application has started, it will open a web page at `http://localhost:8501` in the browser, as shown in the following screenshot:

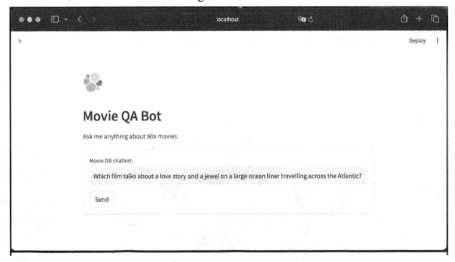

Figure 9.29 – The movie QA application

3. The text area is already populated with the same question we prompted the model with previously, so let's click on **Send** and wait for the results:

Figure 9.30 – The qa_app answer to the movie prompt

The difference between the two answers is immediately noticeable. We receive the correct, or at least the expected, answer when querying through the application. The reason for this, which will be explained in more detail in the *How it works...* section, is primarily due to the use of Elasticsearch to supply the relevant documents to the LLM. The LLM then utilizes these documents to generate the answer. Additionally, note that these documents are also displayed within the application interface, further clarifying the context and sources for the provided answer.

While the previous steps may have given you a fundamental understanding of RAG, let's explore it further with another example. At its core, the concept of RAG is about using LLM in conjunction with confidential data.

To illustrate this, we have a fictional soon-to-be-released movie in our dataset, a movie that is only known to us (authors). As we did earlier, we will first ask the LLM a question directly, and then we will ask the same question about an upcoming movie through the QA application. This approach will demonstrate how RAG leverages both the LLM's capabilities and privately held data to provide responses.

In your terminal, ask the following question to the LLM: `Which movie revolve around a superhero team battling to protect the city of Rennes from the Invader?` The model is quite confused by the question and asks for more context or specific information, as shown in *Figure 9.31*:

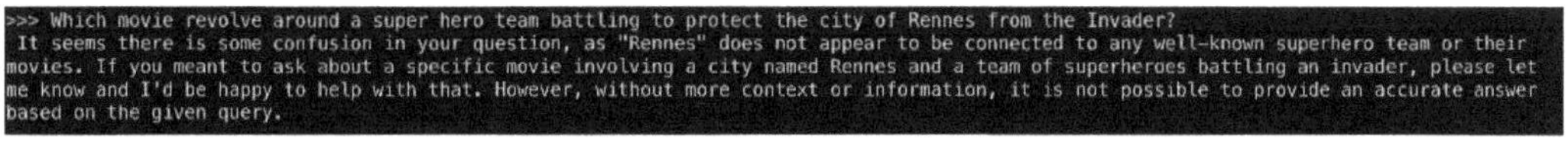

```
>>> Which movie revolve around a super hero team battling to protect the city of Rennes from the Invader?
 It seems there is some confusion in your question, as "Rennes" does not appear to be connected to any well-known superhero team or their
movies. If you meant to ask about a specific movie involving a city named Rennes and a team of superheroes battling an invader, please let
me know and I'd be happy to help with that. However, without more context or information, it is not possible to provide an accurate answer
based on the given query.
```

Figure 9.31 – The LLM answer to the question on the upcoming private movie

This is expected, as the movie we are discussing is still in its early phases and only known to a handful of people.

Now, let's switch to our Q&A application, ask the same question, and observe the results we receive:

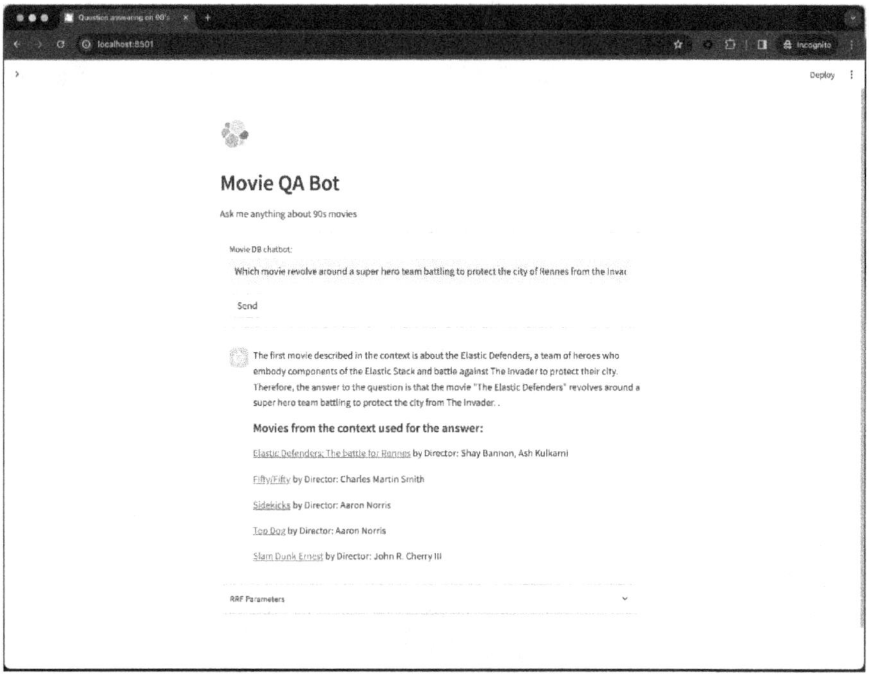

Figure 9.32 – The answer about the private movie from the QA application

Here, we receive a response about an upcoming movie named **Elastic Defenders: The battle for Rennes** right off the bat. This is expected, since the movie is part of the dataset that we've indexed in Elasticsearch, so it emerges as the relevant answer to our query.

How it works...

The concepts covered in this recipe involve many distinct parts, which we will attempt to explain. As discussed in the introduction, our application relies on LangChain, a framework that stitches together the core components when building applications that interact with LLMs. For this application, we used the concepts detailed in the following subsections.

Retrievers

As the name suggests, a retriever's role is to *retrieve* the right data that will be passed into the context of the LLM, performing all the semantic searches. This is the function of Elasticsearch, which is initialized with the following piece of code:

```
def init_retriever(k, db, fetch_k):
    retriever = db.as_retriever(
```

```
        search_type="similarity",
        search_kwargs={
            "k": k,
            "fetch_k": fetch_k,
            "doc_builder": custom_document_builder,
            "custom_query": custom_query_builder,
            "fields": ["title", "director", "plot",
                        "release_year", "wiki_page"],
        }
    )
    return retriever
```

The `init_retriever` function takes a parameter named db. This parameter represents the `ElasticsearchStore` library that we use to connect to our Elastic Cloud deployment. The following line of code demonstrates the initialization of `ElasticsearchStore`:

```
self.db = ElasticsearchStore(
    es_cloud_id=ES_CID,
    index_name=ES_VECTOR_INDEX,
    es_user=ES_USER,
    es_password=ES_PWD,
    query_field="plot",
    vector_query_field="plot_vector",
    distance_strategy="DOT_PRODUCT",
    strategy=ElasticsearchStore.ApproxRetrievalStrategy(
        hybrid=True,
        query_model_id=query_model_id,
    )
)
```

One specific parameter we want to touch on is the `strategy` parameter. As the name implies, it serves to define the search strategy that we'll use to retrieve data from Elasticsearch. In the context of our application, we use a hybrid strategy combining traditional BM25 and k-NN search.

Since we're utilizing documents from an existing Elasticsearch index, their structure is not ideally suited for what LangChain expects. To address this, we have defined a `custom_document_builder` function and a `custom_query_builder` function. `custom_document_builder` is used to determine how a document is constructed, based on data retrieved from Elasticsearch. Conversely, `custom_query_builder` allows for the customization of the query used to retrieve documents.

A prompt template

To make sure that the LLM only generates answers based on the document provided in the context through the retrieval phase, we must set our expectations through some prompt engineering. We use the following prompt:

```
LLM_CONTEXT_PROMPT = ChatPromptTemplate.from_template(
    """Use the following pieces of retrieved context to answer the
question. If the answer is not in the provided context, just say that
you don't know. Be as verbose and educational in your response as
possible..

    {context}
    Question: "{question}"
    Answer:
    """
)
```

To put everything together, LangChain has a concept called a **chain**, which is a sequence of calls to any of the building blocks (e.g., LLM, retrieval step, or tool). Our chain for this application is defined with the following function from the helper.py file:

```
def setup_rag_chain(prompt_template, llm, retriever):
    rag_chain_from_docs = (
        RunnablePassthrough.assign(
            context=(lambda x: qa_format_docs(x["context"])))
        | prompt_template
        | llm
        | StrOutputParser()
    )
    rag_chain_with_source = RunnableParallel(
        {
            "context": retriever,
            "question": RunnablePassthrough()
        }).assign(answer=rag_chain_from_docs)
    return rag_chain_with_source
```

We've explicitly defined in our chain the ability to return and show the sources used to generate the answer.

The LLM is initiated with the setup_chat_model function.

Finally, the `ask` function is called every time the **Send** button is triggered on the interface. It takes the user question, the chain, and chat model and returns the generated answer:

```
resp = ask(
    user_query,
    setup_rag_chain(
        LLM_CONTEXT_PROMPT,
        setup_chat_model(
            st.session_state.llm_base_url,
            st.session_state.llm_model,
            st.session_state.llm_temperature
        ),
        init_retriever(
            st.session_state.k,
            self.db,
            st.session_state.num_candidates
        )
    )
)
```

There's more...

The application comes with a sidebar that is collapsed by default. You can click on the > icon at the top left to expand it:

Figure 9.33 – Expanding the sidebar configuration menu

The following settings are available from the sidebar:

- **Choose your LLM**: The default option is **Ollama/Mistral**, but you can switch to **OpenAI GPT3.5** if you have a developer account with OpenAI API access. This will require you to provide an OpenAI API key and will incur some billings from using the OpenAI API.

- **Ollama base url**: This is the endpoint to reach Ollama models.

- **Temperature**: This controls the randomness of predictions, with lower values producing more predictable text and higher values generating more varied outputs.

The remaining parameters, such as **Number of documents to retrieve**, **Number of candidates**, **RRF window size**, and **RRF rank constant**, have already been covered in previous recipes and are related to vector search configuration.

Experimenting with these settings enables you to observe how they influence the generated answer. This flexibility in configuration is key to optimizing search performance and tailoring results to specific requirements or scenarios.

When using Elasticsearch as a vector store in Langchain, there are indeed interesting alternative approaches you can leverage, such as a self-querying retriever and MultiQueryRetriever:

- **A self-querying retriever**: This approach involves converting the original question to a structured query that can be run against an Elasticsearch index. This can be especially useful, as it allows you to tailor the query to the structure of the documents for better results.

- **MultiQueryRetriever**: This method involves generating multiple queries from the original input query, often by transforming or expanding it in various ways. Each of these generated queries is then used to retrieve information from Elasticsearch. `MultiQueryRetriever` can be particularly beneficial when the original query is ambiguous, broad, or might benefit from different perspectives or angles. Using multiple queries increases the chances of retrieving more diverse and relevant information that can enhance a model's response.

See also

- This recipe serves as a basic introduction to building a RAG application using Elasticsearch and Langchain; you can find many notebooks on the subject in the Elasticsearch labs repository: `https://github.com/elastic/elasticsearch-labs`

- To learn more about Langchain, check out their official documentation: `https://python.langchain.com/docs/get_started/introduction`

Using advanced techniques for RAG applications

In the *Developing a question-answering application with generative AI* recipe, we introduced the concept of RAG and demonstrated it through a simple application, built using Langchain, with Elasticsearch as the primary retriever. In this recipe, we're taking things a step further by transforming our previous QA application into a fully-fledged chatbot. Along the way, we will introduce some very important concepts crucial for developing a production-ready RAG application.

Getting ready

Make sure that you have completed the following recipes:

- *Using hybrid search to build advanced search applications*
- *Developing a question-answering application with generative AI*

This recipe requires the installation of the **Natural Language Toolkit** (**NLTK**) library; run the following command:

```
pip install nltk
```

The snippets of this recipe can be found at `https://github.com/PacktPublishing/Elastic-Stack-8.x-Cookbook/blob/main/Chapter9/snippets.md#using-advanced-technique-for-retrieval-augmented-generation-rag-applications`.

How to do it...

For the chatbot development outlined in this recipe, additional preparation of our movies dataset is necessary. Previously, we connected our QA application directly to an existing `movie-dense-vector` index. While this method is sufficient for straightforward use cases, it's not typically suitable for real-world production scenarios.

Our first steps involve utilizing LangChain to segment the documents before ingestion, as well as employing a more advanced conversational chain to construct the chatbot experience based on the movies dataset.

This process begins with **chunking** the documents, a technique that involves breaking down each document into smaller, more manageable pieces. These chunks can then be indexed individually, allowing the chatbot to retrieve and present information more efficiently and effectively during interactions.

Then, we'll integrate a conversational logic layer, leveraging LangChain's conversational chain to handle more complex dialogues and provide a seamless chatbot experience. This advanced conversational chain will enable the chatbot to understand context, maintain the flow of conversation, and respond accurately to user queries, thereby enhancing the overall user experience with the dataset:

1. Go to the GitHub repository to download the following Python script: `https://github.com/PacktPublishing/Elastic-Stack-8.x-Cookbook/blob/main/Chapter9/rag/indexer.py`. Then, load the documents from the CSV files and ingest them into Elasticsearch. Execute the following command, which may take some time to complete (in our test, it took approximately 10 minutes):

    ```
    python indexer.py
    ```

2. After executing the script, wait for all the documents to be indexed. Upon completion, navigate to Kibana, and then go to **Search | Content | Indices**. You should find an index named `movies-langchain-generated` that contains **9141** documents, as shown in *Figure 9.34*:

movies-langchain-generated

Search Applications ∨

Overview   Documents   Index mappings   Pipelines

Ingestion type
API

Document count
9141

Figure 9.34 – The document count of the LangChain-generated movies index

3. This increased document count may seem surprising at first, but it's important to remember that we chunked the original documents before ingesting them, leading to a higher number of individual documents. Let's create a Kibana data view for this index to explore the results in Discover. For example, filtering on documents with `metadata.title: Titanic` now gives five hits:

Figure 9.35 – Chunking results for the Titanic document

4. We will delve into more details about this process and its implications in the *How it works...* section, explaining the rationale behind document chunking and its benefits for search and retrieval in the context of building a chatbot.

5. Now that we have completed the ingestion phase and have our new vector index ready, let's deploy the chatbot application with the following command:

```
$ streamlit run chatbot_app.py --server.port 8502
```

Once the application is launched, you shall be redirected to your web browser at `http://localhost:8502`.

We will start by asking a question that illustrates the value of chunking in our new index.

6. First, in your navigator, open `http://localhost:8501` to go to the previous QA application (if the application is not running, refer to the steps outlined in the *Developing question-answering applications with generative AI* recipe to launch it). Ask the question `Which film features a love story on a ship that breaks in half and only one of the lovers is alive after?` and check the results, as illustrated in the following screenshot:

Figure 9.36 – A question to the QA app in a non-chunked index

The answer is quite surprising, as the movie *Titanic* cannot be found within the provided context.

Now, switch to the chatbot application we have just deployed to ask the same question and observe the result:

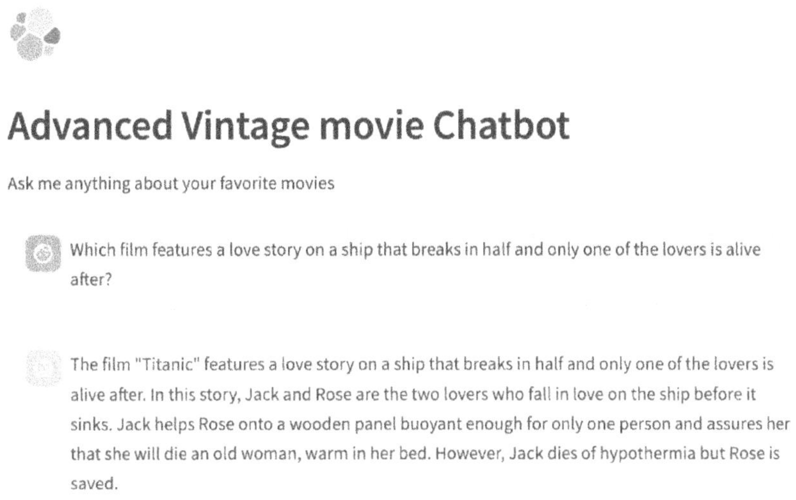

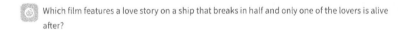

Advanced Vintage movie Chatbot

Ask me anything about your favorite movies

Which film features a love story on a ship that breaks in half and only one of the lovers is alive after?

The film "Titanic" features a love story on a ship that breaks in half and only one of the lovers is alive after. In this story, Jack and Rose are the two lovers who fall in love on the ship before it sinks. Jack helps Rose onto a wooden panel buoyant enough for only one person and assures her that she will die an old woman, warm in her bed. However, Jack dies of hypothermia but Rose is saved.

Figure 9.37 – The answer from the chatbot in the chunked index

> **Note**
> You might have to increase the k parameter to 10, using the configuration available in the sidebar menu.

In this scenario, the relevant context is successfully passed to the LLM, enabling it to provide the expected answer. In the first instance, where the plot was not chunked before indexing, some pieces of information might have been missing due to the plots being truncated. This contrasts our approach with the chunked index, where each part of the plot has been indexed, albeit across different documents. By chunking the plot and indexing each segment as a separate document, we ensure comprehensive coverage and indexing of all relevant details.

Let's now demonstrate the conversation capabilities of our chatbot.

Use the sidebar to locate and click on the **Clear message history** button. Once the previous conversation has been cleared, enter the following query in the textbox – Against who a team of super heroes is fighting to protect the city of Rennes? Wait for the chatbot to answer and ask the follow-up question Who are the heroes in the team?, and then finish by asking the following question about the plot – `How do they stop the Invader?`. At the end of the conversation, your chat window should look something like *Figure 9.38*:

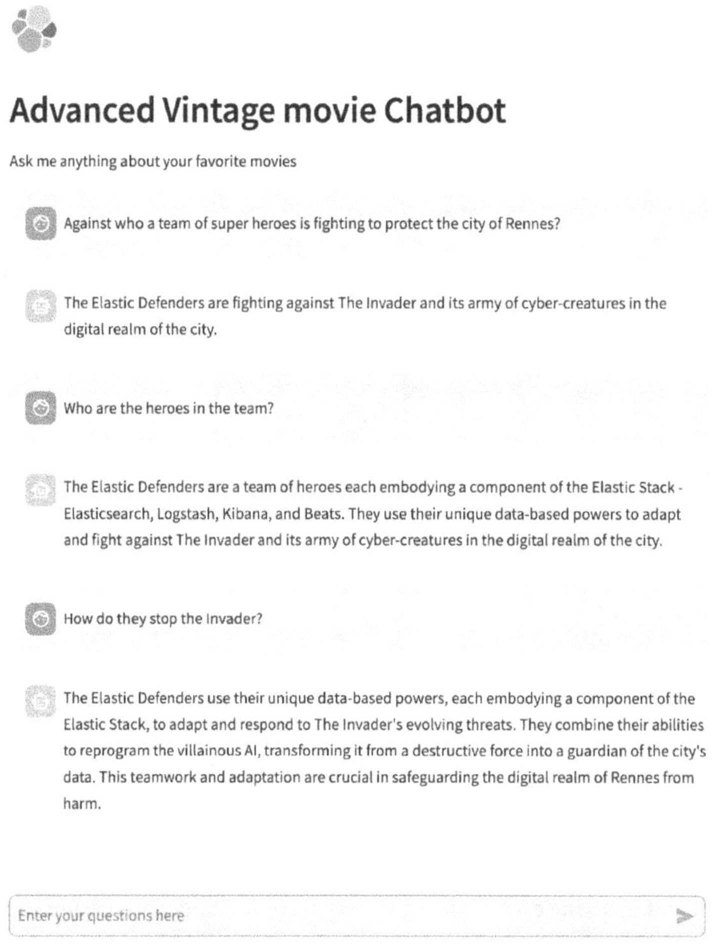

Advanced Vintage movie Chatbot

Ask me anything about your favorite movies

Against who a team of super heroes is fighting to protect the city of Rennes?

The Elastic Defenders are fighting against The Invader and its army of cyber-creatures in the digital realm of the city.

Who are the heroes in the team?

The Elastic Defenders are a team of heroes each embodying a component of the Elastic Stack - Elasticsearch, Logstash, Kibana, and Beats. They use their unique data-based powers to adapt and fight against The Invader and its army of cyber-creatures in the digital realm of the city.

How do they stop the Invader?

The Elastic Defenders use their unique data-based powers, each embodying a component of the Elastic Stack, to adapt and respond to The Invader's evolving threats. They combine their abilities to reprogram the villainous AI, transforming it from a destructive force into a guardian of the city's data. This teamwork and adaptation are crucial in safeguarding the digital realm of Rennes from harm.

Enter your questions here

Figure 9.38 – The chatbot conversation about an upcoming movie

We have successfully demonstrated that our chatbot can engage in conversations and handle follow-up questions, without needing to repeatedly reference a movie's name in each interaction. This capability indicates that our chatbot possesses a form of memory, enabling it to retain context from one exchange to the next. Such contextual awareness is crucial for creating a seamless and natural conversational experience, as it allows the chatbot to understand the flow of the conversation and respond appropriately to queries based on previous interactions, thereby enhancing the user's engagement with the chatbot.

How it works...

In this recipe, we introduced some new concepts to build the chatbot experience. First, we introduced the chunking that happens in the indexer.py script. The actual piece of code where the chunking is performed is in the following extract from the create_vector_db function:

```
def create_vector_db(dataset, index_name):
    loader = CSVLoader(
        file_path=dataset,
        metadata_columns=["release_year", "title", "origin",
                          "director", "cast", "genre", "wiki"],
        csv_args={
            "fieldnames": ["release_year", "title", "origin",
                           "director", "cast", "genre",
                           "wiki", "plot"]
        }
    )
documents = loader.load()
text_splitter = NLTKTextSplitter(chunk_size=1000,
                                 chunk_overlap=200)
docs = text_splitter.split_documents(documents)
```

The function starts by loading the data from the CSV file using LangChain's CSVLoader. While loading, to adhere to LangChain-expected documents structure, we define the metadata fields. Once the documents are loaded, we use the NTLKTextSplitter library to slice them into chunks of 1,000 tokens with a 200-token overlap.

Now, why is it so important to chunk documents, you may ask? Simply because vector models are limited in the number of tokens they can process as input. If you pass a document that exceeds the token limit, it gets truncated. This can be very problematic, as it basically means that we can lose information, which can have a negative impact on our retrieval strategy. For example, the e5-small model we're using has a token limit of 512, as does ELSER. This is also the reason why our new index has almost three times more documents than the original one.

Now, you might be wondering why we chose a chunk size of 1,000. This decision was made to strike a balance between precision and context length. Opting for a 512-token chunk would have resulted in more chunks to feed within the LLM's context, potentially leading to scenarios where some relevant chunks are not included to generate the answer.

Another important thing to note is that we've also defined a specific analyzer in the settings of our index:

```
"analyzer": {
    "my_english_analyzer": {
        "type": "english",
        "stopwords": ["_english_", "where", "which", "how",
```

```
                          "when", "wherever"]
        }
    }
```

As we're using a hybrid search strategy, combining lexical search and vector search, defining this analyzer will help improve the accuracy of lexical search by ignoring some common English stop words.

On the application side, we've implemented the following concepts:

- **Message history**: We've introduced a very basic history for the user query using Streamlit Session State. This is important, as it allows our chat to *remember* the previous conversation.

- **A prompt strategy**: To integrate historical messages, we employ a distinct prompt strategy that condenses the previous conversation and the last question into a single prompt. Using a system prompt, we then ask the LLM to use the resulting context as the ground truth to generate answers:

  - CONDENSE_QUESTION_PROMPT: To condense the previous conversation and the last question into a single prompt

  - SYSTEM_PROMPT: This is a classical RAG prompt that asks the LLM to use the context as ground truth for its answer generation

- **Conversational chain**: Utilizing Langchain's framework, this is a critical component that orchestrates memory management, diverse prompts, and information retrieval to create a cohesive chatbot experience. The get_conversational_rag_chain function from the helper.py file, invoked each time a user poses a query to the chatbot, operates through several high-level steps:

  I.   **Generate a standalone question**: the first step of the chain synthesizes a unified question from the accumulated chat history and the most recent user query. This step ensures that the context of the conversation is maintained and considered when generating responses.

  II. **Retrieve context from Elasticsearch**: Utilizing the standalone question generated in the first step, the system retrieves relevant context from Elasticsearch. This retrieval process is tailored to fetch information that can provide background or support to answer the current query, leveraging the rich data indexed in Elasticsearch,

  III. **Generate the final answer with LLM**: The context obtained from Elasticsearch is then fed into the LLM to generate the final answer. This step allows the chatbot to produce responses that are not only contextually relevant but also informed by the specific details retrieved from the data store.

Overall, RAG is a workflow involving multiple moving parts, as shown in the following figure:

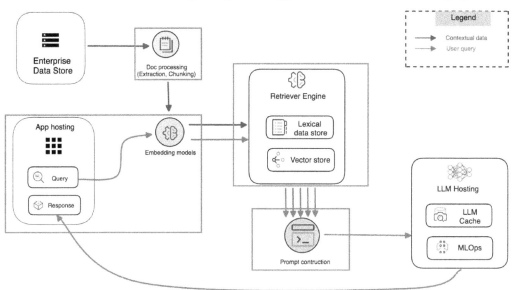

Figure 9.39 – A RAG workflow overview

Throughout this recipe and the previous one, we primarily focused on the components outlined within the rectangles in the preceding diagram, which represent the foundational building blocks essential for creating a successful RAG application:

- **App hosting**: We used Streamlit for the frontend of our application

- **Embedding models**: We relied on the multilingual E5 model included in Elastic Stack to build the vector at ingest and query time

- **Retriever Engine**: The retrieval is performed within Elasticsearch, leveraging both vector and lexical search capabilities.

- **Doc processing**: Langchain uses the CSV Loader and the NTLKTextSplitter library for chunking

These core components include the mechanisms to query and retrieve relevant information from a search engine (such as Elasticsearch), and the integration of this information with an LLM to generate informed and contextually relevant responses.

There's more...

In this recipe, we learned the key concepts of building a RAG application. In the following sections, we will discuss additional techniques relevant to RAG applications.

Chat history management

An important aspect related to chatbots is the management of chat history. In the sample application we provided, we demonstrated how to manage chat history with session messages simply. This allows us to differentiate between new and returning users and provide a personalized experience. To ensure that chat history is not lost between sessions and that it can be used for future analysis, or to offer a continuous user experience, the data must be saved in a persistent storage system.

LangChain offers a specific module to manage chat history; you can find the details in this documentation: `https://python.langchain.com/docs/use_cases/chatbots/memory_management#chat-history`. Elasticsearch can be used as a data store to store the chat history messages in dedicated indices. The implementation of chat history management using Elasticsearch within the LangChain framework is documented at the following address: `https://python.langchain.com/docs/integrations/memory/elasticsearch_chat_message_history#initialize-elasticsearch-client-and-chat-message-history`.

A comprehensive example of chat history management is available at the Elastic Search labs here: `https://www.elastic.co/search-labs/tutorials/chatbot-tutorial/implementation/history`.

Nested k-NN search

In our recipe, we used LangChain to generate chunks of movie plots, with each chunk then stored as a separate document in Elasticsearch. Typically, the retrieved chunks provide the necessary information for the RAG application's prompts. However, due to each chunk's limited length, they may lack the context needed to grasp the full meaning or narrative of the original document. Retrieving the parent document can, therefore, be crucial for providing complete context and ensuring coherence and consistency in the final response.

With the release of Elasticsearch 8.11, a new feature called nested k-NN search was introduced. This feature allows us to retain the chunks within a single document and perform k-NN queries on dense vectors with similar efficiency. Detailed documentation on nested k-NN search is available here: `https://www.elastic.co/guide/en/elasticsearch/reference/current/knn-search.html#nested-knn-search`.

Chain techniques

In this recipe, we used a simple chain for our RAG application, with a basic process involving hybrid search, prompt preparation, and then text generation via LLM. The chain definition can be found in the `get_conversational_rag_chain` function within the `helper.py` module. Note that we defined the chain by using `ConversationalRetrievalChain.from_llm`. The simple chain is sufficient for our scenario, where we deal with straightforward chat tasks with an in-memory chat history and the knowledge base (our 1990s movie dataset) has a narrow and well-defined scope.

In certain situations, more complex or sophisticated chains are required, such as the following:

- Accessing multiple data sources (e.g., a movie dataset and an artist dataset)

- Undertaking multiple rounds of retrieval or refinement (where initial answers guide subsequent retrieval and generation for improved accuracy)

- Engaging in multi-step generation (e.g., a combination of summarization, translation, and generation)

LangChain provides a highly flexible and advanced framework to manage chains with **LangChain Expression Language** (**LCEL**). Detailed documentation is available here: `https://python. langchain.com/docs/modules/chains`.

Cache management

Another important aspect of RAG applications is cache management. Managing the cache helps control costs and latencies associated with API calls to LLMs. LangChain provides mechanisms for caching, including in-memory and SQLite options (`https://python.langchain.com/docs/ modules/model_io/llms/llm_caching`).

Elasticsearch can also be used as a cache layer for RAG applications, taking advantage of both Elasticsearch's vector search and data storage capabilities. You can find a very detailed blog post on this subject here:

`https://www.elastic.co/search-labs/blog/articles/elasticsearch-as- a-genai-caching-layer`.

See also

- If you are interested in building a chatbot experience with Elastic and Langchain, check out this great tutorial: `https://www.elastic.co/search-labs/tutorials/chatbot- tutorial/welcome`.

- Elastic provides a dedicated homepage for LangChain integration at the following link: `https:// www.elastic.co/search-labs/tutorials/integrations/langchain`.

- You can also find other RAG-related integrations on this page: `https://www.elastic.co/search-labs/tutorials/integrations/hugging-face`.

- Langchain also provides a declarative language called LCEL that you can use for more advanced chain setup. Check out the LCEL documentation here: `https://python.langchain.com/docs/expression_language`.

10

Elastic Observability Solution

In addition to its search and data analytics capabilities, the Elastic Stack also serves as a full-stack observability platform, offering a unified view of metrics, logs, traces, user experience, and profiling—commonly referred to as the pillars of observability. *Figure 10.1* shows the reference architecture for **Elastic Observability**, illustrating the different data collectors available for gathering these sources of data:

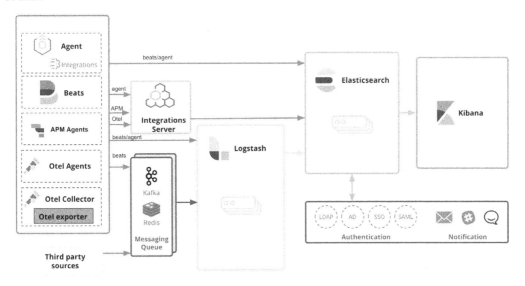

Figure 10.1 – Elastic Observability reference architecture

In this chapter, we will discuss how you can onboard various observability data types into Elastic, and how to use Elastic Observability for troubleshooting and root cause analysis.

In this chapter, we're going to cover the following main topics:

- Instrumenting your application with Elastic Application Performance Monitoring (APM) agents

- Setting up Elastic **Real User Monitoring (RUM)**

- Instrumenting and monitoring with OpenTelemetry

- Monitoring Kubernetes environments with Elastic Agent

- Managing synthetics monitoring

- Gaining comprehensive system visibility with Elastic Universal Profiling

- Detecting incidents with alerting and machine learning

- Gaining insights with the AI Assistant

Technical requirements

To follow the different recipes in this chapter, you will need an Elastic Stack deployment that includes the following:

- Elasticsearch for searching and storing the data

- Kibana for data visualization and stack management (Kibana user with all privileges on Fleet and Integrations)

- Integrations Server (included by default in Elastic Cloud deployment)

- A machine learning node

In addition to the Elastic Stack deployment, you'll also need the following:

- A local machine with an installed Docker engine

- A Kubernetes deployment from a cloud Kubernetes service such as **Google Kubernetes Engine (GKE)**

Instrumenting your application with Elastic APM

Elastic APM, as part of the Elastic Observability solution, provides real-time insights into application performance, allowing developers and operations teams to monitor, troubleshoot, and optimize their applications. In this recipe, we will learn how to instrument your application with Elastic APM agents.

Getting ready

Ensure you have cloned the GitHub repository for the book.

Verify that you have a functioning Docker engine on your local machine.

For your Elastic Cloud deployment, remember the saved password for the default user, `elastic`, and the Elasticsearch endpoint URL (refer to *Chapter 1*). Check that no programs are running on the following local machine ports: `5000`, `5001`, `5002`, `3001`, and `9000`.

The snippets for this recipe can be found at the following link: `https://github.com/PacktPublishing/Elastic-Stack-8.x-Cookbook/blob/main/Chapter10/snippets.md#instrumenting-your-application-with-elastic-apm-agent`.

How to do it...

The goal of this recipe is to instrument a sample microservice application with Elastic APM agents. We will start with setting up the sample application, a movie search application we name **Elastiflix**, and then we will try out the Elastic Java APM agent to instrument one of the services. We will end up having a fully instrumented application with Elastic APM agents for different services. Follow these steps:

1. Let us start with setting up the sample application. Navigate to the local folder of your cloned GitHub repository, `Chapter10/Elastiflix/no-instrumentation`, open the `.env` file, and add your Elastic deployment details and credentials to the `.env` file:

   ```
   ELASTICSEARCH_USERNAME="elastic"
   ELASTICSEARCH_PASSWORD="changeme"
   ELASTICSEARCH_URL="https://foobar.es.us-central1.gcp.cloud.
   es.io"
   ```

2. In your terminal, start the application with the following commands:

   ```
   $ cd no-instrumentation
   $ docker-compose -f docker-compose.yml up --build -d
   ```

3.   Once the application is fully operational, visit `http://localhost:9000` in your browser. You should be able to access the Elastiflix application, as shown in *Figure 10.2*:

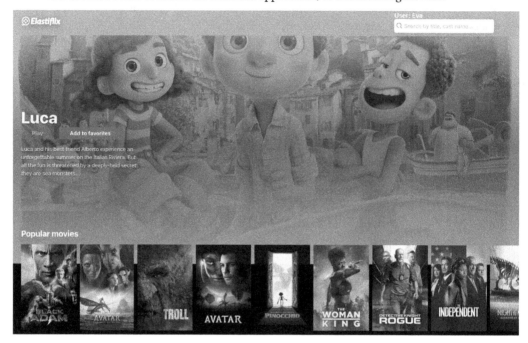

Figure 10.2 – Elastiflix sample application

Stop the application for now (by using the `docker-compose -f docker-compose.yml down` command in your terminal) as we are going to prepare for the instrumentation in the upcoming steps.

4. Now, let us find the APM Server URL that we need to instrument our application. In Kibana, go to **Management | Fleet | Agent policies**, click on **Elastic Cloud agent policy**, and then in the **Actions** column, click on **Edit integration**, as shown in *Figure 10.3*:

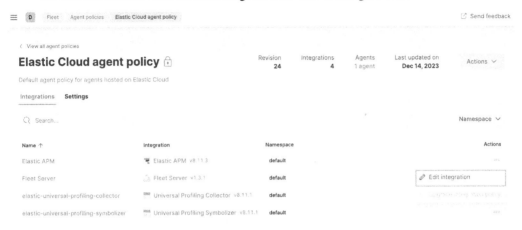

Figure 10.3 – Edit Elastic APM integration

5. Click on the **APM Agents** tab and expand the Java configuration section. Here, you'll find instructions to download the Java APM agent and configure it for your application. Note the Elastic APM Server URL and secret token, as shown in *Figure 10.4*:

Figure 10.4 – Elastic APM agent configuration

6. Return to the local folder of your cloned GitHub repository at `Chapter10/Elastiflix/java-instrumentation`. Update the `.env` file with the APM Server URL and secret token, completing the Elastic deployment details and credentials as done in *Step 1*.

7. Navigate to the `Chapter10/Elastiflix/java-instrumentation/java-favorite` folder and open the Docker file and `start.sh` file to review the snippets prepared for Java instrumentation.

8. In the Docker file, we download the Elastic APM Java Agent with the following snippet:

```
curl -L -o /elastic_apm_agent/elastic-apm-agent.jar https://
repo1.maven.org/maven2/co/elastic/apm/elastic-apm-agent/${AGENT_
VERSION}/elastic-apm-agent-${AGENT_VERSION}.jar
```

9. Additionally, in `start.sh`, we've modified the script to include the `-javaagent` flag and set the system properties according to the example provided in *Step 5*:

```
java \
  -javaagent:/elastic_apm_agent/elastic-apm-agent.jar \
  -Delastic.apm.application_packages=com.movieapi \
  -jar /usr/src/app/target/favorite-0.0.1-SNAPSHOT.jar --server.
port=5000
```

10. After the inspection of the files, navigate to `Chapter10/java-instrumentation` and start the application with the following command in your terminal:

```
$ docker-compose -f docker-compose.yml up --build -d
```

11. In your browser, revisit the application at `http://localhost:9000` and click on **Add to favorite** to generate some load for the Java microservice. Then, in Kibana, go to **Observability | APM | Services**, as shown in *Figure 10.5*, to check whether the Java microservice has been correctly instrumented:

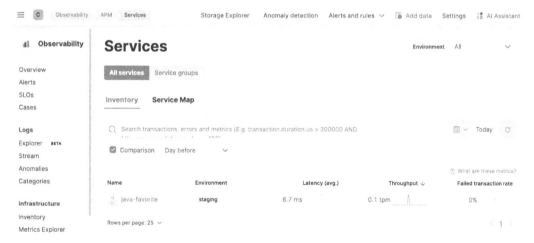

Figure 10.5 – APM instrumentation for a single Java service

12. Let's stop the application for now so you can prepare for full instrumentation. In your terminal, stop the application with the following command:

```
$ docker-compose -f docker-compose.yml down
```

13. We are now ready to extend instrumentation to other services of our sample application. Navigate to the local folder of your cloned GitHub repository, Chapter10/full-instrumentation, and update the .env file as we did in *Step 6* with your APM Server URL, secret token, and the Elastic deployment details and credentials. This folder contains preconfigured instrumentation for the various microservices included in our sample application. Next, we will explore how distributed tracing functions in a microservice environment.

14. Start the application with the following command in your terminal:

```
$ docker-compose -f docker-compose.yml up --build -d
```

15. In Kibana, go to **Observability | APM | Services**, as shown in *Figure 10.6*, to check whether the different microservices have been correctly instrumented:

Figure 10.6 – Elastiflix services

16. Click on the **Service Map** tab to view the visual representation of the application and its interactions between different microservices:

Services

Environment   All   ⌄

[All services] Service groups

Inventory   Service Map

🔍 Search transactions, errors and metrics (E.g. transaction.duration.us > 300000 AND http.response.status_code >= ...)   📅 ⌄   Today   ↻

☑ Comparison   Day before   ⌄

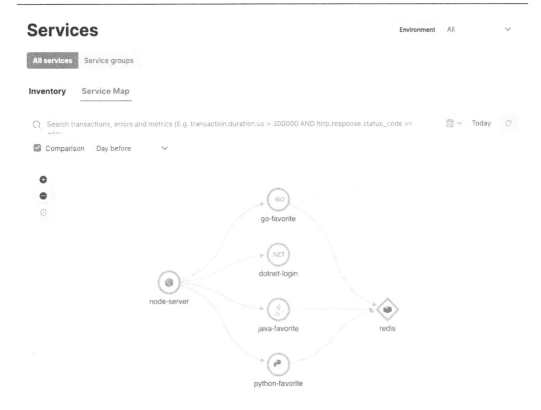

Figure 10.7 – Elastiflix APM service map

Let's stop the application now while you prepare for RUM in the next recipe. In your terminal, stop the application with the following command:

```
$ docker-compose -f docker-compose.yml down
```

How it works...

In our example, we began by identifying the APM Server URL and secret token. This information is necessary to configure Elastic APM agents so that they can securely transmit the collected telemetry data to the APM Server over HTTPS. For cloud deployments, the APM Server is managed by Fleet Server and is preconfigured through the Elastic APM integration within the Elastic Cloud agent policy.

We then integrated the Java APM agent library into our sample application within the Java instrumentation example. Later, we added different APM agents corresponding to other programming languages, enabling distributed tracing for each microservice in the full instrumentation example. During the setup, we defined Agent environment variables, such as service name and service environment, in our docker-compose.yml files.

Once installed, the Elastic APM agents automatically instrument the application by capturing requests, database queries, cache calls, and more depending on the capabilities of the agents and the language/ framework. The agents then transmit the data to the APM Server, which processes and stores it in Elasticsearch. You can then use Kibana to visualize and analyze the APM data through the APM **user interface (UI)**.

There's more...

The APM UI offers a wealth of useful information, such as service maps, detailed performance metrics, transaction samples, and error logs. This enables developers and operations teams to monitor the application, quickly pinpoint issues, and optimize performance. We will explore practical examples later in this chapter in the *Detecting incidents with alerting and machine learning* recipe.

See also

- There are numerous configuration options available for the APM Server. Detailed documentation can be found here: `https://www.elastic.co/guide/en/apm/guide/current/configuring-howto-apm-server.html`

- Besides the Fleet-managed APM Server, it's possible to configure a self-managed APM Server. Detailed documentation is available here: `https://www.elastic.co/guide/en/apm/guide/current/_apm_server_binary.html`

- The complete list of Elastic APM agents and their documentation can be found here: `https://www.elastic.co/guide/en/apm/agent/index.html`

Setting up RUM

In the previous recipe, we saw how to instrument a sample microservice application with Elastic APM agents. This recipe will show us how to extend our monitoring capabilities to not only observe server-side metrics but also directly capture user interactions from within the browser. This approach will provide insights into real user experiences, offering detailed performance data from the user's perspective.

Getting ready

Ensure you have cloned the book's GitHub repository.

Verify that a functional Docker engine is installed on your local machine.

For your Elastic Cloud deployment, note the saved password for our default user, `elastic`, and Elasticsearch endpoint URL (refer to *Chapter 1*).

Confirm that no programs are running on the following ports of your local machine: `5000`, `5001`, `5002`, `3001`, and `9000`.

The snippets for this recipe can be found at the following link: `https://github.com/PacktPublishing/Elastic-Stack-8.x-Cookbook/blob/main/Chapter10/snippets.md#setting-up-real-user-monitoring`.

How to do it...

The goal of this recipe is to continue the instrumentation of our sample Elastiflix application that we used in the previous recipe. We will configure the application with the RUM JavaScript agent, inspect the configurations, and then restart the application to analyze the real user metrics in Kibana:

1. On your local machine, navigate to the local folder of your cloned GitHub repository, `Chapter10/Elastiflix/full-instrumentation-with-rum`. Here, the application is preconfigured with Elastic RUM. Before you start, adjust and inspect the configurations.

2. Update the `.env` file as we did in the previous recipe, with your APM Server URL, secret token, and the Elastic deployment details and credentials.

3. First, let's examine the setup. Navigate to the `Chapter10/Elastiflix/full-instrumentation-with-rum/javascript-client` folder and open the `package.json` file to review Elastic RUM JavaScript dependencies:

```
"dependencies": {
    "@elastic/apm-rum": "^5.12.0",
    "@elastic/apm-rum-react": "^1.4.2",
    ...
}
```

4. Next, navigate to the `Chapter10/Elastiflix/full-instrumentation-with-rum/javascript-client/src` folder and open the `App.js` file to inspect the RUM instrumentation:

```
export const apm = initApm({
    serviceName: "${ELASTIC_APM_SERVICE_NAME}",
    serverUrl: "${ELASTIC_APM_SERVER_URL}",
    serviceVersion: '',
    environment: "${ELASTIC_APM_ENVIRONMENT}",
    distributedTracingOrigins: ['http://localhost:3000','http://localhost:3001','http://localhost:9000']
})
```

5. After inspecting the files, return to `Chapter10/Elastiflix/full-instrumentation-with-rum/` and start the application with the following command in your terminal:

```
$ docker-compose -f docker-compose.yml up --build -d
```

6. In your browser, visit the application at `http://localhost:9000`. After waiting for a few seconds, in Kibana, navigate to **Observability| APM | Services| Service Map**, as shown in *Figure 10.8*, to check whether the **frontend** service has been correctly instrumented, compared to the service map that we have seen in the previous recipe. Click on the **frontend** service in the **Service map** tab, and then click on **Service Details**:

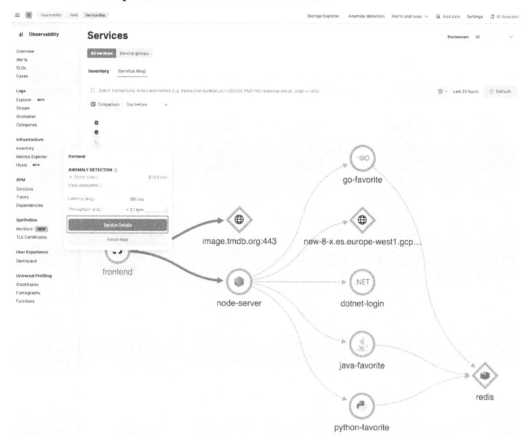

Figure 10.8 – Elastiflix APM service map with RUM

7. On the frontend service detail page, click on **/home** in the transactions sections, then scroll down on the transaction detail page to inspect the distributed tracing, which includes both RUM and backend transactions, as displayed in *Figure 10.9*:

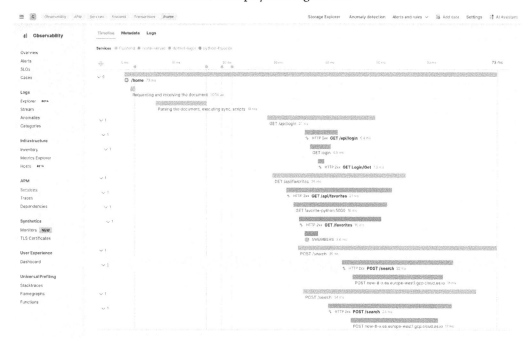

Figure 10.9 – Elastiflix frontend service distributed tracing

8. Finally, let's check the dedicated **User Experience Dashboard** by visiting **Observability| User Experience | Dashboard**. There, you can see the different user experience key performance indicators (KPIs), such as page load distribution, regional page load distribution, and visitor breakdown, as shown in *Figure 10.10*:

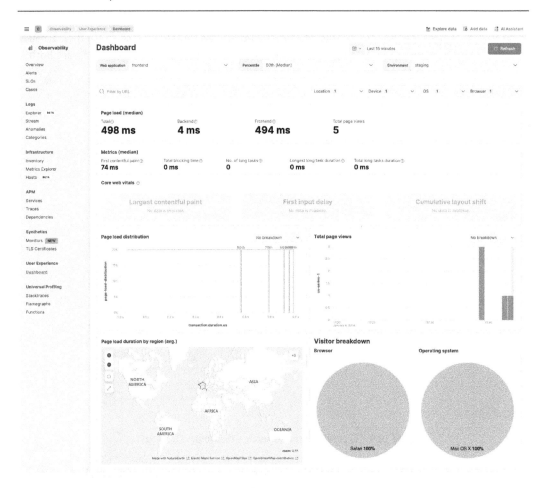

Figure 10.10 – Elastic User Experience Dashboard

How it works...

To set up Elastic RUM, we began by integrating the Elastic RUM JavaScript agent into the **frontend** service of our Elastiflix web application. Adding a small code snippet to the HTML (through App. js) enables the application to be served with the RUM agent. As our frontend is a React application, we also added the JavaScript agent in the package.json file.

Once integrated, the RUM agent begins monitoring the application automatically. It captures a range of performance metrics and user interactions, including page load times, AJAX requests, and **document object model** (**DOM**) loading and rendering duration. It also collects additional information such as the type of browser, the **operating systems** (**OSs**) used by visitors, their geographic locations, and device types.

The collected data is sent to the Elastic APM Server at regular intervals, where it's processed. Once processed, the performance data becomes available in Elasticsearch. We can visualize and analyze it within the Elastic Observability app in Kibana. The Elastic APM app in Kibana offers various views to help us understand aspects of user experience such as throughput, latency, and error rates, along with individual transaction traces. The **Elastic User Experience** app provides important KPIs for performance monitoring, core web vitals, visitor breakdown, JavaScript errors, and more.

There's more...

The following sections will cover two additional topics: custom instrumentation and OpenTelemetry support. These insights will enable you to further refine your observability strategy and ensure compatibility with the evolving landscape.

Custom instrumentation

In our recipe, we used the auto-instrumentation provided by the Elastic RUM agent. However, it's also possible to perform custom instrumentation. This can be particularly beneficial for **single-page applications** (**SPAs**) where the auto-instrumentation of page loads may not be sufficient. Examples of custom transactions are available at this URL: `https://www.elastic.co/guide/en/apm/agent/rum-js/current/custom-transactions.html`.

OpenTelemetry support

OpenTelemetry (`https://opentelemetry.io`) is an open-source initiative that offers a suite of APIs, libraries, agents, and integrations for the standardized generation, collection, and management of observability data—metrics, logs, and traces. At the time these recipes were written, OpenTelemetry supported RUM instrumentation. However, OpenTelemetry-instrumented RUM data had not yet been fully integrated into Elastic's Observability UI. This is why we introduced the Elastiflix sample application and demonstrated how to instrument applications using the Elastic APM agent and the Elastic RUM JavaScript agent. Such instrumentation enables Elastic Observability to correlate frontend performance data with backend monitoring insights, providing a comprehensive view of both client-side and server-side performance.

For the remainder of the chapter, we will switch our focus to the official OpenTelemetry demo and explore how to ingest observability data using the OpenTelemetry protocol and how to broaden data collection across different pillars of observability, including metrics, logs, profiling, and synthetic monitoring.

See also

The complete documentation on the metrics collected by the RUM JavaScript agent is available here: `https://www.elastic.co/guide/en/apm/agent/rum-js/current/supported-technologies.html`.

Our example is based on a React integration. For more detailed documentation on other JavaScript frameworks, such as Angular or Vue, you can refer to the following URL: `https://www.elastic.co/guide/en/apm/agent/rum-js/current/framework-integrations.html`.

Instrumenting and monitoring with OpenTelemetry

In this recipe, we'll focus on how you can easily instrument your application with OpenTelemetry and use Elastic Stack as your observability backend. While the previous two recipes leveraged Elastic APM agents to instrument applications, we will now turn to OpenTelemetry agents for the collection of application traces, logs, and metrics. OpenTelemetry is a framework designed to assist developers in gathering and analyzing telemetry data from their applications. It offers a standardized approach for logging how applications perform, including details related to traces, metrics, and logs. The following figure illustrates how OpenTelemetry integrates with the Elastic Observability architecture:

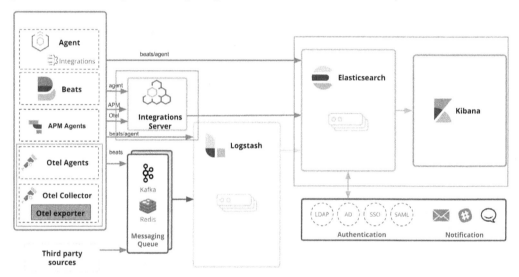

Figure 10.11 – Elastic Observability reference architecture with OpenTelemetry

OpenTelemetry is designed to be user friendly and adaptable, allowing developers to select the most suitable tools and services for their requirements. It also aims to foster collaboration within the industry by creating a unified collection of APIs and libraries dedicated to the monitoring and troubleshooting of distributed systems.

We will be using the official OpenTelemetry demo as our instrumented application.

Getting ready

Ensure that you have an operational Kubernetes cluster with the appropriate requirements as per the official demo documentation:

- Kubernetes 1.24+

- 6 GB of free RAM for the application

- Helm 3.9+

We suggest consulting the OpenTelemetry website for the latest up-to-date requirements: `https://opentelemetry.io/docs/demo/kubernetes-deployment/`.

For your Kubernetes cluster, you may either set up a local one or use a managed Kubernetes service such as Elastic Kubernetes Service (EKS), Azure Kubernetes Service (AKS), or GKE. For this recipe and the remainder of the chapter, we'll be using GKE from Google Cloud to host our Kubernetes cluster. Our Kubernetes setup includes the following:

- Kubernetes GKE 1.27.7

- 3 nodes of type n2-standard-2 with 2 CPUs and 8 GB RAM each

Make sure that you have an Elastic Stack deployment up and running with the integration server enabled.

You'll need to create a Kubernetes secret with the APM Server's URL and secret token. This secret is necessary to forward the data to the APM Server running in Elastic Cloud.

The snippets for this recipe are available at the following link: `https://github.com/PacktPublishing/Elastic-Stack-8.x-Cookbook/blob/main/Chapter10/snippets.md#instrumenting-and-monitoring-with-opentelemetry`.

How to do it...

The goal of this recipe is to instrument an application using OpenTelemetry agents and libraries, with the Elastic Stack serving as the observability backend:

1. Begin by obtaining the URL and secret token of the APM Server to configure OpenTelemetry. Refer to *Step 5* of the *Instrumenting your application with Elastic APM* recipe to retrieve both APM_URL and `secret-token`.

2. Create an environment variable for the APM URL with the following command:

```
$ export APM_URL_WITHOUT_PREFIX=<your-url>
```

3. Next, define an environment variable for the secret token:

    ```
    $ export APM_SECRET_TOKEN=<your-secret-token>
    ```

4. Then, create a Kubernetes secret using the following command:

    ```
    $ kubectl create secret generic elastic-secret --from-
    literal=elastic_apm_endpoint=$APM_URL_WITHOUT_PREFIX --from-
    literal=elastic_apm_secret_token=$APM_SECRET_TOKEN
    ```

5. We will use Helm (https://helm.sh/) to install the official OpenTelemetry demo application. First, add the OpenTelemetry chart repository:

    ```
    $ helm repo add open-telemetry https://open-telemetry.github.io/
    opentelemetry-helm-charts
    $ helm repo update
    ```

6. Next, we will install the demo application. Download and save the custom values.yaml from this address: https://github.com/PacktPublishing/Elastic-Stack-8.x-Cookbook/blob/main/Chapter10/otel-demo/kubernetes/values.yaml

7. Deploy the demo using helm install and pass in the custom values.yaml file that you have just downloaded:

    ```
    $ helm install -f values.yaml cookbook-otel-demo open-telemetry/
    opentelemetry-demo --version 0.29.2
    ```

8. Verify the successful deployment of the application:

    ```
    $ kubectl get pods
    ```

    It may take a few minutes for all pods to enter the running state. Once all are active, you should see the output as shown in *Figure 10.12*:

```
NAME                                                           READY   STATUS    RESTARTS
cookbook-otel-demo-accountingservice-b65c47d77-qd82m           1/1     Running   0
cookbook-otel-demo-adservice-6f86f5b88d-q55hv                  1/1     Running   0
cookbook-otel-demo-cartservice-6bffc64b84-hkdzj                1/1     Running   0
cookbook-otel-demo-checkoutservice-689cc667b6-lhcrt            1/1     Running   0
cookbook-otel-demo-currencyservice-56cb588b74-ljw2d            1/1     Running   0
cookbook-otel-demo-emailservice-b449b9c75-r5ktn                1/1     Running   0
cookbook-otel-demo-featureflagservice-767d66dbb4-ckpbw         1/1     Running   0
cookbook-otel-demo-ffspostgres-669d7867b7-ptvt8                1/1     Running   0
cookbook-otel-demo-frauddetectionservice-d5b6b5598-nrtsq       1/1     Running   0
cookbook-otel-demo-frontend-77f4f46c65-cw8vk                   1/1     Running   0
cookbook-otel-demo-frontendproxy-56cf8d9764-7ghlw              1/1     Running   0
cookbook-otel-demo-kafka-54b858c6b9-p2kx4                      1/1     Running   0
cookbook-otel-demo-loadgenerator-6795788d4-qc42l               1/1     Running   0
cookbook-otel-demo-otelcol-78c8c6554f-9vmzn                    1/1     Running   0
cookbook-otel-demo-paymentservice-78c89f7b89-lwcst             1/1     Running   0
cookbook-otel-demo-productcatalogservice-b6b5dbc65-njc79       1/1     Running   0
cookbook-otel-demo-quoteservice-64f7bdffc8-xvwsg               1/1     Running   0
cookbook-otel-demo-recommendationservice-574d94d775-j5j9l      1/1     Running   0
cookbook-otel-demo-redis-7ffc4b6f5d-f7vfj                      1/1     Running   0
cookbook-otel-demo-shippingservice-79f5bc4745-w679c            1/1     Running   0
```

Figure 10.12 – OpenTelemetry demo application pods in a running state

9. To access the frontend of the demo application, retrieve the external IP address of the frontendproxy service with the following command:

```
$ kubectl -n default get svc cookbook-otel-demo-frontendproxy
```

> **Important note**
>
> In the provided values.yaml file, we've configured the frontendproxy component as a Loadbalancer service type. Depending on your Kubernetes environment, you may need to adjust this setting. For more details, refer to the official OpenTelemetry demo documentation: https://opentelemetry.io/docs/demo/kubernetes-deployment/.

Once you have obtained the IP address of the `frontendproxy` service, open your preferred web browser and go to `http://<frontendproxy-external-IP>:8080`. You should be able to view the page represented in *Figure 10.13*:

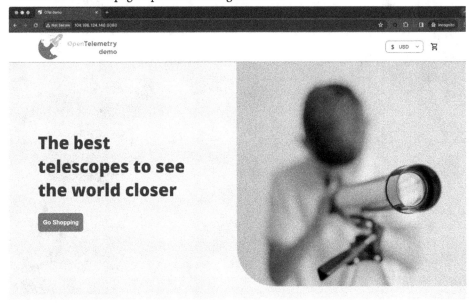

Hot Products

Figure 10.13 – OpenTelemetry Demo home page

10. To confirm that OpenTelemetry data is being forwarded to your Elastic deployment, in Kibana, go to **Observability | APM | Services**, and then click on the **Service Map** tab to observe the interactions between all the instrumented services:

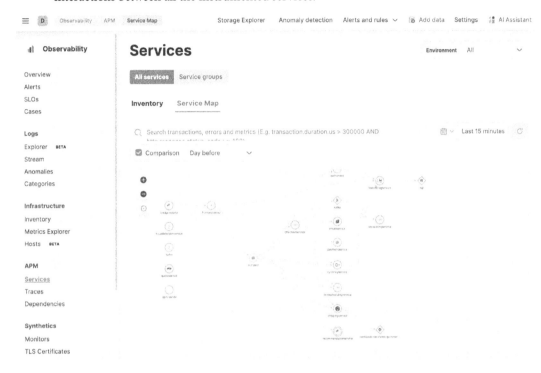

Figure 10.14 – OpenTelemetry demo application service map

How it works...

The Elastic Stack natively supports the **OpenTelemetry Protocol** (**OTLP**), which allows it to receive traces, metrics, and application logs collected by OpenTelemetry agents. These are sent to the collector, which uses the otlp/elastic exporter to forward the data to the APM Server, as shown in *Figure 10.15*:

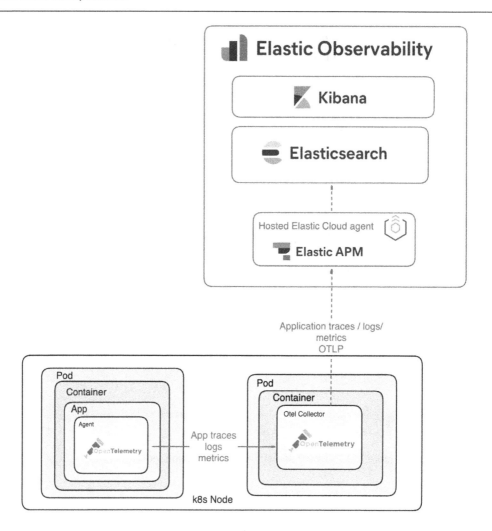

Figure 10.15 – OpenTelemetry into Elastic Observability

The configuration for this collector output is detailed in `values.yaml` under the following section:

```
config:
  exporters:
    otlp/elastic:
      endpoint: ${ELASTIC_APM_ENDPOINT}
      compression: none
      headers:
        Authorization: Bearer ${ELASTIC_APM_SECRET_TOKEN}
```

There's more...

In our recipe, the traces, metrics, and logs collected from the application are sent directly to the Elastic Stack via the OpenTelemetry collector. However, OpenTelemetry agents can also directly send their data to the Elastic Stack; this simplified setup might be useful for initial implementations or when quickly instrumenting a single application. Nonetheless, for production environments, it is considered a best practice to utilize the collector. The collector enhances performance and efficiency by processing and aggregating data before sending it to the backend monitoring systems, which reduces the load on these systems and minimizes network usage.

Elastic also offers an OpenTelemetry bridge—a combination of API/SDK and Elastic APM agents—that translates OpenTelemetry API calls into Elastic APM API calls. Currently, this functionality is available for the following technologies: Java, .NET, Node.js, and Python. The main benefits of this approach include the reusability of existing instrumentation to generate APM data and the ability to integrate various technologies.

See also

- The complete documentation on the native OpenTelemetry integration within Elastic Observability is available here: `https://www.elastic.co/guide/en/apm/guide/current/open-telemetry.html`

- If you're seeking guidance and best practices on using OpenTelemetry alongside Elastic Observability, check out the following article: `https://www.elastic.co/observability-labs/blog/best-practices-instrumenting-opentelemetry`

- For insights into how to both observe and secure OpenTelemetry-instrumented applications in a Kubernetes environment, read this detailed article: `https://www.elastic.co/blog/implementing-kubernetes-observability-security-opentelemetry`

Monitoring Kubernetes environments with Elastic Agent

In this recipe, we'll focus on implementing infrastructure monitoring within a Kubernetes environment where our application, instrumented with OpenTelemetry, is running. Our objective is to collect logs from containers and infrastructure metrics to gain insights into both application performance and the health of the underlying infrastructure. To accomplish this, we will install and configure an Elastic Agent with the Kubernetes integration. This agent will be tasked with collecting Kubernetes monitoring data and forwarding it to our Elastic Stack deployment. Once the data is gathered and transmitted, we will utilize the preconfigured Kubernetes dashboards provided by Elastic to better understand our environment.

Getting ready

Make sure that you've followed all the steps in the previous recipe, *Instrumenting and monitoring with OpenTelemetry*.

The snippets for this recipe can be found at the following link: `https://github.com/PacktPublishing/Elastic-Stack-8.x-Cookbook/blob/main/Chapter10/snippets.md#monitoring-kubernetes-environments-with-elastic-agent`.

How to do it...

The goal of this recipe is to install an Elastic Agent in the Kubernetes cluster where our application is running to collect infrastructure logs and metrics:

1. Let's start by installing Elastic Agent into the Kubernetes cluster. In **Kibana**, go to **Management | Fleet** and click on **Add agent**.

2. In the flyout, click on **Create new agent policy**. Then, name the policy `kubernetes-policy` and click on **Create policy**. Under **Install Elastic Agent on your host**, select the **Kubernetes** tab. Click on **Download Manifest** to save the manifest file to your local machine.

3. In your terminal, execute the following command to deploy the manifest in the Kubernetes cluster:

    ```
    $ kubectl apply -f elastic-agent-managed-kubernetes.yml
    ```

4. In Kibana, wait for confirmation that the agent has enrolled and is transmitting data. Check that your Elastic Agent pod is running in the `kube-system` namespace:

    ```
    $ kubectl get pods -n kube-system | grep elastic-agent
    ```

    Once the Elastic Agent is successfully enrolled and starts sending data, you will receive a confirmation on the **Add Agent** flyout page. At this point, you're ready to proceed to the next step, which involves adding the Kubernetes integration.

    Before installing the Kubernetes integration, we need to deploy `kube-state-metrics` in the cluster:

> **Important note**
> `kube-state-metrics` is a simple service that listens to the Kubernetes API server and generates metrics about the state of the objects. The Elastic Kubernetes integration depends on these metrics to populate its dashboards.

    I.   To check whether `kube-state-metrics` is running, use the following command:

    ```
    $ kubectl get pods --namespace=kube-system | grep kube-state
    ```

II.   If `kube-state-metrics` pods are not present, you can deploy them with the following commands:

```
$ git clone https://github.com/kubernetes/kube-state-metrics.git
kube-state-metrics
$ kubectl apply -f kube-state-metrics/examples/standard
```

III. After that, verify whether the `kube-state-metrics` pods are running:

```
$ kubectl get pods --namespace=kube-system | grep kube-state
```

5. Now, return to **Kibana| Management| Integrations** and search for Kubernetes. Click on the **Kubernetes** integration:

Integrations

Choose an integration to start collecting and analyzing your data.

Browse integrations   **Installed integrations**

All categories	368	Q kubernetes	
APM	1		
AWS	40	**Kubernetes**	**Kubernetes Security Posture Management (KSPM)**
Azure	24	Collect logs and metrics from Kubernetes clusters with Elastic Agent.	
Cloud	9		Identify & remediate configuration risks in Kubernetes
Containers	15		
Custom	40		

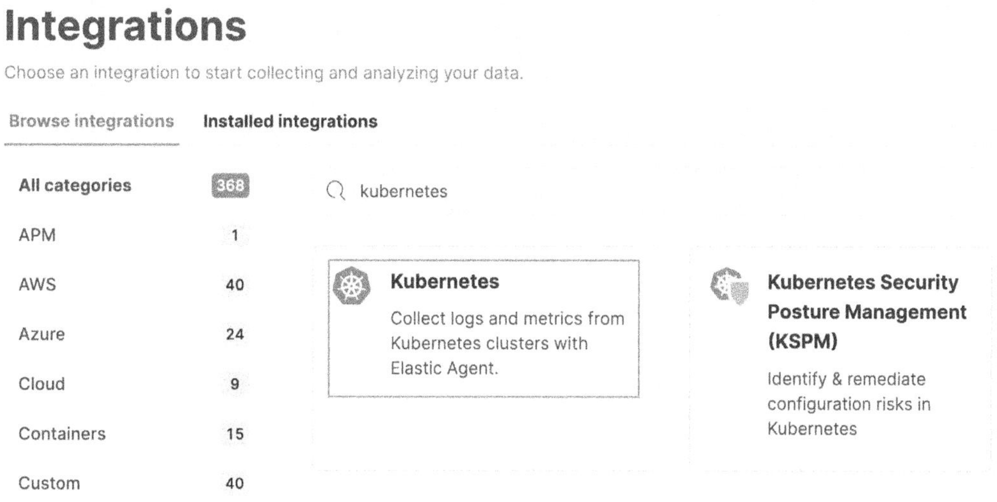

Figure 10.16 – The Elastic Agent Kubernetes integration

6. Click on the **Add Kubernetes** option that you will find in the top-right corner of the UI as shown in the following screenshot:

Figure 10.17 – Adding the Kubernetes integration to the policy

7. You can leave the default settings and scroll down to the **Where to add this integration?** section. Select **Existing hosts** and make sure the **Agent policy** field is set to kubernetes-policy, which the Elastic Agent is currently running.

Figure 10.18 – Selecting Existing hosts for Kubernetes Integration

8. Click on **Save and continue**, and then confirm by clicking on **Save and deploy changes**.

   We have now deployed the Kubernetes integration into the cluster. This integration comes with an extensive set of dashboards that provide a comprehensive view and detailed information about the cluster's activity. You can monitor metrics for clusters, pods, DaemonSets, services, and more.

9. To verify that the dashboards are populating correctly, in Kibana, go to **Analytics| Dashboard**. In the search bar, type `kubernetes overview` and select the **[Metrics Kubernetes] Cluster Overview** dashboard.

Dashboards

Q kubernetes overview

Name, description, tags

[Metrics Kubernetes] Cluster Overview
Overview of Kubernetes cluster metrics

Kubernetes   Managed

Rows per page: 20 ⌄

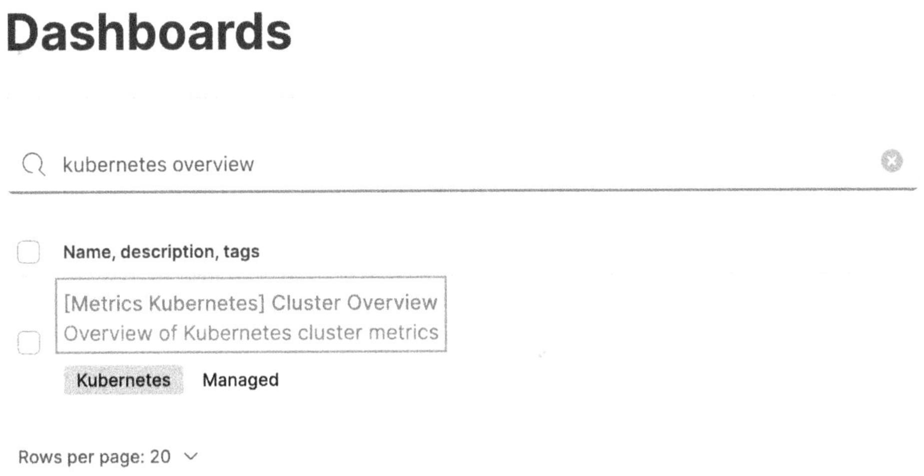

Figure 10.19 – Locating the Kubernetes Cluster Overview dashboard

Opening the dashboard will bring you to the **Cluster Overview** page with information regarding nodes, memory used and objects currently running in the cluster.

> **Note**
>
> The following screenshot shows what the Kubernetes dashboard overview looks like, giving you an idea of the type of visualization to expect when following this recipe.

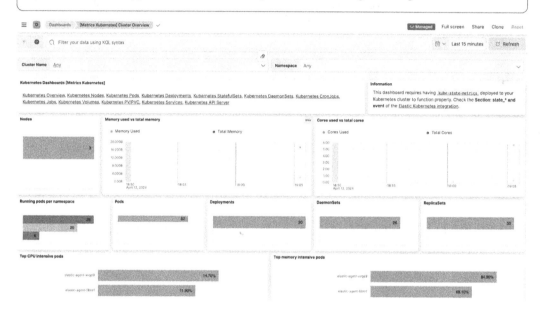

Figure 10.20 – An overview of the populated Kubernetes dashboard

Another especially useful dashboard is [**Metrics Kubernetes**] **Pods**. It gives a detailed overview of pods resource usage (CPU, memory, network) as illustrated in *Figure 10.21*:

Figure 10.21 – Kubernetes pods metrics overview

10. Finally, the Kubernetes integration also gathers container logs. To check logs streaming into Elasticsearch, head to **Observability| Logs| Stream**. You should find entries from the `kubernetes.container_logs` dataset, as shown in *Figure 10.22*:

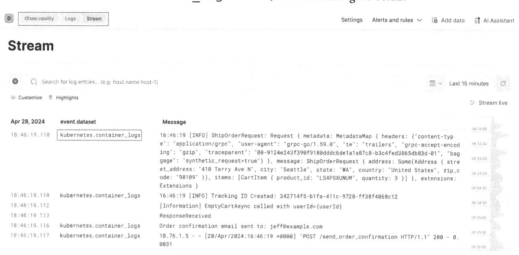

Figure 10.22 – Kubernetes container logs stream

How it works...

To collect logs and metrics from Kubernetes clusters, we deploy the Elastic Agent as a DaemonSet on each node. The agent executes the Kubernetes integration, which retrieves metrics from various components such as `kube-state-metrics`, proxy, kubelet, and the API Server. Logs are collected by accessing the default container log path used by Kubernetes. *Figure 10.23* provides an overview of the various components involved and their interactions:

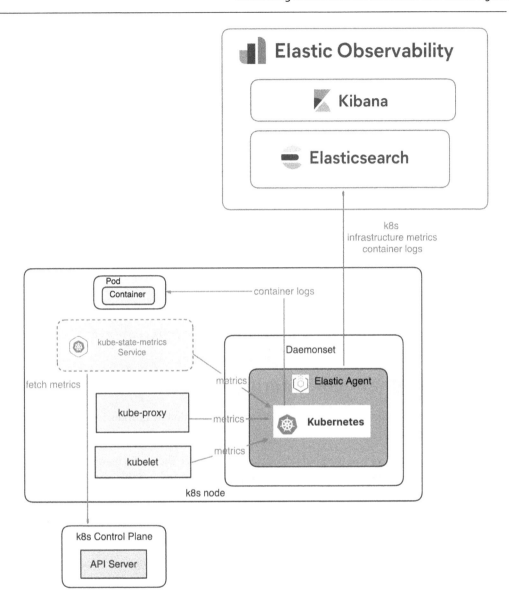

Figure 10.23 – An overview of the Kubernetes Elastic integration

The Kubernetes integration is highly configurable, allowing you to select which metrics and events you wish to gather by toggling the corresponding configurations. You can also tailor the log collection path to suit your specific environment.

There's more...

When operating on cloud-managed Kubernetes services, be aware that the Elastic Agent may not be able to access certain data sources, such as audit logs and metrics from the Kubernetes control plane (this applies to components such as `kube-scheduler` and `kube-controller-manager`).

On self-managed Kubernetes clusters, you should have access to this data if your configuration permits.

You may also opt to run the Elastic Agent in standalone mode. Although you would forgo the central management features provided by Fleet, this approach affords greater flexibility to manage and deploy your Elastic Agent as code. Moreover, it enables the Kubernetes auto-discovery feature.

See also

- The complete documentation for running the Elastic Agent on self-managed Kubernetes can be found here: `https://www.elastic.co/guide/en/fleet/current/running-on-kubernetes-managed-by-fleet.html`

- For more details on auto-discovery with the Elastic Agent on Kubernetes, see the following link: `https://www.elastic.co/guide/en/fleet/current/elastic-agent-kubernetes-autodiscovery.html`

Managing synthetics monitoring

In this recipe, we'll keep improving the observability of our application by introducing another feature: **synthetic monitoring**.

Synthetic monitoring allows you to simulate, track, and visualize the performance of critical user journeys. We will set up a browser monitor that runs periodically and use the synthetic application to analyze the results.

Getting ready

Make sure that you've followed all the steps in the previous recipe, *Monitoring Kubernetes environments with Elastic Agent*.

How to do it...

The goal of this recipe is to set up a browser monitor that simulates a user browsing for a random product on the website and placing an order:

1. In Kibana, navigate to **Observability | Synthetics | Monitors** to configure our browser monitors. Since this is our first monitor, click on **select a different monitor type** in the create monitor popup:

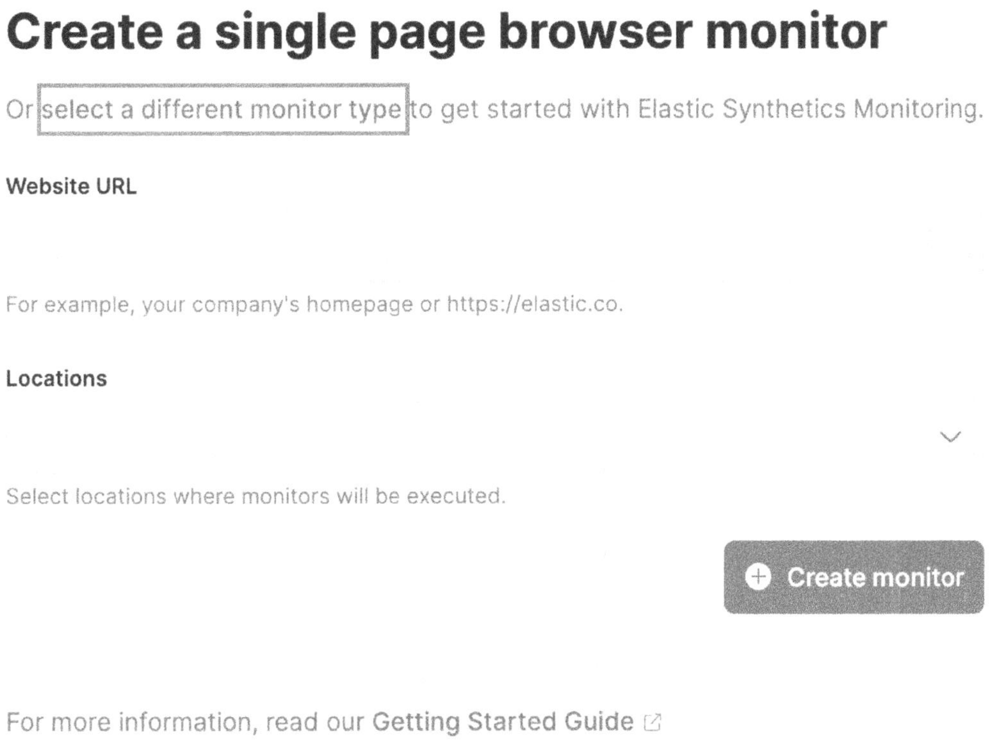

Create a single page browser monitor

Or select a different monitor type to get started with Elastic Synthetics Monitoring.

Website URL

For example, your company's homepage or https://elastic.co.

Locations

⌄

Select locations where monitors will be executed.

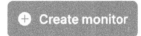 Create monitor

For more information, read our Getting Started Guide ☐

Figure 10.24 – Selecting a different monitor type

2. On the **Create monitor** page, select the **Multistep** option as the monitor type. In the **Monitor details** section, name your monitor and select some hosted locations where it will run. We'll keep the default frequency of every 10 minutes:

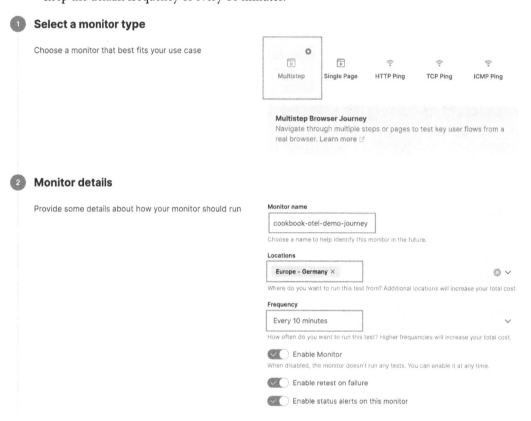

Figure 10.25 – Configuring monitor

3. In the third section of our recipe, you will incorporate the Playwright script into the editor. This script outlines the sequence of actions that the monitor will execute to simulate the behavior of a real user interacting with your application. You can copy the script source code in our GitHub repository: `https://github.com/PacktPublishing/Elastic-Stack-8.x-Cookbook/blob/main/Chapter10/snippets.md#sample-playwright-script`.

4. To add the script, select the **Script editor** tab and paste the content:

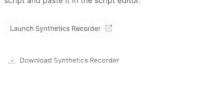

Figure 10.26 – Monitor script editor

5. The script references a my_ip variable for the frontend application's IP address. Add the corresponding value in the script parameters, as shown in the following figure's { "my_ip":"<replace-with-your-external-ip>"} script parameters:

Parameters

```
1    {"my_ip": "104.19        8"}
```

Use JSON to define parameters that can be referenced in your script with params.value

Figure 10.27 – Monitor script parameters

> **Reminder**
> To retrieve your corresponding frontend external IP, execute kubectl -n default get svc cookbook-otel-demo-frontendproxy and note the EXTERNAL-IP value.

6. Before completing the monitor setup, it's wise to run a test to confirm the configuration is accurate. To do this, click on **Run Test** and wait for the test to be completed. If the configuration is set up correctly and the test runs successfully, you should observe positive results, indicating that the monitor is ready to be deployed:

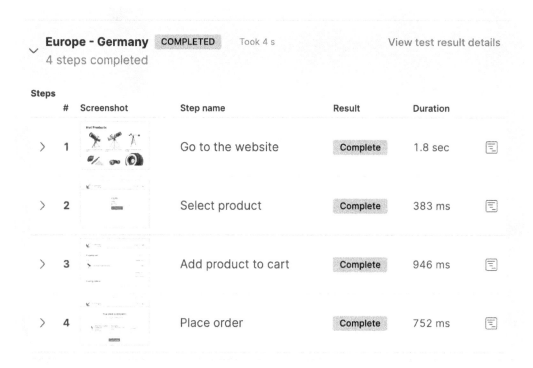

Figure 10.28 – Run test results

Go ahead and click **Create monitor**.

7. In **Monitors | Overview**, your new monitor should now be listed:

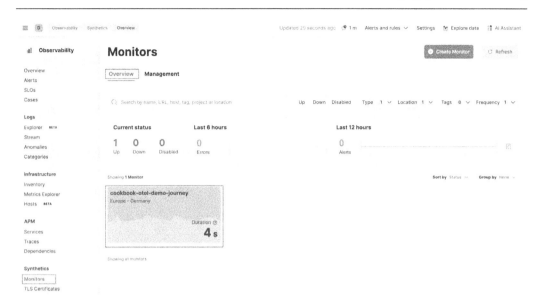

Figure 10.29 – The Monitors overview page

8.   Clicking on the monitor will open a flyout menu; from there, click on **Go to Monitor** to access the detailed page. This page provides a comprehensive overview of all the steps executed by the script, along with associated performance metrics. You'll have the opportunity to analyze duration trends and examine the breakdown of each step's duration. This detailed insight allows you to understand how each part of the process contributes to the overall performance, enabling you to identify potential bottlenecks or areas for optimization in the user journey simulated by the script:

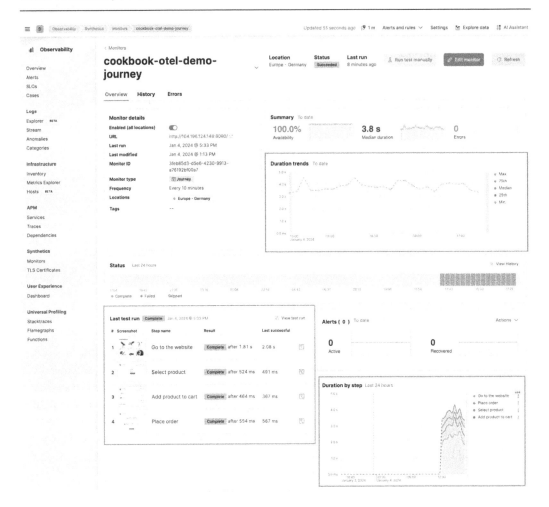

Figure 10.30 – An overview of the Monitor details page

9. If you want to view the performance breakdown of a specific step, you can go to the **Last test run** field and click on the icon next to the steps:

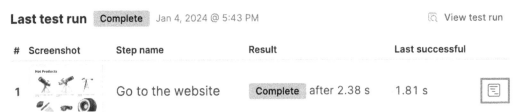

Figure 10.31 – Breakdown of a specific test

The following screenshot shows the extensive metrics captured by the synthetic monitoring agent:

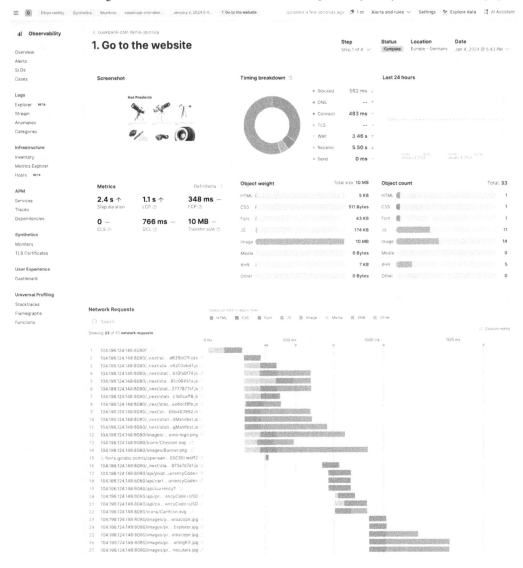

Figure 10.32 – An overview of the Step breakdown

As you will notice, some metrics such as **core web vitals**, **Largest Contentful Paint (LCP)**, **First Contentful Paint (FCP)**, and **Cumulative Layout Shift (CLS)** are also captured by the RUM agent that was covered in the *Setting up RUM* recipe. However, the synthetic monitor captures additional information such as **timing breakdown**, **object weight**, **object count**, and **network requests** associated with the waterfall listing of assets loading.

How it works...

Synthetic browser monitoring in the Elastic Stack is a powerful feature designed to simulate user interactions with web applications to monitor their performance and availability. This is achieved using Playwright scripts that mimic end-user actions such as clicking links, filling out forms, and navigating through a site. These scripts are executed at scheduled intervals from various locations, ensuring a broad and realistic overview of user experience.

Under the hood, Elastic's synthetic browser monitoring uses a **headless browser**, which is a browser without a graphical UI. This enables the scripts to run in an environment such as a real browser but is optimized for automation and monitoring tasks. The data collected by these synthetic transactions includes metrics such as page load times, response times for various web elements, and transaction paths. This data is then ingested into the Elastic Stack, where it can be visualized and analyzed using tools such as Kibana.

One of the key benefits of this approach is that it allows for the detection of issues before they impact real users. It can identify slow-loading pages, broken links, or malfunctioning features, providing insights not just into whether a site is up or down, but also into the quality of the user experience. Furthermore, because this testing is automated and can be run at regular intervals, it provides continuous monitoring, offering a more consistent and reliable way to ensure web application performance and availability.

When you're using Elastic deployments, synthetic monitors run as fully managed services in Elastic global infrastructure, meaning you don't have to provision machines and deploy agents to take advantage of it. However, in case your website operates in a non-publicly reachable location, you can set up private locations to run monitors from your own premises.

There's more...

In our recipe, we've configured the monitor by using the Synthetics application in Kibana but there is another more powerful approach called **Project monitor**. To create monitors with the Project monitor feature in the Elastic Observability solution, you can follow a streamlined process. Firstly, you define your synthetic monitors as code in a version-controlled repository. Then, by leveraging the observability integration with the Elastic Agent, you can easily deploy these monitors across your infrastructure. This approach allows for the efficient management and scaling of synthetic monitoring and seamless integration with existing continuous integration (CI)/ continuous delivery (CD) workflows.

We've created a multistep browser journey monitor, to reproduce user action, but there are other additional types of monitors available:

- **Page monitor**: This is used for loading and rendering a single page
- **Lightweight monitors**: This is used to check the status and performance of network endpoints using the following methods:

  - **HTTP**: This is used to oversee your website. It ensures that chosen endpoints are returning the appropriate status codes.

  - **Internet Control Message Protocol** (**ICMP**): This method is useful for verifying the presence of your hosts. It employs ICMP (v4 and v6) Echo requests to assess whether the hosts you're querying are reachable within the network. This informs you about the host's network availability, but it won't indicate whether a specific service on the host is operational.

  - **Transmission Control Protocol** (**TCP**): This is designed to check the services operating on your hosts. By examining individual ports, the TCP monitor verifies that a service is not only reachable but also actively running.

When creating monitors, you have alerting enabled by default. You just need to set up a connector of your choice to start receiving alerts every time your monitors are down.

The data from a synthetic monitor is stored in Elasticsearch data streams; thanks to flexible data life cycle management, you can easily reduce the amount of storage required or customize the retention for each data stream.

See also

- For detailed instructions and more information on setting up and using Project monitors, you can refer to the Elastic guide at `https://www.elastic.co/guide/en/observability/current/synthetics-get-started-project.html`

- To manage the retention of synthetic monitoring data, check the documentation `https://www.elastic.co/guide/en/observability/current/synthetics-manage-retention.html`

- To learn how you can apply synthetic monitoring on the Kibana dashboard, check out this excellent blog: `https://www.elastic.co/blog/what-can-elastic-synthetics-tell-us-about-kibana-dashboards`

Gaining comprehensive system visibility with Elastic Universal Profiling

So far in this chapter, we've seen how by unlocking key features of Elastic Observability such as APM with native OpenTelemetry support, Kubernetes infrastructure monitoring, and Synthetics monitoring, we've improved our monitoring posture. However, the complexity of this type of environment can often obscure critical insights into performance and resource utilization. This is where **Universal Profiling** becomes a game changer. This new feature in Elastic Observability marks a significant leap forward, offering deep, always-on insights into both system and application performance without the typical overhead associated with traditional profiling tools.

In this recipe, we will delve into the practical application of Universal Profiling in our Kubernetes cluster. We will guide you through setting up Universal Profiling in your Kubernetes environment, illustrating how it seamlessly integrates with Elastic Observability to offer real-time visibility into the performance deep into kernel-level code.

Getting ready

Make sure that you've followed all the steps in the previous recipe, *Monitoring Kubernetes environments with Elastic Agent*.

For Universal Profiling, there are additional system requirements:

- The workloads you are profiling must be running on Linux machines with x86_64 or ARM64 CPUs

The minimum supported kernel version is 4.15 for x86_64 or 5.5 for ARM64. As always, we recommend checking the latest documentation for the most current information on system requirements. You can find the prerequisites for Universal Profiling here: `https://www.elastic.co/guide/en/observability/current/profiling-get-started.html#profiling-prereqs`.

How to do it...

The goal of this recipe is to set up Universal Profiling in our Kubernetes cluster, providing system-wide visibility into methods, classes, threads, and containers. This will help us identify the most CPU resource-intensive workloads. Follow these steps:

1. Navigate to **Kibana| Observability| Overview**. Scroll down to **Universal Profiling** and click on **Stacktraces**; then, click on **Set up Universal Profiling**. This will bring you to the **Add profiling data** page:

Select an option below to deploy the Universal Profiling Agent.

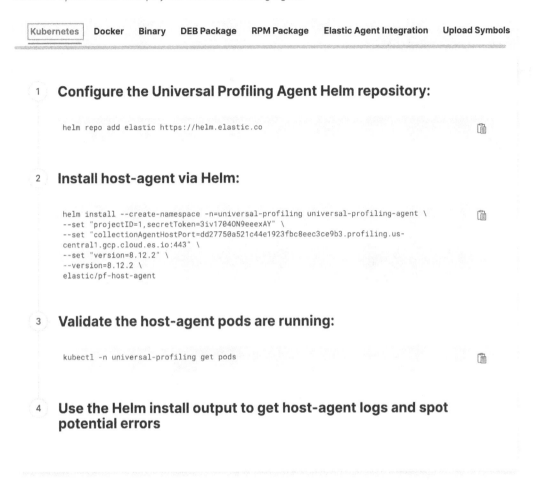

Kubernetes   Docker   Binary   DEB Package   RPM Package   Elastic Agent Integration   Upload Symbols

1 Configure the Universal Profiling Agent Helm repository:

```
helm repo add elastic https://helm.elastic.co
```

2 Install host-agent via Helm:

```
helm install --create-namespace -n=universal-profiling universal-profiling-agent \
--set "projectID=1,secretToken=3iv1704ON9eeexAY" \
--set "collectionAgentHostPort=dd27750a521c44e1923fbc8eec3ce9b3.profiling.us-
central1.gcp.cloud.es.io:443" \
--set "version=8.12.2" \
--version=8.12.2 \
elastic/pf-host-agent
```

3 Validate the host-agent pods are running:

```
kubectl -n universal-profiling get pods
```

4 Use the Helm install output to get host-agent logs and spot potential errors

Figure 10.33 – Adding Profiling data

The **Kubernetes** tab should be selected by default. Simply follow the outlined steps. Note that the install command includes the token associated with your deployment and the relevant profiling host.

2. After running the `helm install` command, you should observe the following output:

```
NAME: universal-profiling-agent
LAST DEPLOYED: Fri Jan  5 16:40:50 2024
NAMESPACE: universal-profiling
STATUS: deployed
REVISION: 1
TEST SUITE: None
NOTES:
#######################################################################################


88         88                  88                                                                         88
88         88                  ""                                                                         88
88         88                                                                                             88
88         88  8b,dPPYba,   88  8b       d8   ,adPPYba,  8b,dPPYba,  ,adPPYba,  ,adPPYYba,   88
88         88  88P'   `"8a  88  `8b     d8'  a8P_____88  88P'   "Y8  I8[    ""  ""     `Y8   88
88         88  88       88  88   `8b   d8'   8PP"""""""  88          `"Y8ba,   ,adPPPPP88   88
Y8a.     .a8P  88       88  88    `8b,d8'    "8b,   ,aa  88          aa    ]8I  88,    ,88   88
 `"Y8888Y"'   88       88  88      "8"       `"Ybbd8"'  88          `"YbbdP"'  `"8bbdP"Y8   88




88888888ba                           ad88  88  88  88
88      "8b                          d8"    ""  88  ""
88      ,8P                          88         88
88aaaaaa8P'  8b,dPPYba,   ,adPPYba,  MM88MMM  88  88  88  8b,dPPYba,    ,adPPYb,d8
88""""""""'  88P'   "Y8  a8"     "8a   88     88  88  88  88P'   `"8a  a8"    `Y88
88           88          8b       d8   88     88  88  88  88          8b       88
88           88          "8a,   ,a8"   88     88  88  88  88          "8a,   ,d88
88           88           `"YbbdP"'    88     88  88  88  88           `"YbbdP"Y8
                                                                       aa,    ,88
                                                                       "Y8bbdP"


Elastic Universal Profiling host agent is installed and configured to send data to
                          .profiling.eu-west-1.aws.cloud.es.io:443 on project ID 1.

You can check that the Pods from the DaemonSet are running properly with:

kubectl -n universal-profiling get pods -l app=profiling-agent

You can get the agent logs with:

kubectl -n universal-profiling logs -l app=profiling-agent

#######################################################################################
```

Figure 10.34 – Profiling agents install output

3. Next, run the `kubectl` command to verify that `host-agent` pods are up and running:

```
$ kubectl -n universal-profiling get pods
```

You should get the following output:

```
NAME                               READY   STATUS    RESTARTS   AGE
universal-profiling-agent-brj2z    1/1     Running   0          8m51s
universal-profiling-agent-hlmh6    1/1     Running   0          8m51s
universal-profiling-agent-q7m2z    1/1     Running   0          8m51s
```

Figure 10.35 – Universal Profiling pods up and running

4. Return to **Kibana | Observability| Universal Profiling | Stacktraces** to see the dashboard being populated by the data sent from the profiling agents. Universal Profiling offers multiple perspectives for data analysis. As we are profiling a Kubernetes environment, visit the **Containers** tab to begin. The stacktrace view is especially useful in our case, for example, to find which container is using the most CPU. To find a specific container, type this query in the search bar: `orchestrator.resource.name: cookbook-*`.

Figure 10.36 shows the stacktraces captured by the Universal Profiling agent, grouped by container:

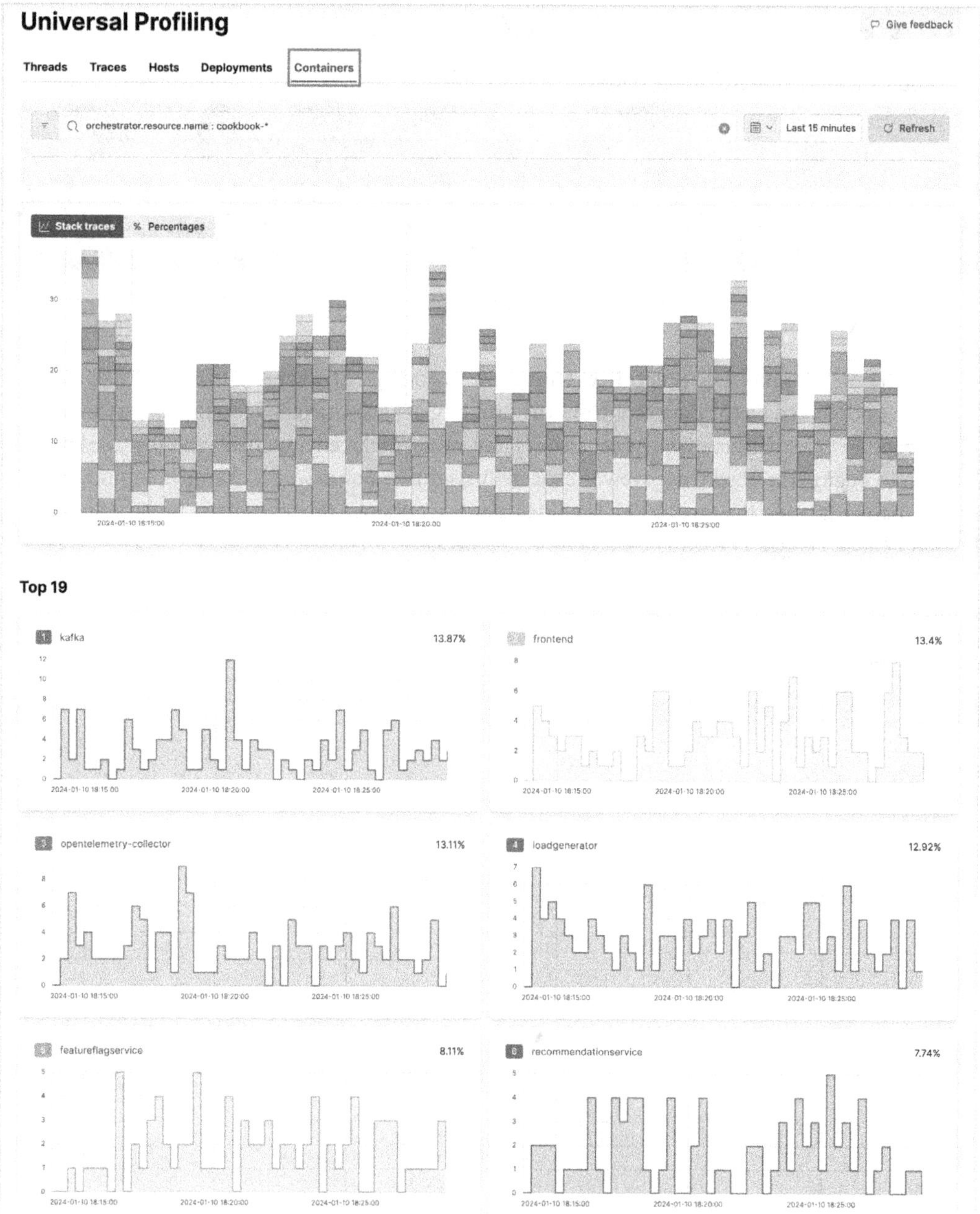

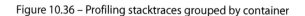

Figure 10.36 – Profiling stacktraces grouped by container

It turns out that Kafka is the most CPU-intensive container in our application, as it is displayed at the top of the results. Click on the Kafka container to view the details of the associated traces:

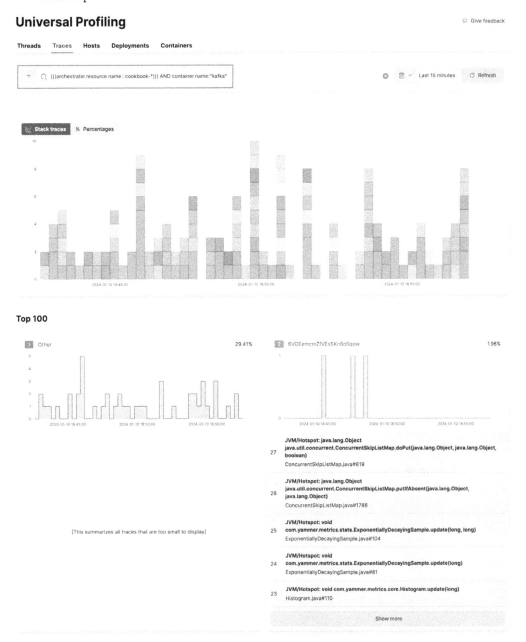

Figure 10.37 – Stacktraces – Kafka container

The percentage shown in the top-right corner of each individual graph represents the relative frequency of occurrences for every time over the total number of samples in the group. You can click on the **Show more** button at the bottom to display the full stacktrace.

5. Stack traces can sometimes be very difficult to interpret, which is why the Universal Profiling application includes a **Flamegraph** feature. Flamegraphs provide an easily understandable visual representation of hierarchical data, allowing for quick identification of the most frequent and critical code paths. Click on **Flamegraph** on the left menu:

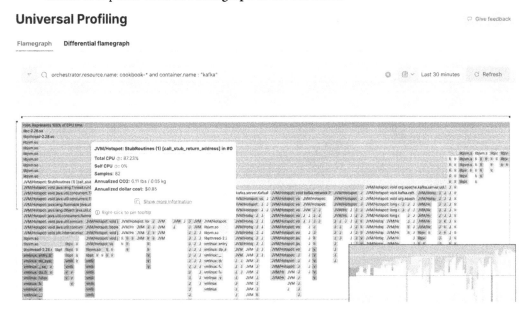

Figure 10.38 – The Flamegraph view for Kafka container stacktraces

6. To focus on a specific frame within a flamegraph, you can right-click on the tooltip associated with that frame to "pin" it in place and then click on **Show more information**. This action will provide detailed insights about the selected frame:

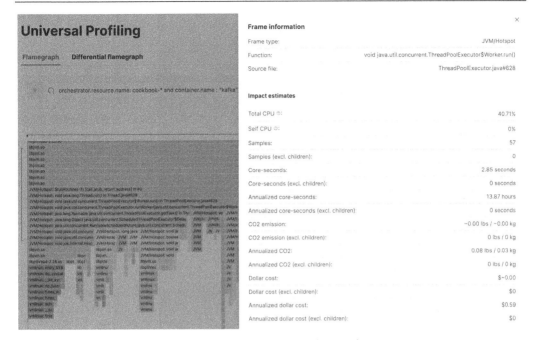

Figure 10.39 – Flamegraph detailed frame information

As you can see, valuable pieces of information are displayed, such as total CPU impact, CO_2 emissions, and dollar cost. Annualized dollar cost is particularly important as it gives an idea of what are the costliest pieces of software running in your infrastructure and where your engineers could spend time optimizing.

> **Important note**
>
> In **Flamegraph**, you might notice when you expand certain frames a `Missing symbols error` message. In the context of profiling, "symbols" refer to the human-readable names of functions and variables in your code. When you profile an application, the profiler captures raw data such as memory addresses, which are not very meaningful on their own. Symbolization is the process of converting these raw memory addresses back into the readable names of the functions and variables they represent

7.   Under the **Universal Profiling** menu located in the left sidebar, click on **Functions**:

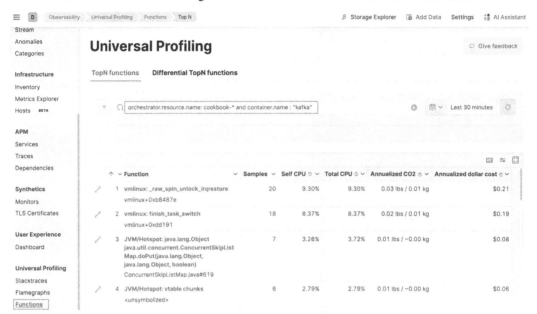

Figure 10.40 – TopN functions for Kafka

The **TopN functions** view in Universal Profiling displays the functions that are most frequently sampled, categorizing them by their CPU usage time, estimated annual carbon dioxide emissions, and projected annual costs. This feature helps in pinpointing the functions that incur the most significant resource consumption across your entire system. Also, it allows filters to zoom in on areas for an in-depth evaluation. By selecting a function name, you will be guided to the flamegraph, where you can analyze the sequence of function calls in detail.

How it works...

Universal Profiling operates through a host agent that is deployed on servers, or as a DaemonSet in Kubernetes environments. Elastic leverages the **Extended Berkeley Packet Filter (eBPF)** at the data collection level for the Universal Profiling feature by utilizing eBPF's ability to capture data directly safely and efficiently from the kernel space. This agent intelligently recognizes its deployment context, without needing any manual configuration by the user. It also seamlessly profiles all system processes without altering the functioning of any applications, interpreters, compilers, or runtimes present on the server.

What sets Universal Profiling apart from many existing solutions is its comprehensive visibility. For instance, if a Python application utilizes a C library, Universal Profiling extends its analysis beyond just the Python app to include the C library and even the OS kernel. This multi-layered approach helps uncover performance issues across the entire software stack. Universal Profiling is distinguished by its minimal system impact, requiring less than 1% CPU usage and about 250 MB of RAM. The following figure gives an overview of how Universal Profiling operates:

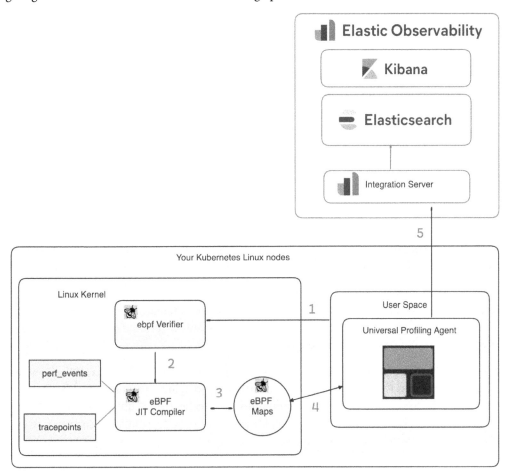

Figure 10.41 – Profiling overview architecture

Briefly, we can explain how it works in the following steps:

1. **Deployment of the unwinder eBPF bytecode**: The unwinder eBPF bytecode is dispatched to the kernel.

2. **Kernel verification and execution**: The kernel conducts a safety check on the eBPF program. Once it passes verification, the program is linked to specific probes. It then activates upon the occurrence of designated events.

3. **Data transmission to userspace**: The eBPF program gathers data and relays it to the userspace utilizing maps.

4. **Data retrieval by the Elastic Universal Profiling Agent**: The Elastic Universal Profiling agent accesses and retrieves the accumulated data, which includes stacktraces, metrics, and metadata, from the maps.

5. **Data integration into the Elastic Stack**: The collected data, encompassing stacktraces, metrics, and metadata, is then integrated into the Elastic Stack for further processing and analysis.

There's more...

In our recipe, we've learned how to deploy the Universal Profiling agent and use it in the Kubernetes environment. Profiling can be used in conjunction with APM and starting from version 8.11, you now have a **Universal Profiling** tab for each service, which shows you the flamegraph from the service's hosts so you can easily correlate with transactions and traces.

To measure the impact of changes made to the code once you've spotted some inefficiency or performance issue, you can leverage differential flamegraphs or differential TopN functions to help you identify regressions and measure your change impact not only in terms of performance but also in terms of carbon emissions and cost savings.

To view function names and line numbers in traces from applications coded in languages that compile into native code (such as C, C++, Rust, Go, etc.), it's necessary to upload symbols to the Elastic Stack. You can do that by using the `symbtool` **command-line interface** (**CLI**) utility. Elastic also hosts a public service containing symbol information for popular Linux distributions. You can find the complete list on the official Elastic documentation page.

See also

*   The full documentation for Universal Profiling is available on the official Elastic products guide page: `https://www.elastic.co/guide/en/observability/current/universal-profiling.html`

*   If you want to delve deeper into how you can leverage Universal Profiling to optimize your applications, check out this blog: `https://www.elastic.co/blog/whole-system-visibility-elastic-universal-profiling`

Detecting incidents with alerting and machine learning

In the previous recipes, we deployed the official OpenTelemetry Demo on GKE and collected different observability data from the application such as traces, metrics, logs, profiling, and Synthetics. In this recipe, we will discover how Elastic Observability can assist in detecting incidents in real time using its correlation capabilities, along with alerting and machine learning features within the Elastic Observability app.

Getting ready

Make sure that you have completed the previous four recipes:

- *Instrumenting and monitoring with OpenTelemetry*
- *Monitoring Kubernetes environments with Elastic Agent*
- *Managing synthetics monitoring*
- *Gaining comprehensive system visibility with Elastic Universal Profiling*

How to do it...

The goal of this recipe is to first set up alerting and machine learning jobs within the Elastic Observability app and simulate application incidents on the deployed official OpenTelemetry Demo. We will then see how we can detect those incidents and find the root cause of the incidents:

1. Let's start with configuring alerting rules for the APM services. In Kibana, go to **Observability | APM | Services**, and then on the top right menu bar, click on **Alerts and rules | Create threshold rule | Latency**. This will bring up a flyout for the rule creation form as shown in *Figure 10.42*. You can keep the default values and can add optionally an action such as **Email** at the bottom of the flyout. Create the rule by clicking on the **Save** button:

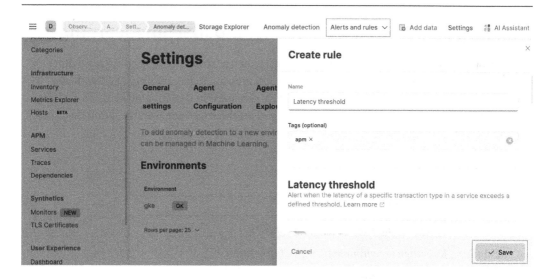

Figure 10.42 – Creating APM latency threshold rule

2. From the top right menu bar, click on **Alerts and rules | Create threshold rule | Failed transaction**, and then create the rule following the same steps as for the latency rule.

3. Now, let's prepare for the machine-learning-based anomaly detection job for the service latencies. From the top-right menu bar, click on **Anomaly detection** and then click on the **Create job** button. Complete the job creation by choosing **gke** from the environments list and clicking on the **Create Jobs** button as shown in *Figure 10.43*:

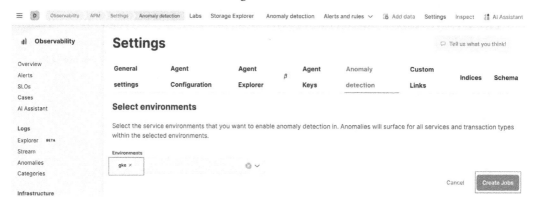

Figure 10.43 – Creating APM latency anomaly detection job

4. Once the anomaly detection job is created, from the top-right menu bar, click on **Alerts and rules | Create anomaly rule**, leave the default values, and create the rule following the same steps as for the latency rule.

5. Now, let's create rules and anomaly detection jobs related to logs. Go to **Observability | Logs | Stream**, and then on the top-right menu bar, click on **Alerts and rules | Create rule**; this will bring up a rule creation flyout. Name the rule `Log errors threshold` then adjust the error count condition to **more than 5** and group by `service.name`, as shown in *Figure 10.44*. You can optionally create an action such as **Email** and click on the **Save** button to create the rule:

Figure 10.44 – Creating a Log errors threshold rule

6. Once the static alerting rule based on **Log errors threshold** is created, let's also activate the machine learning anomaly detection for logs: go to **Observability | Logs | Anomalies**, then enable both the **Log rate** anomaly detection and **Categorization** anomaly detection by clicking on the **Enable anomaly detection** button, as shown in *Figure 10.45*, and then follow the wizard by leaving the default values to create both jobs.

> **Note**
> You will need to enable one of them first and then return to enable the second one.

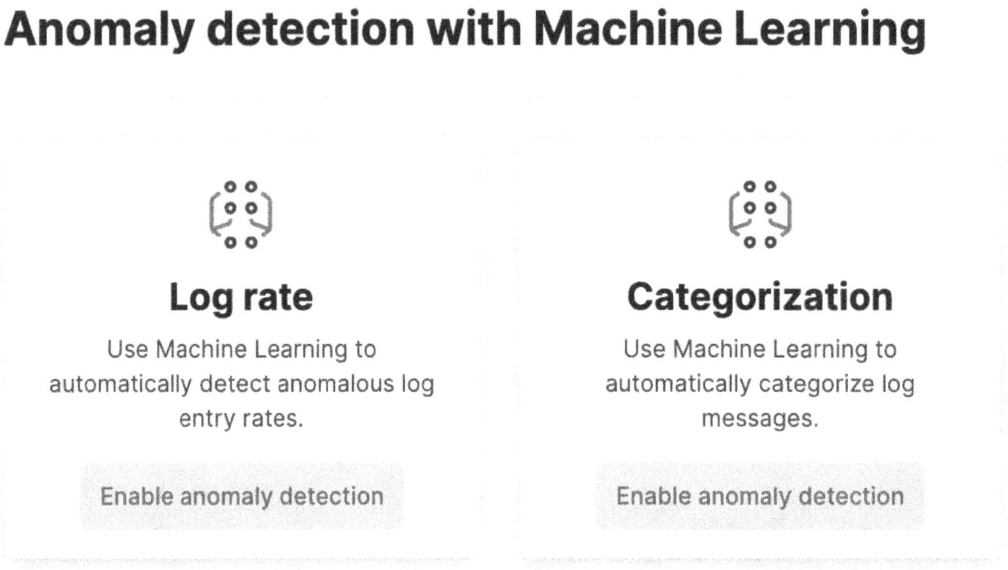

Anomaly detection with Machine Learning

Log rate

Use Machine Learning to automatically detect anomalous log entry rates.

Enable anomaly detection

Categorization

Use Machine Learning to automatically categorize log messages.

Enable anomaly detection

Figure 10.45 – The Log rate and Categorization anomaly detection

7. We can now simulate some application errors by using the OpenTelemetry Demo's feature flags. In your browser, go to `http://{serviceip}:8080/feature/featureflags/productCatalogFailure/edit` (in the *Instrumenting and monitoring with OpenTelemetry* recipe, you can find guidance on how to locate `serviceip`), set **Enabled** to `1.0`, and then click on the **Save** button:

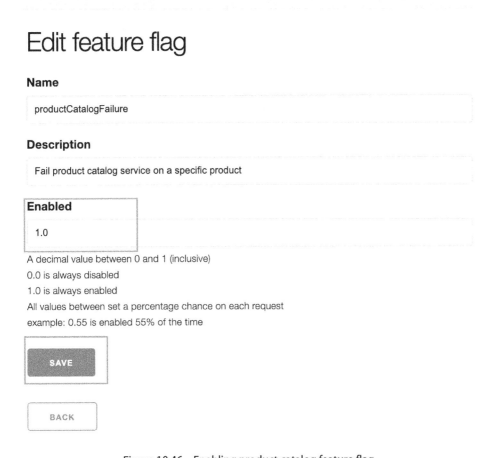

Figure 10.46 – Enabling product catalog feature flag

8. The feature flag is now periodically generating an error for a specific product from the product catalog. Wait 10 minutes so that the data can be populated, and then, in Kibana, go to **Observability | Alerts**, where you can see that an alert has been generated from the frontend service. Click on the error so that you can inspect the alert detail in the flyout, as shown in *Figure 10.47*:

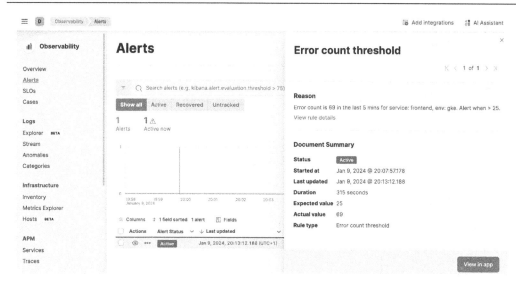

Figure 10.47 – Alert inspection

9. Let's try to have a better understanding of the alert. Go to **Observability | APM | Services** and choose the time range to **Last 15 minutes**. From this view, we can see the correlation between different services (`frontend`, `frontend-proxy`, `productioncatalogservice`, and `loadgenerator`) on **Failed transaction rate**, as shown in *Figure 10.48*:

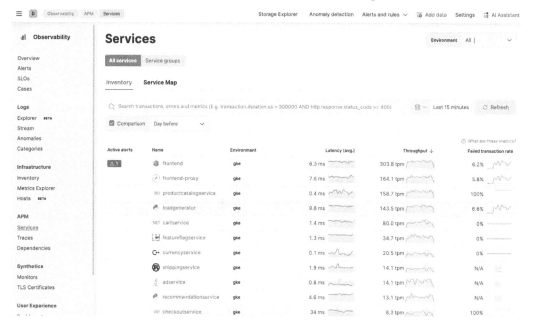

Figure 10.48 – Service list view correlation

10. Now, you can click on the **frontend** service – this brings you to the service home page for the frontend service. Click on **GET** from the **Transactions** panel, as shown in *Figure 10.49*:

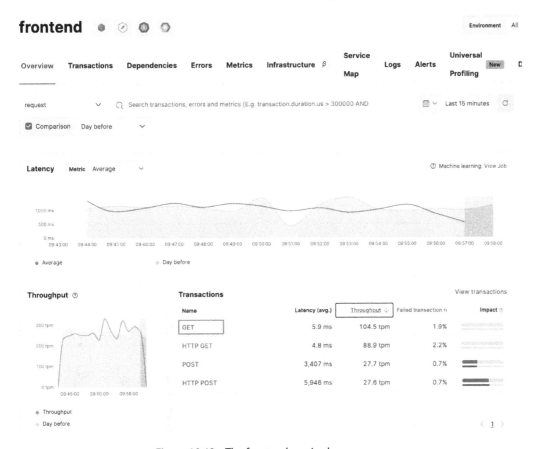

Figure 10.49 – The frontend service home page

11. Once on the transaction's details page, scroll down and click on the **Failed transactions correlations** tab, as shown in *Figure 10.50*. We can see that the failed product is identified:

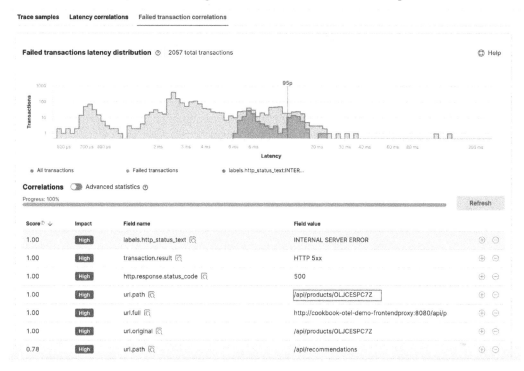

Figure 10.50 – The frontend service – Failed transaction correlations

12. We can now try a more complex scenario with another feature flag from the OpenTelemetry Demo, which simulates the memory leak due to an exponentially growing cache from the recommendation service. In your browser, go to `http://{serviceip}:8080/feature/featureflags/recommendationCache/edit`, set **Enabled** to 1, and then click on the **Save** button:

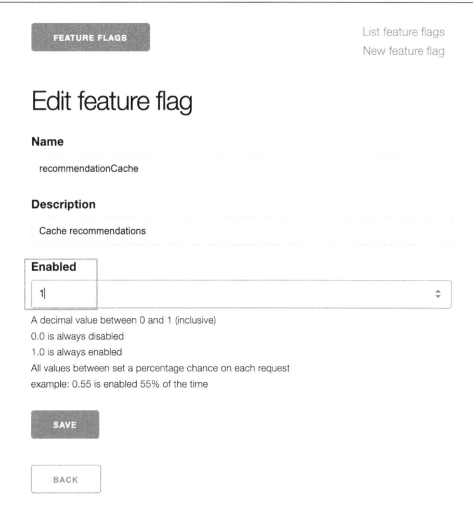

Figure 10.51 – Enabling the recommendation cache feature flag

13. This time, wait for a longer time (more than 30 minutes) so that the memory leaks can take their effects and impact other services. Then, in Kibana, go to **Observability | Alerts**. Set the time range to **Last 1 hour**. You can notice that there are multiple alerts generated by the APM service threshold rule, APM anomaly detection rule, and the monitor status rules that have been created, along with the synthetic monitor (the synthetic monitor has been created in the previous recipe, *Managing synthetics monitoring*, of this chapter):

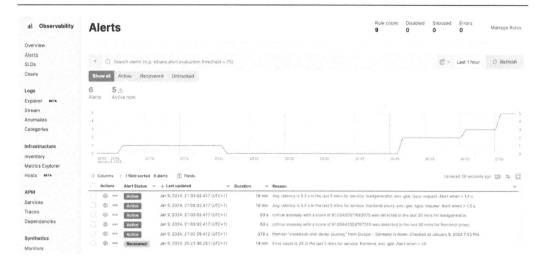

Figure 10.52 – Recommendation feature-flag-generated alerts

14. To get a better understanding of the root cause, let's go to **Observability | APM | Services**. You can see the latency trend correlation among different services. In the **Health** column, service latency machine learning anomaly detection assigns the health indicator for different services within the time range:

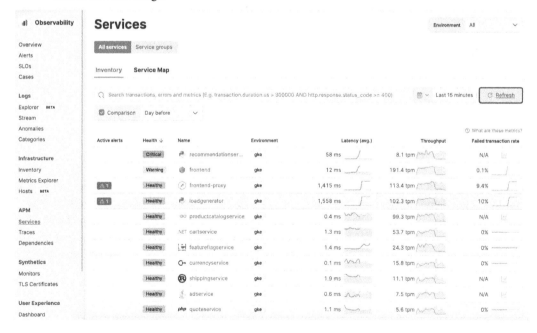

Figure 10.53 – Root cause analytics with service view

15. You can click on the **frontend** service again and then click on **GET** from the **Transactions** panel, as we did for *Step 10*. Once on the transaction's details page, scroll down and click on the **Latency correlations** tab, as shown in *Figure 10.54*. We can observe that the root cause of the latency has been clearly identified, with the recommendation service being pinpointed as having a significant impact on latencies:

Figure 10.54 – Latency correlations

16. Let's now try to invest more in the recommendation service. Go to **Observability | APM | Service** and click on **recommendationservice**. Click on the **Infrastructure** tab, and then click on the **cookbook-otel-demo-recommendationservice-5cb...** pod, as shown in *Figure 10.55*:

Figure 10.55 – The Infrastructure view for recommendation service

17. You can see the detailed metrics for the recommendation service, which clearly show the increased CPU and memory usage that caused the pod to restart, as shown in *Figure 10.56*:

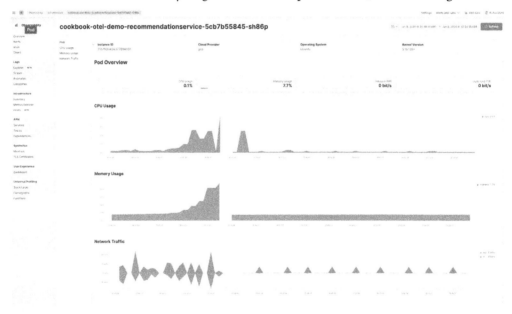

Figure 10.56 – Recommendation service metrics

18. For the correlation with logs, navigate to **Observability | Logs | Categories**. Here, you will find that similar log entries have been categorized, and the machine learning anomaly detection job indicates the trends and anomaly scores for each category. By setting the time range to **Last 1 hour**, we can observe that during the recommendation service feature flag, a considerable number of cache misses and cache hits have occurred, as shown in *Figure 10.57*:

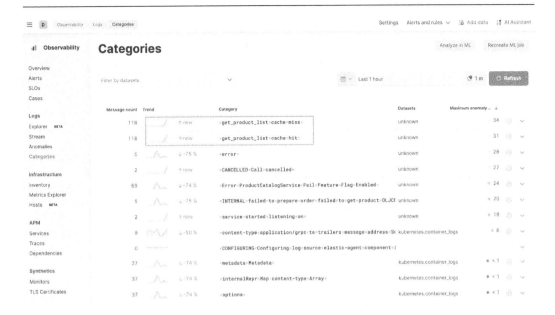

Figure 10.57 – Log entry correlations

How it works...

In this recipe, we leveraged three key features of Elastic Observability that helped us with troubleshooting. Let's discuss them in detail.

Integrated alerting

Alerts and rules are integrated directly into the **Observability** app. In our recipe, we set up rules in APM and logs. The other native apps, including Infrastructure, Uptime, and service-level objective (SLO), also provide native integration with the alerting framework. Elastic Observability provides a unified alerting view and lists all the alerts on one page. This centralized view facilitates the monitoring of alert patterns and the correlation of various alerts.

Integrated machine learning

Elastic Observability integrates machine learning anomaly detection. In our recipe, we configured machine learning to monitor service latency and analyze the rate of log entries. That enabled us to understand the behavior of our observability data and easily identify anomalies while reducing false positives.

Additionally, we also leveraged log categorization, which helped us identify similar log patterns and track their trends to reduce the mean time to resolution.

In-app correlation

For both feature flags in our recipe, we started our investigation by identifying the most impacted services and transactions. Elastic Observability provides built-in correlation functionality that compares the attributes associated with high latencies and errors against the complete set of transactions, pointing out the metadata that is notably prevalent in failing or slow transactions. It helps us to easily narrow down to specific groups of transactions; we can then jump to relevant UI to drastically reduce the investigation time.

There's more...

While the features highlighted in this recipe provide powerful tools for addressing immediate issues, Elastic Observability does not stop there. Its capabilities extend to include strategic planning and collaborative problem-solving functionalities, enabling teams to not only respond to current problems but also to prevent future ones.

Service level objectives

In addition to alerts, Elastic stack 8.x introduces **service-level objectives** (**SLOs**), which allow us to define precise goals for our service's performance, considering aspects such as uptime, latency, error rate, and other custom metrics. This can be useful for determining the severity of incidents and helps continuously improve the service's quality. The full documentation for SLOs is available at this address: `https://www.elastic.co/guide/en/observability/current/slo.html`.

Cases

Troubleshooting usually requires teamwork; this is the reason why creating cases and enriching them with alerts, comments, and visualizations can be useful to track the investigation process and ensure clear communication and documentation of each step taken toward resolution. Elastic Observability natively provides case management; the detailed documentation on case management is available at the following link: `https://www.elastic.co/guide/en/observability/current/create-cases.html`.

See also

- The whole list of the alerting integration for Elastic Observability can be found here: `https://www.elastic.co/guide/en/observability/current/create-alerts.html#create-alerts-rules`

- To explore more on APM correlation in Elastic Observability, you can check out this blog post: `https://www.elastic.co/blog/apm-correlations-elastic-observability-root-cause-transactions`

Gaining insights with the AI Assistant

In the previous recipe, we explored methods for detecting incidents and correlating different data points to facilitate investigation and resolution. In this recipe, we will see how we can leverage the Elastic AI Assistant to connect with **large language models** (**LLMs**), allowing us to get additional insights and go deeper into the investigation.

Getting ready

Make sure that you have completed the previous recipe, *Detecting incidents with alerting and machine learning*.

For this recipe, you will need either an OpenAI GPT-4 account or an Azure account with GPT-4 or GPT-4-32k with the minimum API version 2023-07-01-preview. The dialog results with the AI Assistant can vary depending on the API version and Elastic Stack version that you are using.

How to do it...

In this recipe, we will first configure the AI Assistant connector. Following that, we will resume the investigation of the feature flags of the previous recipe (product catalog error and recommendation memory leak), this time using the AI Assistant:

1. We will begin by creating the AI Assistant connector. In Kibana, navigate to **Management | Stack Management | Connectors**, click on **Create connector**, and opt for **OpenAI** from the list of available connectors. This will bring up a flyout – give a name to the connector, choose the OpenAI provider of your choice (either OpenAI or Azure OpenAI Service), and then enter the URL and API key. Click on the **Save and test** button to save and validate the connector configuration.

2. Let's now focus on using the AI Assistant to further examine the feature flags from our earlier recipe. We'll start by reviewing the errors caused by the feature flag product catalog. Go to **Observability | APM | Services**. Click on the **frontend** service, choose the **Errors** tab, and then locate and click on the `13 INTERNAL: Error. ProductionCatalogService Fail Feature Flag Enable` error, as shown in *Figure 10.58*:

Figure 10.58 – Frontend service errors

3. Once on the error details page, click on **What's this error?**. The AI Assistant will start generating insights, as shown in *Figure 10.59*. Interestingly, the AI Assistant gives extremely useful explanations on feature flags and offers in-depth code-level observations and troubleshooting suggestions:

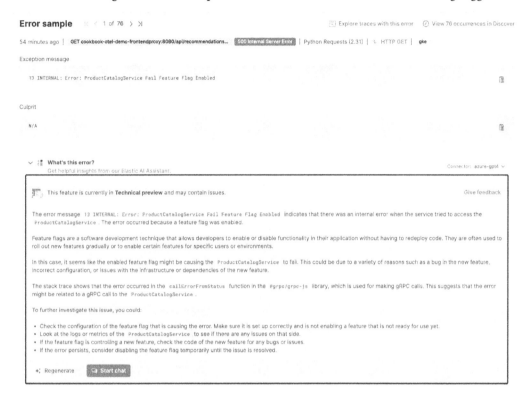

Figure 10.59 – Code error explanation with the AI Assistant

4. Now, let's switch to the second feature flag scenario we enabled for the recommendation service. At the end of the previous recipe, we investigated the *cache miss* log category identified by the machine learning job. In Kibana, go to **Observability** | **Logs** | **Categories**, find the category labeled get_product_list*cache*miss, and click on the downward arrow symbol on the right. This action will display the log entries associated with this category. From here, you can locate the log entry from the kubernetes.container_logs dataset, click on ... on the right, and then click on **View in stream** from the drop-down menu, as indicated in *Figure 10.60* (you can also find these log entries by navigating to **Observability** | **Logs** | **Stream**, and then searching for cache miss):

Figure 10.60 – The log Categories contextual menu

5. You are now redirected to the log stream UI and the log entry is highlighted. From here, click on ... on the right and choose **View details** from the drop-down menu, as shown in *Figure 10.61*:

Figure 10.61 – Viewing the log message in detail

6. This action brings up the log entry's detail flyout. You can use the AI Assistant to provide an explanation for the log message by clicking on **What's this message?**. As shown in *Figure 10.62*, the AI Assistant gives us details on the metadata and indicates that if there is a high rate of cache misses, further investigation would be necessary. Click on **Start chat** to do so.

Details for log entry TkCy74wBjRyPPR_Kej60
From index .ds-logs-kubernetes.container_logs-default-2024.01.03-000001

Figure 10.62 – Log message explanation with the AI Assistant

7. Once the chat is started, let's ask the question: how much did it impact gke services?, as you can see in *Figure 10.63*. You will see that the AI Assistant can extract relevant information from Elastic Observability. This includes service health indicators derived from the service latency machine learning job. It gives us valuable insight, highlighting latency issues within the recommendation service and establishing a link between cache misses and these latency problems.

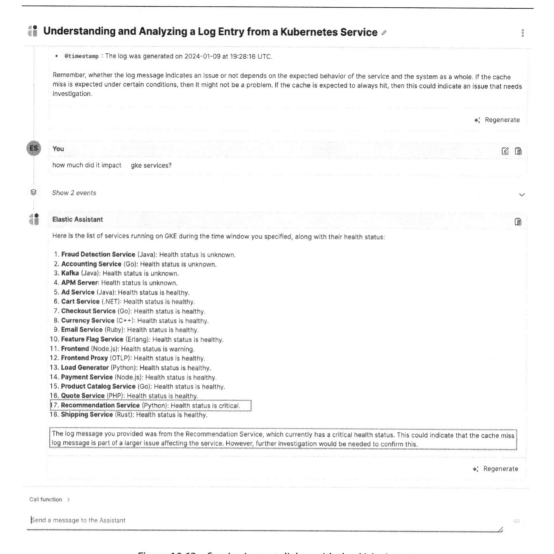

- **@timestamp** : The log was generated on 2024-01-09 at 19:28:16 UTC.

Remember, whether the log message indicates an issue or not depends on the expected behavior of the service and the system as a whole. If the cache miss is expected under certain conditions, then it might not be a problem. If the cache is expected to always hit, then this could indicate an issue that needs investigation.

+ Regenerate

You

how much did it impact   gke services?

Show 2 events

Elastic Assistant

Here is the list of services running on GKE during the time window you specified, along with their health status:

1. **Fraud Detection Service** (Java): Health status is unknown.
2. **Accounting Service** (Go): Health status is unknown.
3. **Kafka** (Java): Health status is unknown.
4. **APM Server**: Health status is unknown.
5. **Ad Service** (Java): Health status is healthy.
6. **Cart Service** (.NET): Health status is healthy.
7. **Checkout Service** (Go): Health status is healthy.
8. **Currency Service** (C++): Health status is healthy.
9. **Email Service** (Ruby): Health status is healthy.
10. **Feature Flag Service** (Erlang): Health status is healthy.
11. **Frontend** (Node.js): Health status is warning.
12. **Frontend Proxy** (OTLP): Health status is healthy.
13. **Load Generator** (Python): Health status is healthy.
14. **Payment Service** (Node.js): Health status is healthy.
15. **Product Catalog Service** (Go): Health status is healthy.
16. **Quote Service** (PHP): Health status is healthy.
17. **Recommendation Service** (Python): Health status is critical.
18. **Shipping Service** (Rust): Health status is healthy.

The log message you provided was from the Recommendation Service, which currently has a critical health status. This could indicate that the cache miss log message is part of a larger issue affecting the service. However, further investigation would be needed to confirm this.

+ Regenerate

Call function >

Send a message to the Assistant

Figure 10.63 – Service impact dialog with the AI Assistant

8.   Proceed with the conversation by asking the AI Assistant, do you have any suggestions for further investigation?. As shown in *Figure 10.63*, the AI Assistant begins by clarifying the cause of the latency anomaly and then recommends examining related alerts and service metrics. These suggestions are pertinent in our context.

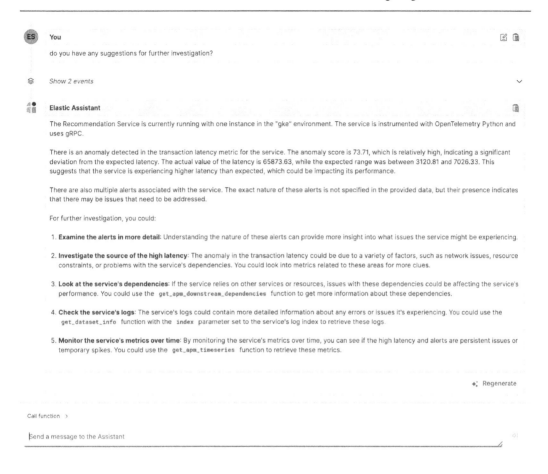

Figure 10.64 – Investigation dialog with the AI Assistant

How it works...

In this recipe, we used the AI Assistant, which is capable of providing insights within the context of the observability data from our OpenTelemetry Demo application. Behind the scenes, it uses Elastic's semantic search engine to retrieve information from its knowledge base index, enabling it to generate responses through **retrieval-augmented generation** (**RAG**), as explored in *Chapter 9*. *Figure 10.65* illustrates how it works.

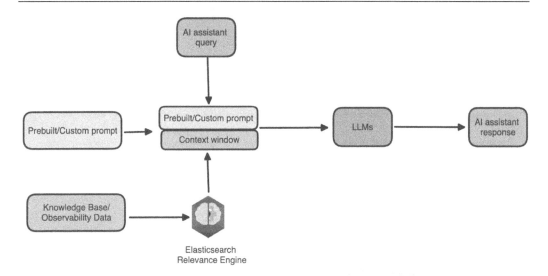

Figure 10.65 – Investigation dialog with the AI Assistant

There's more...

We explored some of the AI Assistant integrations with Elastic Observability, such as explaining errors in APM and interpreting log messages.

The additional integrations include the following:

- Explaining the processes in infrastructure UI
- Interpreting log rate changes in alerting
- Identifying resource-intensive libraries and functions in Universal Profiling

The AI Assistant can leverage other knowledge bases such as GitHub issues, and Site Reliability Engineering (SRE) runbooks. The full documentation of the data ingestion to the knowledge base can be found here: `https://www.elastic.co/guide/en/observability/current/obs-ai-assistant.html#obs-ai-add-data`.

See also

More examples of the Elastic AI Assistant and Observability can be found in this blog post: `https://www.elastic.co/blog/context-aware-insights-elastic-ai-assistant-observability`.

11

Managing Access Control

In previous chapters, we explored various capabilities of the Elastic Stack through different use cases, including search, data analytics, and observability. For simplicity's sake, we have utilized the default **elastic** admin user for all recipes performed in this book so far. However, such practice is not recommended in real-world scenarios. In this chapter, we will delve into managing authentication and authorization within the Elastic Stack and demonstrate how to implement these security practices in concrete scenarios.

In this chapter we're going to cover the following recipes:

- Using built-in roles
- Defining custom roles
- Granting additional privileges
- Managing and securing access to Kibana spaces
- Managing access with API keys
- Configuring single sign-on
- Mapping users and groups to roles

Technical requirements

To follow the recipes in this chapter, you'll need an Elastic Stack deployment that includes the following:

- Elasticsearch for searching and storing the data
- Kibana for data visualization and Sack Management (Kibana user with **All privileges** granted on Fleet and Integrations)

Additionally, it's essential to have data ingested for different use cases within your deployment. Make sure you have completed the recipes in *Chapters 5, 6, and 10*.

Snippets for different recipes in this chapter are available at the following link: `https://github.com/PacktPublishing/Elastic-Stack-8.x-Cookbook/blob/main/Chapter11/snippets.md`.

Using built-in roles

As an essential aspect of any robust data management system, **role-based access control** (**RBAC**) in the Elastic Stack is pivotal for regulating access and ensuring the integrity and confidentiality of your data. Throughout this book, we have ingested various types of data into Elasticsearch, designed dashboards and visualizations, and configured machine learning jobs and alerts. Now, imagine you want to share this productive setup with friends or colleagues to showcase the Elastic Stack's capabilities. It is essential to give them access to Kibana and your data without compromising or altering the existing configuration.

In this recipe, we will be looking at how you can take advantage of the built-in Elasticsearch roles to easily give access while maintaining a strong security posture.

Getting ready

Make sure you have completed the following recipes:

- *Monitoring Apache HTTP logs and metrics using the Apache integration* in *Chapter 4*
- *Creating and using Kibana dashboards* in *Chapter 6*

How to do it...

The objective is to utilize a predefined role within our Elastic Stack deployment to efficiently grant a third-party user access to our data. The chosen role should be restrictive, ensuring that the user's permissions are limited to prevent any unintentional modifications or disruptions. While the concept of adding security measures might seem daunting to some users, the Elastic Stack's user-friendly interface simplifies this process, demonstrating that enforcing security does not have to be a complex task.

Follow these steps:

1. Head to **Kibana | Stack Management | Security | Users** to create a new user. Click on **Create user** to open the user creation form, and enter the following details:

   - **Username**: `cookbook_reader`
   - **Full name**: `Cookbook Reader`
   - **Email address**: (you can leave it blank)

- **Password**: Choose a secure password that meets the minimum length of 6 characters, and then confirm the password

- **Roles**: Select the built-in **viewer** role

Make sure the information is correct and it resembles the example shown in *Figure 11.1*:

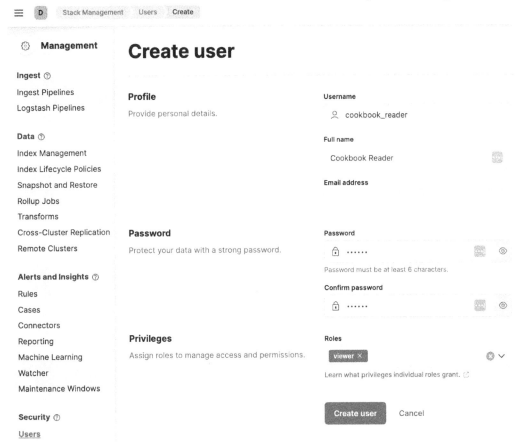

Figure 11.1 – Creating a user

Once you've reviewed the form and everything looks correct, click on **Create user**.

2.   After creating the user, they will be listed in the **Users** section, as illustrated in *Figure 11.2*:

Figure 11.2 – Verifying the created user

3.   Now, let's log in as the new user, `cookbook_reader`, to verify that the **viewer** role satisfies our requirements. To switch users, click on your initial in the upper-right corner and select **Log out** from the pop-up menu:

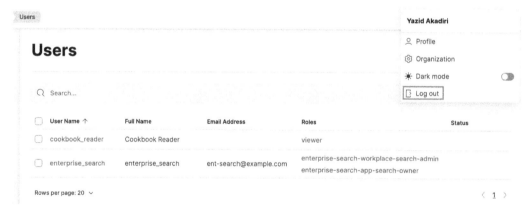

Figure 11.3 – Logging out to switch users

4.   Upon logging out, you will be redirected to the initial welcome screen, where two authentication options are presented: **Log in with Elasticsearch** and **Log in with Elastic Cloud**. We will use the **Log in with Elasticsearch** method, as the `cookbook_reader` user credentials are stored within Elasticsearch. The **Log in with Elastic Cloud** option, which incorporates single sign-on mechanisms, will be discussed in the *Configuring single sign-on* recipe.

5. After selecting **Log in with Elasticsearch**, enter the credentials of the `cookbook_reader` user to log back in. You're now connected as a user with a **viewer** role. What does that imply? Let's find out.

6. To quickly spot the difference and confirm that you're effectively connected as a user with a more restrictive profile, navigate to **Stack Management**. You'll observe significantly fewer options than those available to a superuser:

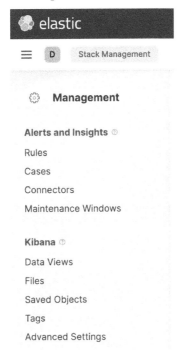

Figure 11.4 – Stack management with the viewer role

7. This is reassuring, but let's examine actual data and dashboards. Open a simple dashboard by typing `[Metrics System] Overview` in the search bar at the top. Click on the first result to access the dashboard, as shown in *Figure 11.5*:

Figure 11.5 – Searching for the Metrics Overview dashboard

8. On the dashboard page, you can click on a hostname to see an overview and various visualizations. Notice that some actions are absent, such as the **Clone** or **Edit** buttons, in the top-right menu. *Figure 11.6* shows the difference between a superuser (upper screenshot) and a user with the **viewer** role (lower screenshot):

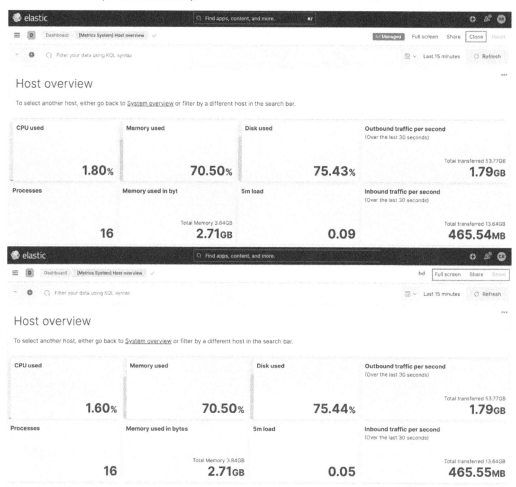

Figure 11.6 – Dashboard menu differences between superuser and viewer

We have successfully created a user and assigned them the built-in **viewer** role. This role's restrictive nature has been observed, yet it still permits access to the data and the various assets developed in Kibana and Elasticsearch. This serves as an effective demonstration of how built-in roles can be employed to enhance the security of your cluster.

How it works...

In Elastic Stack, built-in roles are predefined sets of permissions that govern user access to various features and data within the cluster. These roles, integral to the security framework, determine what users can see and do, ranging from read-only access to full administrative privileges. Each role encompasses a specific scope of permissions for accessing indices, managing cluster operations, or utilizing Kibana features.

By assigning these roles to users, administrators can effectively control access to the cluster, ensuring that each user has the appropriate level of access aligned with their responsibilities, thereby maintaining the overall security and integrity of the Elastic Stack environment.

Leveraging built-in security roles in Elastic Stack 8 offers several key advantages:

- **Enhanced security and compliance**: Built-in roles provide a structured and secure way to manage access, ensuring that only authorized users can perform specific actions. This approach aligns well with compliance requirements, offering a clear audit trail of who has access to what.

- **Simplified access management**: With predefined roles, administrators can quickly assign appropriate access levels to users based on their job functions. This simplifies the process of managing user permissions and reduces the risk of human error in assigning granular access rights.

- **Scalability and efficiency**: Using built-in roles allows for easier scaling of security protocols as the organization grows. It streamlines user onboarding, as assigning a role is quicker than configuring individual permissions, thus enhancing overall efficiency in managing a large user base.

While built-in roles offer some clear advantages, keep in mind that they cannot be updated and the set of privileges that constitute them is fixed.

See also

The complete list of built-in roles is available on the official Elastic documentation page here: `https://www.elastic.co/guide/en/elasticsearch/reference/current/built-in-roles.html`.

Defining custom roles

Defining a custom role in the Elastic Stack is necessary to establish fine-grained access control according to your security and operational requirements. In this recipe, we will see how to define custom roles by specifying what users can see and do, to make sure that access rights are in sync with each user's permissions.

Getting ready

Make sure that you have completed the recipes in *Chapter 6* and *Chapter 10* specifically because, in this recipe, we will explore how to create different roles for various use cases. For example, one role could permit access solely to business data (such as the Rennes traffic data that we explored in previous chapters), while another role might be configured to view only **observability data** (such as the different logs, metrics, and trace data we examined in *Chapter 10*).

How to do it...

The upcoming steps will guide you through the process of creating and assigning custom roles within Kibana. You'll learn how to set up one role specifically for business data access and another for observability data, ensuring that users have access only to the relevant datasets they need for their roles. By the end, you will see how these roles impact user access by logging in and verifying the permissions against specific dashboards and datasets.

Follow these steps:

1. Let's start with creating a custom business user role. In Kibana, switch to the **elastic** admin user first.

2. Once logged in with the **elastic** user, go to **Stack Management | Security | Roles**, click on **Create role**, and set `cookbook_business_reader` as the **Role** name. Then, as shown in *Figure 11.7*, in the **Index privileges** section, set `metrics-rennes_traffic-raw` for **Indices** and set **read** and **view_index_metadata** for **Privileges**:

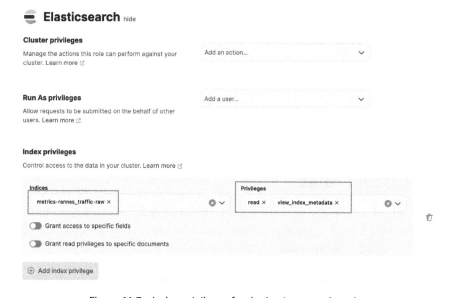

Figure 11.7 – Index privileges for the business reader role

3. Scroll down the page and click on the **Add Kibana privilege** button. This action opens a flyout to set the Kibana privileges, as shown in *Figure 11.8*. In the **Spaces** input box, select * **All Spaces** and then click on the **Read** option to apply the preconfigured set of privileges for a read-only role (we will provide examples of how to customize the privileges in a later recipe of this chapter: *Granting additional privileges*). Click on **Create global privilege** to save the privileges and close the flyout. Finally, click on the **Create role** button:

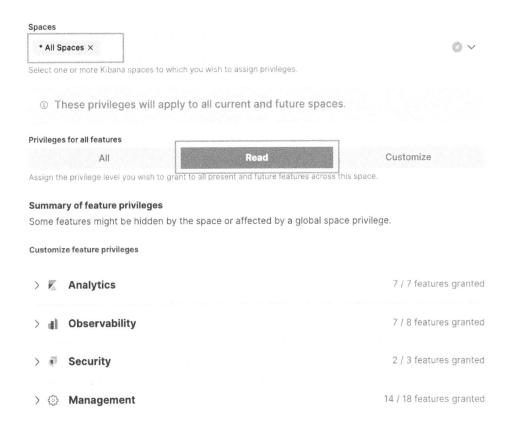

Figure 11.8 – Setting global Kibana privileges

4. Once the role is created, we can create a user and assign the role. Go to **Stack Management |
    Security | Users** and click on the **Create user** button. This opens a form, as shown in *Figure 11.9.*
    Set **Username** to cookbook_business_reader1, **Full name** to Business Reader1,
    enter the password of your choice and confirm it, and set **Roles** to **cookbook_business_reader**.
    You can then click on the **Create user** button:

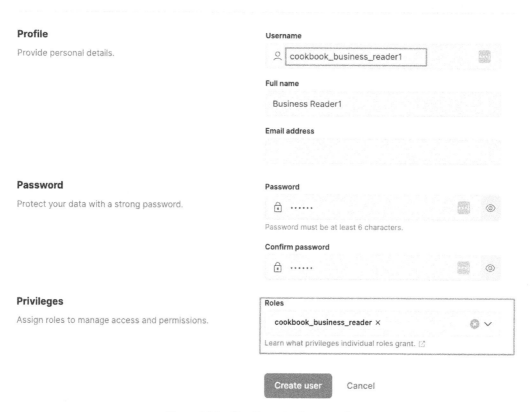

Figure 11.9 – Creating a business reader

5. We can now log in with the new user to test the granted access. Switch to the **cookbook_business_reader1** user, as we did in *Step 4* of the previous recipe. Then, navigate to **Analytics | Dashboards**, locate **[Rennes Traffic] Overview Dashboard**, and click to open it. As illustrated in *Figure 11.10*, you should observe that the newly created user indeed has read-only access to the dashboard and the data. Also, note that the user only has read access, as indicated by the *eye* icon in the menu bar:

Figure 11.10 – Verifying Rennes traffic data access

6. Now, let's go to **Observability | Overview**. We can notice that the user does not have access to the observability data, as shown in *Figure 11.11*:

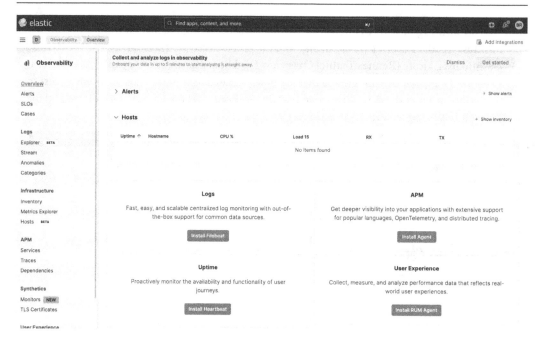

Figure 11.11 – Verifying observability data restriction

7. Now, let's create another role dedicated to observability data viewing only. Log back in with your **elastic** admin user. Go to **Stack Management** | **Security** | **Roles** and click on **Create role**. Set **Role** name to `cookbook_observability_reader`. As shown in *Figure 11.12*, in the **Index privileges** section, add the following indices:

    - **.profiling-***
    - **profiling-***
    - **logs-***
    - **apm-***
    - **traces-apm***
    - **metrics-kubernetes.***
    - **metrics-apm-***
    - **metrics-system-***
    - **metrics-apache-***
    - **metricbeat-***
    - **metrics-elastic_agent***

As you can see, we add observability-related indices for logs, metrics, traces, and profiling. Then, add **read** and **view_index_metadata** for the privileges:

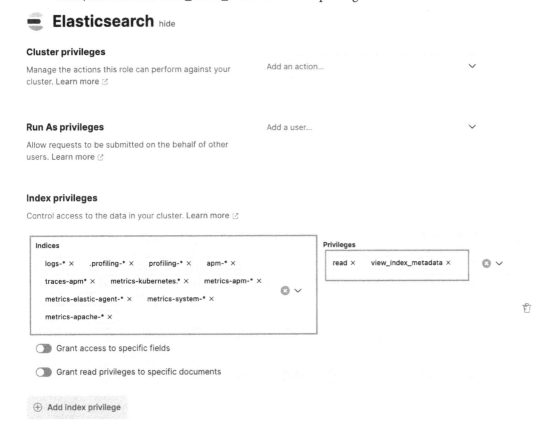

Figure 11.12 – Index privileges for the observability reader role

8. Scroll down the page and click on the **Add Kibana privileges** button. Then, set the Kibana privileges for all spaces using the predefined **Read** privilege. Save the Kibana privileges and create the new role as we did in *Step 2*.

9. Once the new role is created, we can create another user and assign the role as we did in *Step 3*. Go to **Stack Management | Security | Users** and click on the **Create user** button. This opens a form as shown in *Figure 11.13*. Set **Username** to `cookbook_observability_reader1`, **Full name** to `Observability Reader1`, enter the password of your choice and confirm it, and set **Roles** to **cookbook_observability_reader**. You can then click on the **Create user** button:

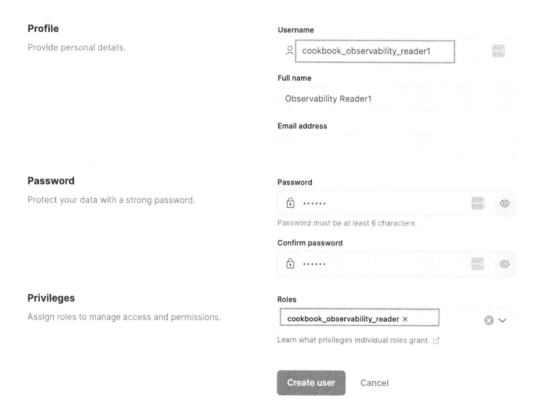

Figure 11.13 – Creating the observability reader

10. We can now log in with the new user to test the granted access. Switch to the **cookbook_ observability_reader1** user, then navigate to **Observability | Overview**. We can notice that this time, we can access various observability data:

Figure 11.14 – Verifying observability data access

11. Navigate to **Analytics | Dashboards**, locate **[Rennes Traffic] Overview Dashboard**, and click to open it. As illustrated in *Figure 11.15*, this time, as an observability reader, the user does not have access to the business data. The popup in the lower-right corner of the screen indicates that the user does not have the required role to access the index.

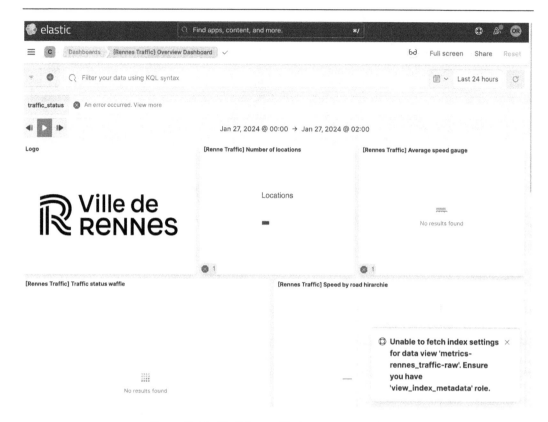

Figure 11.15 – Verifying traffic data access restriction

How it works...

Kibana's RBAC system enables fine-grained control over user access to data and features within the Elastic Stack:

- **Roles**: Administrators define roles that encapsulate a set of permissions for accessing data and Kibana features.

- **Cluster privileges**: These define the level of access users have to cluster-wide operations and settings. Cluster privileges can include the ability to monitor the health and performance of the cluster, manage users and roles, execute administrative tasks, and manage features such as index lifecycle management (ILM) policies, snapshots, and more.

- **Index privileges**: These control what actions a user can perform on Elasticsearch indices, such as reading or writing data, managing index settings, or executing index-level actions.

- **Kibana space privileges**: These determine which Kibana spaces a user can enter and what they can do within those spaces, such as creating and managing dashboards and visualizations or using dev tools.

- **Feature privileges**: These manage access to specific features in Kibana such as visualizations, dashboards, and so on.

- **User Assignment**: Users are assigned one or more roles that correspond to their effective permissions.

- **Advanced Security**: Additional settings allow for restrictions at the document, field, and sub-feature levels; we will see examples in the following recipes of this chapter.

Figure 11.16 shows how RBAC for Kibana roles works and the relationship between different entities:

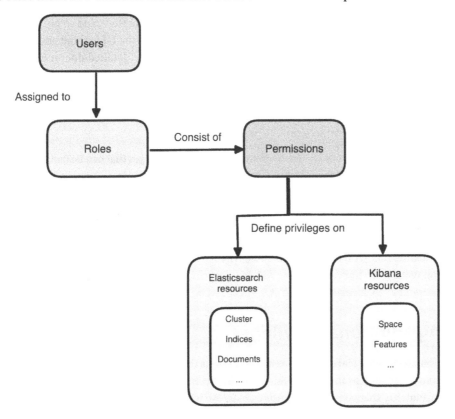

Figure 11.16 – RBAC for Kibana roles

RBAC ensures that users have access only to the data (indices) and features necessary for their roles, maintaining security and compliant data practices.

There's more...

In our recipe, we created custom roles that grant Elasticsearch and Kibana privileges. We call this type of role **Kibana roles**. These roles can be created either via the UI, as we did in this recipe, or via the Kibana REST API (the detailed documentation can be found here: `https://www.elastic.co/guide/en/kibana/current/role-management-api.html`).

Kibana roles and Elasticsearch roles

In real-world use cases, there are situations where the user does not need access to Kibana, such as search applications, third-party applications using Elasticsearch as a data store, and so on. In these use cases, we can create **Elasticsearch roles**. These roles can be created via the UI without adding Kibana privileges or via a dedicated Elasticsearch role management API (the detailed documentation can be found here: `https://www.elastic.co/guide/en/elasticsearch/reference/current/security-api-put-role.html`).

See also

- For a comprehensive list of cluster-level and index-level privileges that can be assigned to roles, refer to the Elasticsearch security privileges documentation: `https://www.elastic.co/guide/en/elasticsearch/reference/current/security-privileges.html`

- For a detailed list of Kibana privileges, consult the Kibana privileges documentation: `https://www.elastic.co/guide/en/kibana/current/kibana-privileges.html`

- For documentation on the full suite of security APIs, including role management, visit the following link: `https://www.elastic.co/guide/en/elasticsearch/reference/current/security-api.html#security-role-apis`

Granting additional privileges

In the previous recipe, we learned how to create various custom roles with different privilege sets to manage multi-tenancy in the Elastic Stack. In this recipe, we will extend these roles to provide additional granularity that goes beyond basic RBAC. We will use field-level security and document-level security to demonstrate how to achieve more precise management of who can view and interact with different pieces of data.

Getting ready

Make sure you have completed the recipes in *Chapter 6* and *Chapter 10*.

Make sure that you have completed the previous recipe, *Defining custom roles*.

The snippets in this recipe are available at the following link: `https://github.com/PacktPublishing/Elastic-Stack-8.x-Cookbook/blob/main/Chapter11/snippets.md#granting-additional-privileges`

How to do it...

We'll now tighten security for the two roles that we created in the previous recipe: firstly, by applying field-level security to **cookbook_business_reader** to limit field access within an index, and secondly, by setting document-level restrictions for **cookbook_observability_reader** to control access to Kubernetes logs.

Follow these steps:

1. Log in to Kibana with the default admin user, **elastic**. Navigate to **Stack Management | Security | Roles** and click on the **cookbook_business_reader** role. We can now add extra field-level security in the **Index privileges** section. Under **metrics-rennes_traffic-raw**, toggle **Grant access to specific fields** to the *on* position, then set * in **Granted fields** and select **insee** in **Denied fields**, as shown in *Figure 11.17*. Click on the **Update role** button:

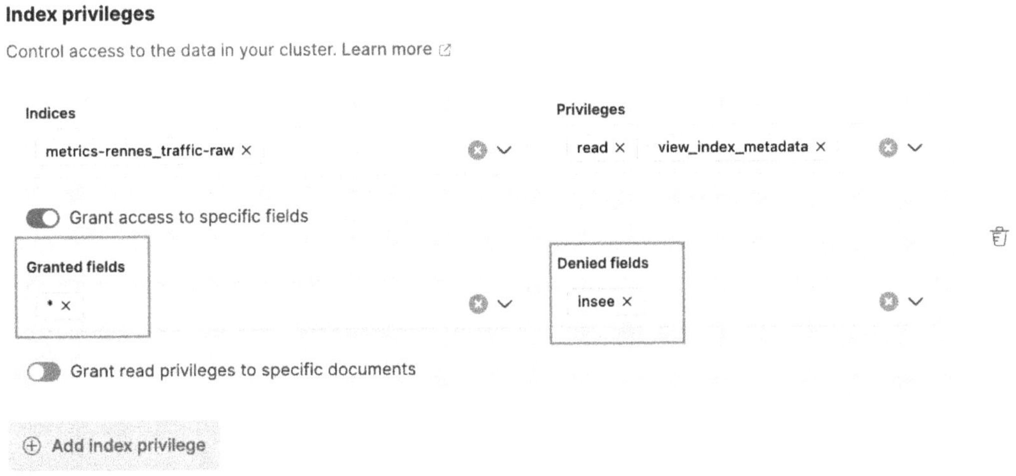

Index privileges

Control access to the data in your cluster. Learn more

Indices	Privileges
metrics-rennes_traffic-raw ×	read × view_index_metadata ×

○ Grant access to specific fields

Granted fields

* ×

Denied fields

insee ×

○ Grant read privileges to specific documents

⊕ Add index privilege

Figure 11.17 – Managing field restrictions

2. Once the role is updated, switch to the **cookbook_business_reader1** user and then navigate to **Analytics | Discover**. Select the **metrics-rennes_traffic-raw** data view and click to expand on any document to see the document details. You can see in *Figure 11.18* that both on the left sidebar and the document detail flyout, the user cannot see the **insee** field:

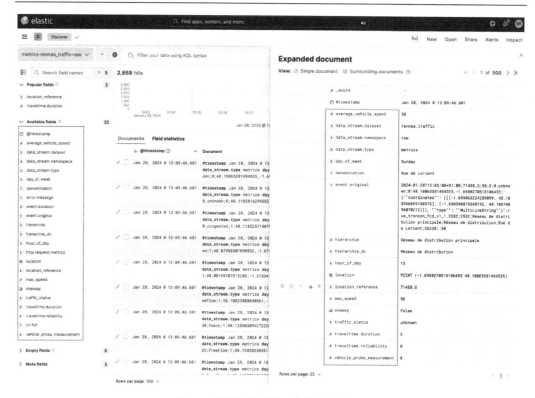

Figure 11.18 – Verifying field visibility

3. Now, let's try to update the **cookbook_observability_reader** role by limiting access just to Kubernetes logs. Instead of using field-level security, we will leverage document-level security. Log back in with the admin user, **elastic**. As the document-level security requires a query compatible with the Query DSL, we can first test the following query in Kibana's Dev Tools to verify that it returns results including only Kubernetes logs (the snippet is available at https://github.com/PacktPublishing/Elastic-Stack-8.x-Cookbook/blob/main/Chapter11/snippets.md#dev-tools-command-to-return-kubernetes-logs):

```
GET /logs-*/_search
{
  "query": {
    "bool": {
      "must": {
        "exists": {
          "field": "kubernetes.namespace"
        }
      }
    }
}
```

```
        }
    }
```

4. Now, we can update the observability role. Navigate to **Stack Management | Security | Roles** and click on the **cookbook_observability_reader** role. In the **Index privileges** section, let's first remove **logs-*** from the existing indices list and click on **Add index privilege** to configure dedicated document-level security for **logs-***.

    In the newly created **Index privileges** section, as shown in *Figure 11.19*, set **logs-*** for **Indices** and set **read** and **view_index_metadata** for **Privileges**. Then, enable the **Grant read privileges to specific documents** option, and in the **Granted documents query** box, insert the query object taken from the Query DSL example that we used in the previous step (the snippet can be found at `https://github.com/PacktPublishing/Elastic-Stack-8.x-Cookbook/blob/main/Chapter11/snippets.md#granted-read-privileges-to-specific-documents`). To finalize the updates, click on the **Update role** button:

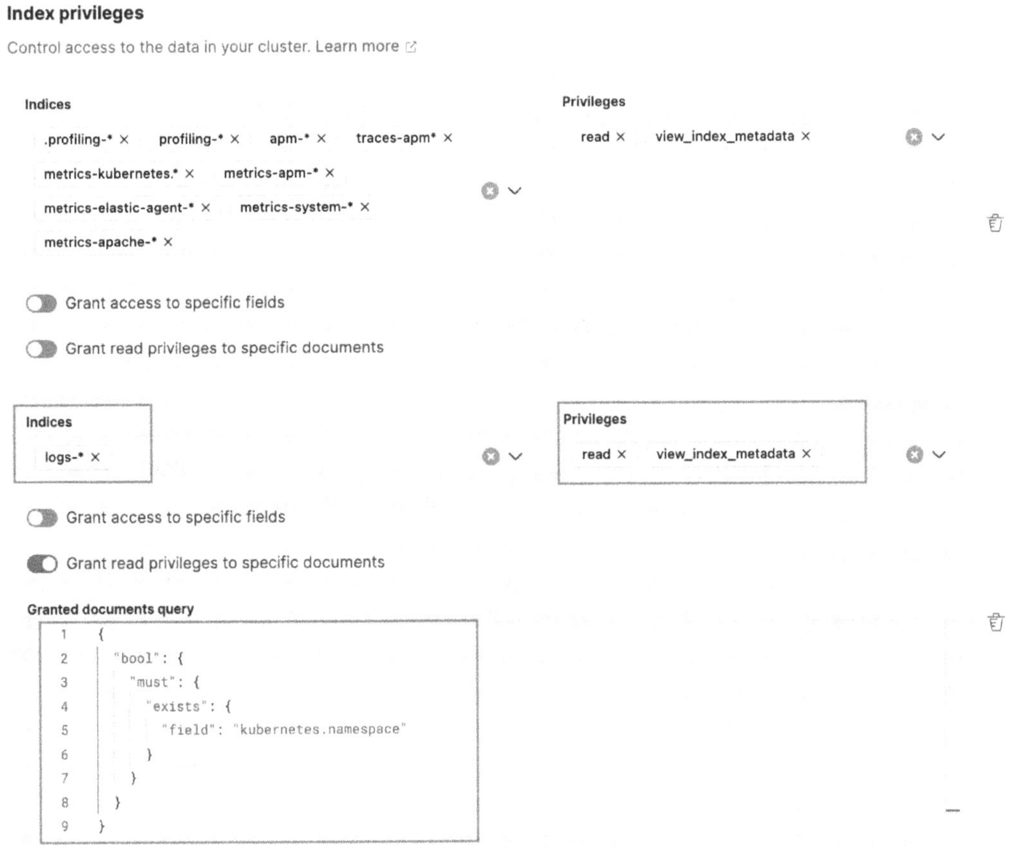

Index privileges

Control access to the data in your cluster. Learn more ☑

Indices	Privileges
.profiling-* ✕ profiling-* ✕ apm-* ✕ traces-apm* ✕	read ✕ view_index_metadata ✕ ✕ ⌄
metrics-kubernetes.* ✕ metrics-apm-* ✕	
metrics-elastic-agent-* ✕ metrics-system-* ✕ ✕ ⌄	
metrics-apache-* ✕	

🔘 Grant access to specific fields

🔘 Grant read privileges to specific documents

Indices	Privileges
logs-* ✕ ✕ ⌄	read ✕ view_index_metadata ✕ ✕ ⌄

🔘 Grant access to specific fields

🔘 Grant read privileges to specific documents

Granted documents query

```
1  {
2    "bool": {
3      "must": {
4        "exists": {
5          "field": "kubernetes.namespace"
6        }
7      }
8    }
9  }
```

Figure 11.19 – Adding document restrictions for the business reader role

5. Once the observability reader role is updated, switch to the **cookbook_observability_reader1** user and then navigate to **Observability | Logs | Stream**. You can see, as shown in *Figure 11.20*, that only Kubernetes-related log entries are displayed:

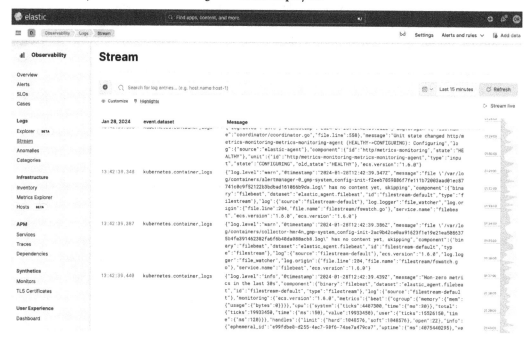

Figure 11.20 – Verifying the log access

How it works...

As you can see in our recipe, field-level security and document-level security in the Elastic Stack offer granular controls to limit the visibility of specific data within Elasticsearch indices.

Field-level security

Field-level security restricts the fields of a document that a user can see. When set up for a particular role, excluded fields do not appear in search results, visualizations, or Kibana Discover and it's possible to include or exclude fields explicitly.

Document-level security

Document-level security restricts access to specific documents within an index. It allows administrators to define a query in a role definition that matches only the documents that a role's users are authorized to see:

- When a user performs a search, the DSL query is automatically applied, and only the matching documents are returned

- Non-matching documents are not visible in search results, visualizations, or dashboards

- It is transparent to the user; the user sees an index as if it only contains the allowed documents

There's more...

As the DSL query is automatically applied, implementing document-level security may have a performance impact; it's preferred to choose simple queries rather than complex queries to avoid performance overhead.

See also

There are some limitations related to field- and document-level security; details can be found in this documentation: `https://www.elastic.co/guide/en/elasticsearch/reference/current/security-limitations.html#field-document-limitations`.

Managing and securing access to Kibana spaces

In the two previous recipes, we created custom roles to manage data access through index privileges. In this recipe, we're taking a step further: we will learn how to control user access to Kibana's personalized spaces and how to combine Kibana privileges with custom roles for a more efficient approach to multi-tenancy management in the Elastic Stack.

Getting ready

Make sure you have completed the recipes in *Chapter 6* and *Chapter 10*.

Make sure that you complete the previous recipes, *Defining custom roles* and *Granting additional privileges*.

How to do it...

In this recipe, we will first create dedicated spaces for traffic analysis and DevOps to provide cleaner access and better data segmentation for our **cookbook_business_reader** and **cookbook_observability_reader** roles. We will then allocate the existing saved objects (such as dashboards, visualizations, and data views) to the appropriate spaces, and finally, update the space-level privileges in the custom roles.

Follow these steps:

1. First, let's create a space dedicated to traffic analysis. Log in to Kibana with the default admin user, **elastic**. Navigate to **Stack Management | Kibana | Spaces** and click on **Create space**. In the form, set **Name** as `Traffic analysis` and **URL identifier** as `traffic-analysis`:

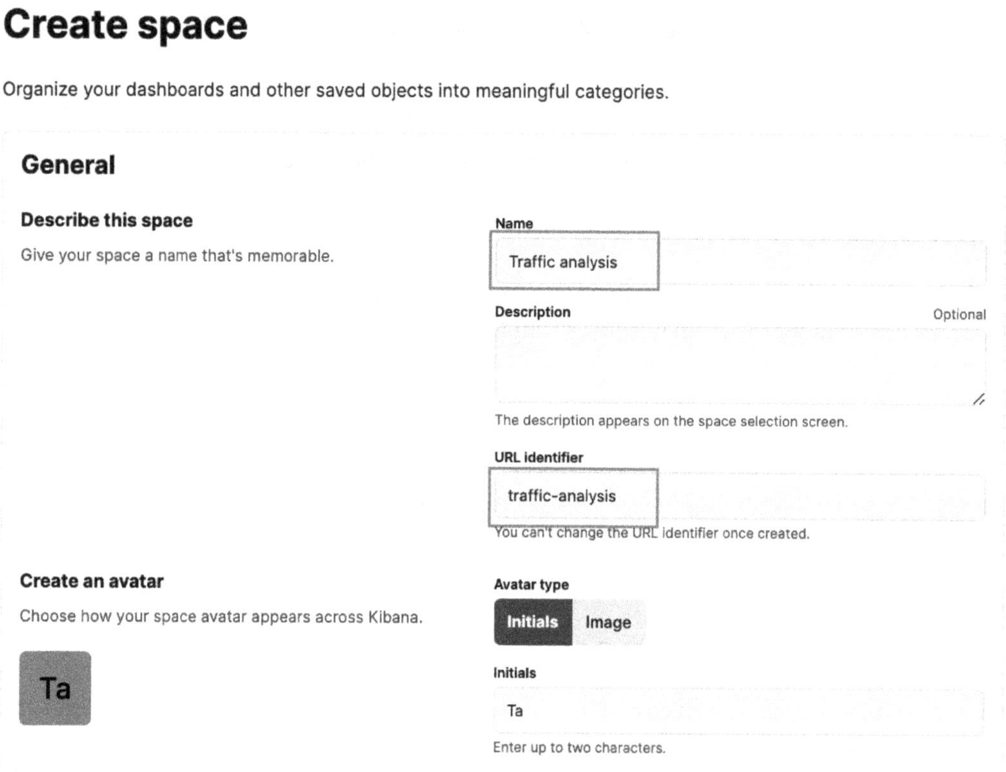

Create space

Organize your dashboards and other saved objects into meaningful categories.

General

Describe this space

Give your space a name that's memorable.

Name

Traffic analysis

Description   Optional

The description appears on the space selection screen.

URL identifier

traffic-analysis

You can't change the URL identifier once created.

Create an avatar

Choose how your space avatar appears across Kibana.

Ta

Avatar type

Initials   Image

Initials

Ta

Enter up to two characters.

Figure 11.21 – Creating the Traffic analysis space

2.   Scroll down to the **Features** section and set **Feature visibility** as shown in *Figure 11.22*, then click on the **Create space** button. As the space is dedicated to **Traffic analysis**, we can remove the access to **Search**, **Observability**, and **Security**, as well as **Graph** under **Analytics** because we didn't create any resources for **Graph**:

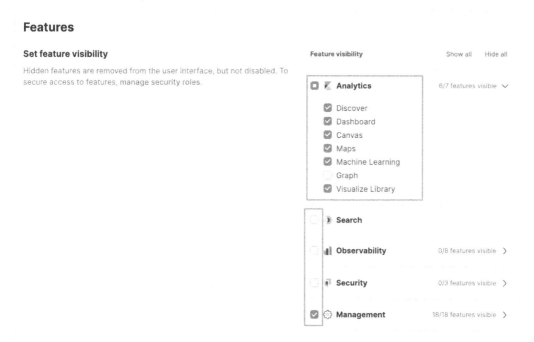

Figure 11.22 – Adjusting the feature visibility for the Traffic analysis space

3. Once the new space is created, we need to move the Rennes traffic-related saved objects to the new space. Go to **Kibana | Saved Objects**, and filter the saved objects by typing `rennes` in the search bar. Once you get the filtered list of the saved objects, bulk select all the objects, then click on the **Export** button. Finally, export all the objects by clicking on the **Export** button in the popup, as shown in *Figure 11.23*. Save it to a local file:

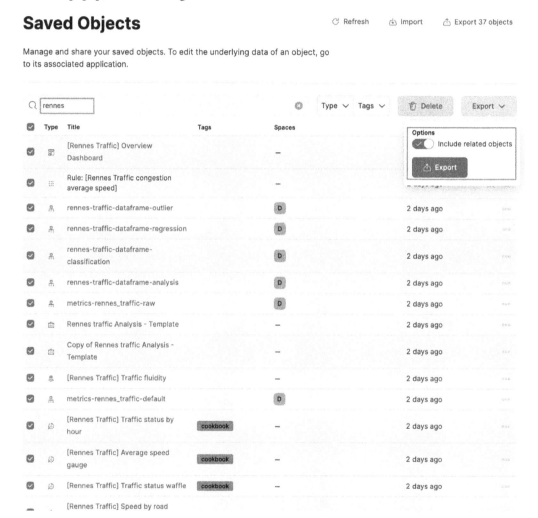

Figure 11.23 – Exporting the Rennes traffic-related saved objects

4. Switch to the new space now by clicking on the space switcher at the top left of the menu bar, and then select the freshly created **Traffic analysis** space, as shown in *Figure 11.24*:

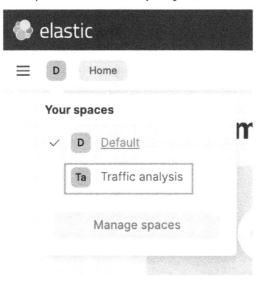

Figure 11.24 – Switching to the new space

5. Now, go to **Stack Management | Saved Objects** and click on **Import**, upload the file that contains the saved objects that we exported in *Step 3*, leave the default options, and click on the **Import** button, as shown in *Figure 11.25*.

    If you get a warning during the import process, you will need to create the metrics-rennes_traffic-raw data view in this new space and restart the import process:

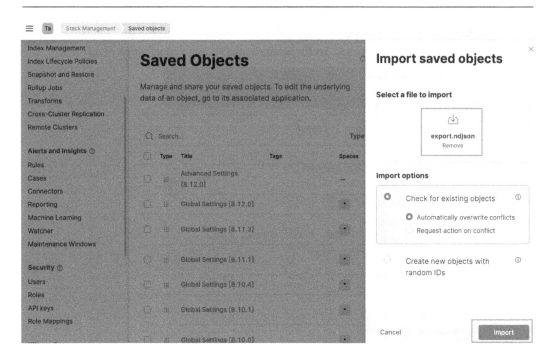

Figure 11.25 – Importing saved objects

6.   We can now update the **cookbook_observability_reader** role by adjusting the space-level privileges. In Kibana, navigate to **Stack Management | Security | Roles** and click on the **cookbook_observability_reader** role. On the role edit form, scroll down to the **Kibana privileges** section and click on the modify button for * **All Spaces**, as shown in *Figure 11.26*:

Figure 11.26 – Editing space privileges

7. On the **Kibana privileges** flyout, in the **Spaces** field, remove * **All Spaces** and select **Traffic analysis**. Then, click on **Customize** and adjust the privileges by granting read access only to all the features in **Analytics** except for **Graph**. We can then save the changes:

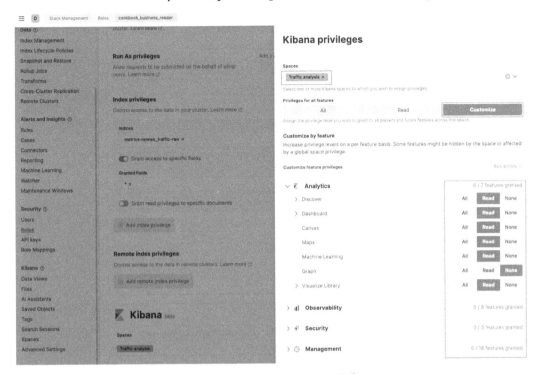

Figure 11.27 – Customizing space privileges

8. Now, let's log in with the **cookbook_business_reader1** user. We can see that the user is automatically logged in to the **Traffic analysis** space and now has a much cleaner navigation menu on the left, as shown in *Figure 11.28*. We can explore different features such as **Dashboards**, **Discover**, and so on by navigating within the space. Verify that we have access to the saved objects such as [**Rennes Traffic**] **Overview Dashboard** among the accessible dashboards, the **metrics-rennes_traffic-raw** data view in **Discover**, and so on:

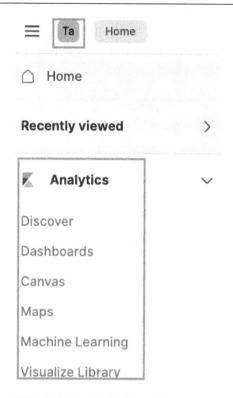

Figure 11.28 – Verifying the Traffic analysis space privileges

9. At this stage, let's create a space dedicated to observability. Log in to Kibana with the default admin user, **elastic**, and select the **Default** space. Navigate to **Stack Management | Kibana | Spaces** and click on **Create space**. In the form, set **Name** as DevOps and **URL identifier** as devops.

    In the **Feature visibility** section, uncheck **Search** and **Security**, as shown in *Figure 11.29*:

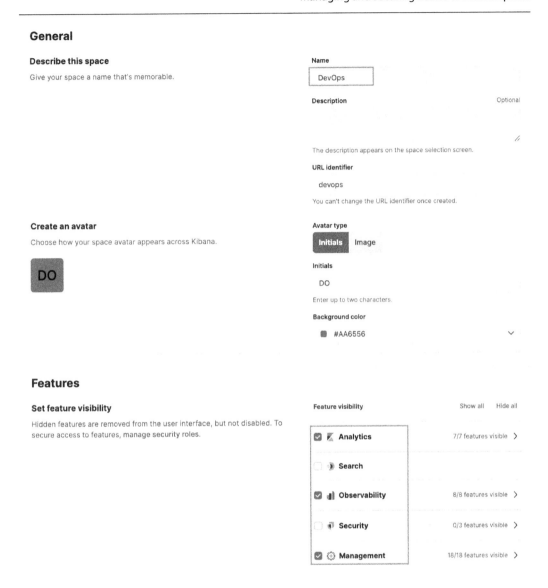

Figure 11.29 – Creating the DevOps space

10. This time, we will skip the saved objects for now – export and import – and instead test the access to the native **Observability** application. Go to **Stack Management | Roles** and click on the **cookbook_observability_reader** role. Then, modify the **Kibana privileges** section as shown in *Figure 11.30*:

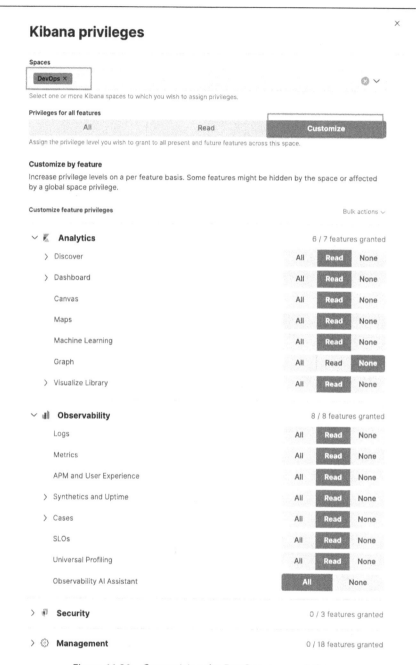

Figure 11.30 – Customizing the DevOps space privileges

Log in with the **cookbook_observability_reader1** user. We can see that the user is automatically logged in to the **DevOps** space. Go to **Observability | Logs | Stream** to verify the access.

How it works...

As you can see in this recipe, Kibana spaces are like workspaces or containers in Kibana that enable you to organize your dashboards, visualizations, rules, and other saved objects. Because saved objects are not shared across spaces by default, sensitive information is better protected. Users only interact with and have visibility into the saved objects within the spaces they have been granted access to, aligning with the principle of least privilege.

When combined with Kibana's RBAC, spaces can control user access at a granular level. You assign roles to users that specify which spaces they can access and what they can do within those spaces via Kibana privileges.

There's more...

As we noticed in this recipe, managing saved objects is key to controlling access in Kibana. Often, we need to transfer saved objects between spaces via exporting, importing, or copying. When establishing these objects, using easily understandable and descriptive names, descriptions, and tags is essential. This practice not only simplifies the process of identifying and sharing the correct saved objects but also aids users in comprehending their functionality and content.

See also

Similar to Kibana role management, it is also possible to manage Kibana spaces and saved objects via a REST API; the detailed documentation for each can be found here:

- `https://www.elastic.co/guide/en/kibana/current/spaces-api.html`
- `https://www.elastic.co/guide/en/kibana/current/saved-objects-api.html`

Managing access with API keys

In the previous recipes, we learned how to manage access control in Kibana with roles. This is very useful for the interactive usage of Kibana. API keys are ideal for service-to-service authentication such as Logstash-to-Elasticsearch communication. This approach aligns with security best practices by separating human user access from service accounts and automating the seamless rotation and revocation of keys where necessary. In this recipe, we will learn how to create and use API keys and how to apply correct access control to them.

Getting ready

Make sure that you have completed the *Installing self-managed Logstash* and *Creating a Logstash pipeline* recipes in *Chapter 5*.

Make sure that you have admin access to the virtual machine that hosts the Logstash installation.

The snippets in this recipe are available at the following link: `https://github.com/PacktPublishing/Elastic-Stack-8.x-Cookbook/blob/main/Chapter11/snippets.md#managing-access-with-api-key`.

How to do it...

In *Chapter 5*, we installed a self-managed Logstash instance and configured a Logstash pipeline to ingest the Rennes traffic data into our Elastic deployment. We used basic authentication with our admin user for the Elasticsearch output in the Logstash pipeline. In this recipe, we will learn how to create an API key and use it to replace the basic authentication in the Logstash pipeline.

Follow these steps:

1. First, let's start by creating an API key via the API. We create this key by granting the `monitor` and `read_ilm` cluster-level privileges and `view_index_metadata` and `create_doc` index-level privileges for indices matching the `metric-rennes_traffic-*` pattern. In Kibana, go to **Dev Tools** and execute the following command (the code snippet can be found at `https://github.com/PacktPublishing/Elastic-Stack-8.x-Cookbook/blob/main/Chapter11/snippets.md#dev-tools-command-to-create-api-key`):

```
POST /_security/api_key
{
  "name": "rennes_traffic_writer_key",
  "role_descriptors": {
    "rennes_traffic_writer": {
      "cluster": ["monitor","manage_ilm","read_ilm"],
      "index": [
        {
          "names": ["metric-rennes_traffic-*"],
          "privileges": ["view_index_metadata", "create_doc"]
        }
      ]
    }
  }
}
```

2. Once the command is executed, you should see the following response. Note `api_key_id` and `api_key_value` in the response. We will use them to update the Logstash configuration file:

```
1 ▾ {
2      "id": yVk9Uo0BpgyC1Aehchm4",          ◄─────────  [api_key_id]
3      "name": "rennes_traffic_writer_key",
4      "api_key": "hS0Q0BD0Q9uYYvzNpoS-Hg",  ◄──────────[api_key_value]
5      "encoded": "eVZr0VVvMEJwZ3lDMUFlaGNobTQ6blMwUTBCRDBROXVZWXZ6TnBvUy1IZw=="
6 ▴ }
```

Figure 11.31 – Output of API key creation

3. Now, let's prepare a new `logstash.conf` file (you can find an example file at `https://github.com/PacktPublishing/Elastic-Stack-8.x-Cookbook/blob/main/Chapter11/logstash/rennes_traffic-raw.conf`). Update the file with your `CLOUD_ID`, `api_key_id`, and `api_key_value` in the output block of the configuration file:

```
output {
    elasticsearch {
        cloud_id => "CLOUD_ID"
        api_key => "api_key_id:api_key_value"
        data_stream => true
        data_stream_type => "metrics"
        data_stream_dataset => "rennes_traffic"
        data_stream_namespace => "raw"
    }

    stdout { codec => rubydebug }
}
```

4. You can now log in to the virtual machine hosting Logstash; stop the Logstash service with the following command:

```
$ sudo systemctl stop logstash.service
```

5. Replace the existing `logstash.conf` file with the one we just uploaded (the `logstash.conf` file is located at `/etc/logstash/conf.d/` on Debian-based systems, as it was in *Chapter 5*). Then, start the service again with the following command:

```
$ sudo systemctl start logstash.service
```

6. Once the service is started, wait for 10 minutes for the data to come in. Then, in Kibana, go to **Analytics | Discover** and select **metrics-rennes_traffic-raw** in the data view. We should see the latest data, as shown in *Figure 11.32*:

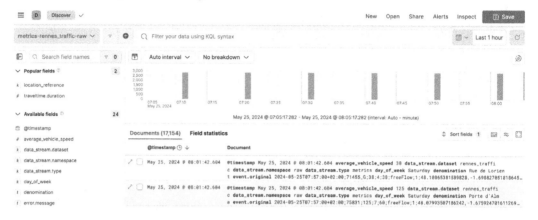

Figure 11.32 – Verifying Rennes traffic ingestion

7. We can also verify the API key status in Kibana. Go to **Stack Management | API keys**, click **Locate**, and click on the key that we created in this recipe. As shown in *Figure 11.33*, it is possible to update the privileges directly in the console:

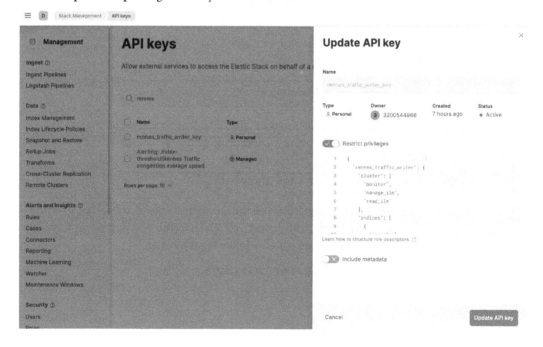

Figure 11.33 – Viewing API key details

How it works...

In our recipe, using an API key for Logstash as an authentication method consists of four main parts:

- **API key creation**: We used the Kibana **Dev Tools** console to generate the key. For the Elasticsearch output plugin of Logstash, only the format `id:apikey` is supported for the API key. That is the reason why we couldn't create the API key via **Kibana | Stack Management | API keys**, as API keys created via the UI would only generate an encoded API key value.

- **Assigning privileges**: The privileges define what actions the API key bearer can perform. In our example, they include writing data to the Rennes traffic data stream and monitoring index lifecycles.

- **Expiration time**: The API keys can either be created with or without an expiration time.

- **Using the API key in Logstash**: The API key is then configured in the `logstash.conf` file, specifically within the output section. Instead of using a username and password (basic authentication), the API key is used to authenticate requests from Logstash to the Elasticsearch cluster.

There's more...

There are some additional use cases and considerations related to API keys that we will explore in the following subsections.

Using API keys for reading and filtering

In our recipe, we used our API key for Logstash ingestion to Elasticsearch. Logstash also has an Elasticsearch input plugin and filter plugin; both plugins provide authentication via API keys. Detailed documentation can be found at the following links:

- `https://www.elastic.co/guide/en/logstash/current/plugins-inputs-elasticsearch.html#plugins-inputs-elasticsearch-api_key`

- `https://www.elastic.co/guide/en/logstash/current/plugins-filters-elasticsearch.html#plugins-filters-elasticsearch-api_key`

Cross-cluster API key

While creating API keys, it's also possible to define cross-cluster API keys for **cross-cluster search** (**CCS**) or **cross-cluster replication** (**CCR**). We will learn more CCS/CCR concepts in *Chapter 12*.

Audits and tracking

Since API keys can be created for specific uses or applications, it provides better tracking and audit logs. You can see which key was used to perform which actions, thus improving traceability over using a generic user account. We will have a dedicated recipe in *Chapter 13: Enabling audit logging*.

See also

- Documentation to create API keys via a REST API can be found at the following link: `https://www.elastic.co/guide/en/elasticsearch/reference/current/security-api-create-api-key.html`

- Documentation to create a cross-cluster API key can be found at the following link: `https://www.elastic.co/guide/en/elasticsearch/reference/current/security-api-create-cross-cluster-api-key.html`

Configuring single sign-on

In this recipe, we delve into the technical intricacies of setting up **single sign-on** (**SSO**) using the **OpenID Connect** option within the Elastic Stack environment. SSO is an essential authentication process that enables users to access multiple applications or services with a single set of credentials. This streamlined authentication approach not only enhances user experience by reducing password fatigue but also bolsters security by centralizing the authentication process.

OpenID Connect is an authentication layer built on top of the OAuth 2.0 protocol. It extends OAuth 2.0 capabilities with ID tokens, which provide additional user information. OpenID Connect is widely adopted due to its simplicity and extensibility, making it a preferred choice for implementing SSO. During this recipe, we will guide you through integrating OpenID Connect with Elastic Stack, providing a step-by-step approach to harness the power of SSO for enhanced security and streamlined user access management.

Getting ready

The process described in this recipe applies to deployment on Elastic Cloud. During configuration, you'll need to provide specific information, such as the Kibana URL. To locate it, log in to the **Cloud** console at `https://cloud.elastic.co`, click on **Manage** to the right of your deployment name, and on the deployment information page, click **Copy endpoint** right beside **Kibana**, as shown in *Figure 11.34*:

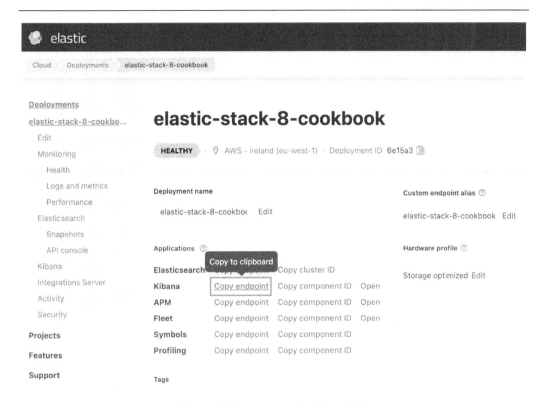

Figure 11.34 – Copying the Kibana URL

Remember, this URL will be referred to as `<KIBANA_URL>`.

Make sure that you have completed the earlier recipe, *Defining custom roles.*

Setting up SSO with OpenID Connect requires an **OpenID provider** (**OP**). We will use an Okta developer account, which provides free access to Okta security features. Set up your account by heading to `https://developer.okta.com/signup/` and signing up for the free Developer Edition of **Workforce Identity Cloud**, as shown in *Figure 11.35*:

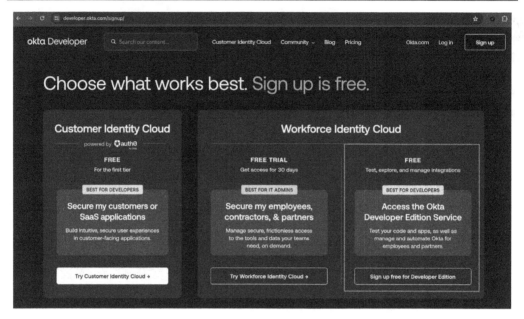

Figure 11.35 – Workforce Identity Cloud developer sign-up

Once you complete your registration and verify your Okta developer account, log in to access it. You will arrive at the **Getting started** page. Click on your username in the upper-right corner and copy the Okta domain, as illustrated in *Figure 11.36*. This domain, referred to as <YOUR-OKTA-DOMAIN>, will be used later in this recipe:

Figure 11.36 – Okta developer account ID

With these prerequisites met, you are prepared to move forward.

The snippets for this recipe can be found at `https://github.com/PacktPublishing/Elastic-Stack-8.x-Cookbook/blob/main/Chapter11/snippets.md#configuring-single-sign-on`

How to do it...

Buckle up, as we're going to embark on the setup; while this might seem intimidating, the overall process is quite simple. *Figure 11.37* sums up the flow we are about to follow:

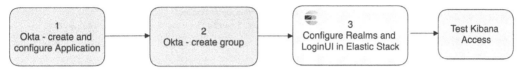

Figure 11.37 – SSO with OIDC setup flow

Let's start by creating and configuring an application integration in Okta:

1. In the Okta console, go to **Applications** from the left-hand menu, then click **Create App Integration**. Okta app integrations allow administrators to connect users to external applications and systems securely, managing SSO options and user lifecycle operations, such as onboarding and device/license management. Our integration will connect with Kibana.

2. In the **Create a new app integration** popup, select **OIDC - OpenID Connect** as the sign-in method and **Web Application** as the application type, as illustrated in *Figure 11.38*:

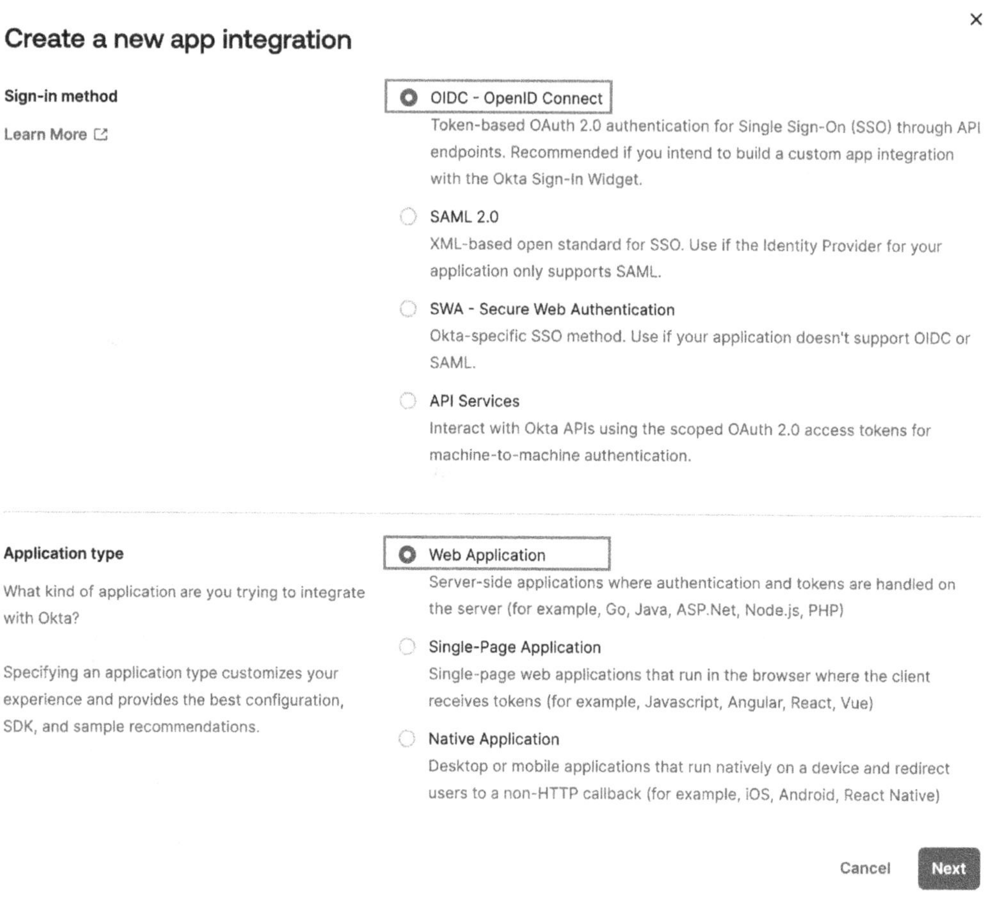

Create a new app integration

×

Sign-in method

Learn More ⤢

- ○ **OIDC - OpenID Connect**
  Token-based OAuth 2.0 authentication for Single Sign-On (SSO) through API endpoints. Recommended if you intend to build a custom app integration with the Okta Sign-In Widget.

- ○ **SAML 2.0**
  XML-based open standard for SSO. Use if the Identity Provider for your application only supports SAML.

- ○ **SWA - Secure Web Authentication**
  Okta-specific SSO method. Use if your application doesn't support OIDC or SAML.

- ○ **API Services**
  Interact with Okta APIs using the scoped OAuth 2.0 access tokens for machine-to-machine authentication.

Application type

What kind of application are you trying to integrate with Okta?

Specifying an application type customizes your experience and provides the best configuration, SDK, and sample recommendations.

- ○ **Web Application**
  Server-side applications where authentication and tokens are handled on the server (for example, Go, Java, ASP.Net, Node.js, PHP)

- ○ **Single-Page Application**
  Single-page web applications that run in the browser where the client receives tokens (for example, Javascript, Angular, React, Vue)

- ○ **Native Application**
  Desktop or mobile applications that run natively on a device and redirect users to a non-HTTP callback (for example, iOS, Android, React Native)

Cancel   **Next**

Figure 11.38 – Creating new app integration part 1

3. Click on **Next** to continue to the configuration. On the **New Web App Integration** page, enter the following information:

- **App integration name:** `Kibana Cookbook`

- **Grant type: Authorization Code** (checked by default)

- **Sign-in redirect URIs:** `<KIBANA_URL>/api/security/oidc/callback`

- **Sign-out redirect URIs:** `<KIBANA_URL>/security/logged_out`

- **Controlled access: Skip group assignment for now**

4. Click on **Save**. Our app integration is now created, but we still have configuration steps to follow. Copy the client ID and client secret of the application as we will need them in the coming steps. These will be referred to as `<CLIENT-ID>` and `<CLIENT-SECRET>`, respectively:

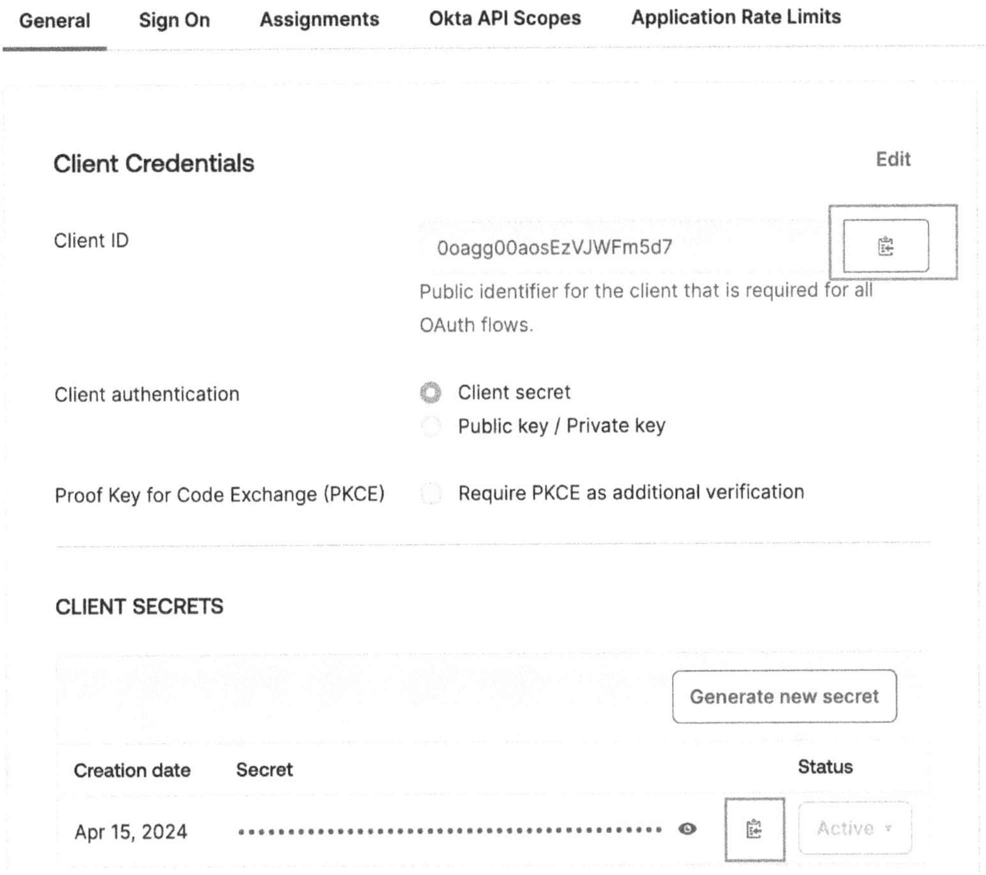

General	Sign On	Assignments	Okta API Scopes	Application Rate Limits

Client Credentials

Edit

Client ID

Ooagg00aosEzVJWFm5d7

Public identifier for the client that is required for all OAuth flows.

Client authentication
- ⦿ Client secret
- ◯ Public key / Private key

Proof Key for Code Exchange (PKCE)
- ☐ Require PKCE as additional verification

CLIENT SECRETS

Generate new secret

Creation date	Secret		Status
Apr 15, 2024	•• 👁		Active ▾

Figure 11.39 – Copying the client ID and secret

5. Scroll down to **General Settings** and select **Edit**. Scroll to locate the **LOGIN** section. Here, the focus is on modifying the initiation of the login process. Find the **Login initiated by** option and, using the drop-down menu on the right, choose **Either Okta or App**. Then, under **Application visibility**, opt for **Display application icon to users**. This will adjust the visibility settings and login initiation method. Finally, in **Initiate login URI**, enter your `<KIBANA_URL>/` URL. The **LOGIN** section should look as follows:

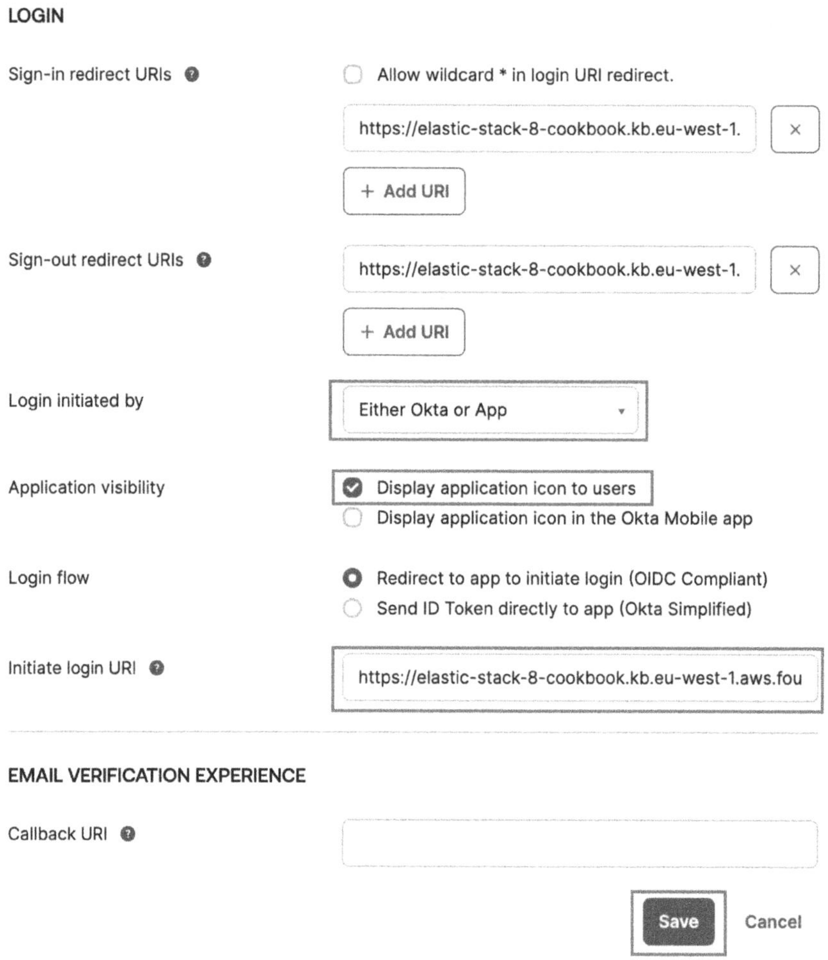

LOGIN

Sign-in redirect URIs ❔   ⃝ Allow wildcard * in login URI redirect.

                                      https://elastic-stack-8-cookbook.kb.eu-west-1.   ✕

                                      + Add URI

Sign-out redirect URIs ❔   https://elastic-stack-8-cookbook.kb.eu-west-1.   ✕

                                      + Add URI

Login initiated by   Either Okta or App   ▾

Application visibility   ✅ Display application icon to users
                                      ⃝ Display application icon in the Okta Mobile app

Login flow   ⦿ Redirect to app to initiate login (OIDC Compliant)
                                      ⃝ Send ID Token directly to app (Okta Simplified)

Initiate login URI ❔   https://elastic-stack-8-cookbook.kb.eu-west-1.aws.fou

EMAIL VERIFICATION EXPERIENCE

Callback URI ❔

                                                                    Save   Cancel

Figure 11.40 – Login section for the app integration

6. Go to the **Sign On** tab to configure how the **groups** claims will be filtered from the OpenID Connect token.

> **Important note**
>
> Group claims in OpenID Connect are data elements that assert the group memberships of an authenticated user, such as roles or teams they belong to. These claims are important because they enable the efficient management of access rights and permissions in applications, allowing for tailored authorization based on the user's group memberships.

7. Scroll down to the **OpenID Connect Token** section and click **Edit**. In the **Groups claim filter** parameter, replace **Starts with** with **Matches regex**, and enter . *, as shown in *Figure 11.41*:

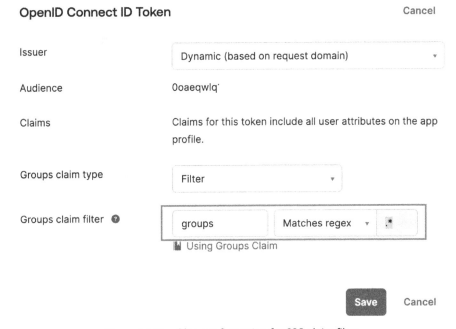

Figure 11.41 – Okta configuration for SSO claim filter

With this last configuration complete, save all changes.

We will now proceed to create a group and assign both our user and the application we created:

1. In your Okta developer console, select **Directory** from the left menu, then **Groups**, and click **Add Group**. In the pop-up window, enter `Elastic Group` for **Name** and `SSO Group for Elastic Stack` for **Description**, and click **Save**:

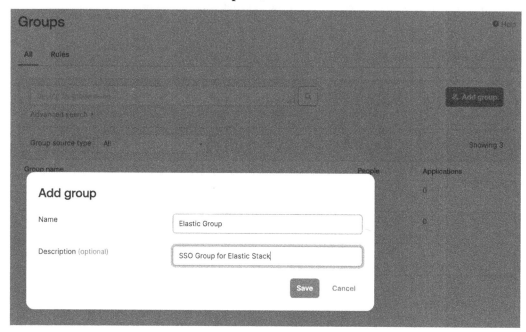

Figure 11.42 – Creating a group in Okta

2. Next, add our user to this group. Select the group from the list and click **Assign people**. Use the + symbol to assign your user to the newly created group.

3. Go to the **Applications** tab and click **Assign applications**. In the popup, select **Assign** beside the **Kibana Cookbook** application to link it with our group. During authentication, it's critical for Okta to verify the user's permission to use the application. Since our user is part of **Elastic Group**, the application must also be linked to this group. Your group configuration should now match that shown in *Figure 11.43*:

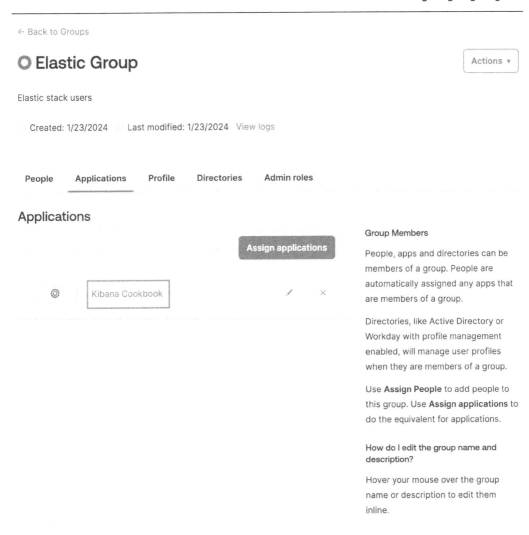

Figure 11.43 – Assigning the Kibana application to Elastic Group

With that, the Okta setup is complete. Let's configure SSO in the Elastic Stack:

1. Log in to Elastic Cloud (`https://cloud.elastic.co`). Find your deployment and click **Manage** beside it.

2. First, add the client secret from your Okta app integration to the keystore. From the **Deployment** menu, navigate to **Security | Elasticsearch keystore | Add setting**. In the **Create settings** window, enter the settings name as `xpack.security.authc.realms.oidc.oidc-okta.rp.client_secret` and choose **Single string** as the type. In the **Secret** field, paste the `<CLIENT-SECRET>` value that we copied from the application integration setup in Okta in the previous step.

3. Let's set up the Elasticsearch cluster to authenticate users via the OpenID Connect realm and correctly assign roles. In the **Deployment** menu, click **Edit** under the deployment name and select **Manage user settings and extensions** to update Elasticsearch's settings. The snippet for this can be found at the following GitHub link: `https://github.com/PacktPublishing/Elastic-Stack-8.x-Cookbook/blob/main/Chapter11/snippets.md#elasticsearch-user-settings-for-oidc-realms`.

    To add the right values, you will need to update the snippet with the following values that you retrieved from the early steps of this recipe:

    *   `<CLIENT_ID>`

    *   `<KIBANA_URL>`

    *   `<YOUR-OKTA-DOMAIN>`

    Following is the code snippet where the preceding values should be added:

    ```
    xpack.security.authc.realms.oidc:
        oidc-okta:
            order: 3
            rp.client_id: <CLIENT_ID>
            rp.response_type: code
            rp.redirect_uri: "<KIBANA_URL>/api/security/oidc/
    callback"
            op.issuer: "https://<YOUR-OKTA-DOMAIN>"
            op.authorization_endpoint: "https://<YOUR-OKTA-DOMAIN>/
    oauth2/v1/authorize"
            op.token_endpoint: "https://<YOUR-OKTA-DOMAIN>/oauth2/
    v1/token"
            op.jwkset_path: "https://<YOUR-OKTA-DOMAIN>/oauth2/v1/
    keys"
            op.userinfo_endpoint: "https://<YOUR-OKTA-DOMAIN>/
    oauth2/v1/userinfo"
            rp.post_logout_redirect_uri: "<KIBANA_URL>/security/
    logged_out"
            rp.requested_scopes: ["openid", "groups", "profile",
    "email"]
            claims.principal: email
            claims.name: name
            claims.mail: email
            claims.groups: groups
    ```

4. *Figure 11.44* shows the updated OIDC realm settings on the **Elasticsearch user settings and extensions** page. Click on **Back** at the bottom of the flyout:

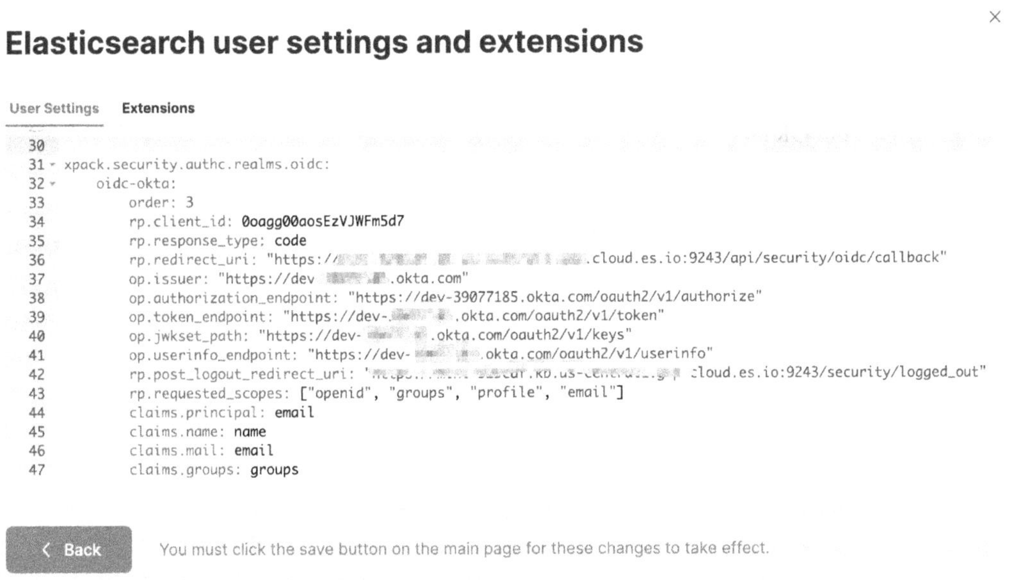

Elasticsearch user settings and extensions

User Settings   **Extensions**

```
30
31 ▾ xpack.security.authc.realms.oidc:
32 ▾     oidc-okta:
33           order: 3
34           rp.client_id: 0oagg00aosEzVJWFm5d7
35           rp.response_type: code
36           rp.redirect_uri: "https://          .cloud.es.io:9243/api/security/oidc/callback"
37           op.issuer: "https://dev     .okta.com"
38           op.authorization_endpoint: "https://dev-39077185.okta.com/oauth2/v1/authorize"
39           op.token_endpoint: "https://dev-      .okta.com/oauth2/v1/token"
40           op.jwkset_path: "https://dev-      .okta.com/oauth2/v1/keys"
41           op.userinfo_endpoint: "https://dev-      .okta.com/oauth2/v1/userinfo"
42           rp.post_logout_redirect_uri: '          .cloud.es.io:9243/security/logged_out"
43           rp.requested_scopes: ["openid", "groups", "profile", "email"]
44           claims.principal: email
45           claims.name: name
46           claims.mail: email
47           claims.groups: groups
```

< Back   You must click the save button on the main page for these changes to take effect.

Figure 11.44 – Example of Elasticsearch OIDC realm settings

5. Next, update the Kibana configuration to utilize OpenID Connect with Okta as the authentication provider. Scroll down to **Kibana**, click **Edit user settings**, and add the following code (https://github.com/PacktPublishing/Elastic-Stack-8.x-Cookbook/blob/main/Chapter11/snippets.md#kibana-user-settings-for-oidc-provider):

```
xpack.security.authc.providers:
    oidc:
        oidc1:
            order: 2
            realm: oidc-okta
            description: SSO with Okta via OIDC
    basic:
        basic1:
            order: 3
```

Note that with this configuration, we're also allowing native realm users to authenticate. This configuration also maintains login capability for native realm users.

6. Click **Back** (located at the bottom), then scroll to find and click the **Save** button. After saving, confirm the changes by clicking **Confirm**. A summary popup will appear, outlining all the new settings to be implemented. Review and validate these settings, and then wait for the new configuration to be applied to your deployment.

While the changes are being rolled out to our deployment, let's take a step back and recap what we just went through. We started in Okta to configure a group and an application integration to manage the SSO with Kibana; then, we configured the OpenID Connect realm in Elasticsearch armed with the information from OKTA.

Now, we can test our OIDC integration and access to Kibana:

1.   Open a new browser window in *incognito* or *private* mode and enter the `<KIBANA_URL>` value. You'll be greeted with the Kibana login page as shown:

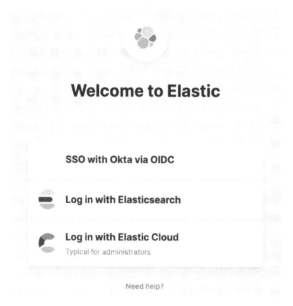

Figure 11.45 – Kibana login page with OIDC provider

2.   You can see that a new login option has been added. Click on **SSO with Okta via OIDC** and you'll be redirected to the Okta login page:

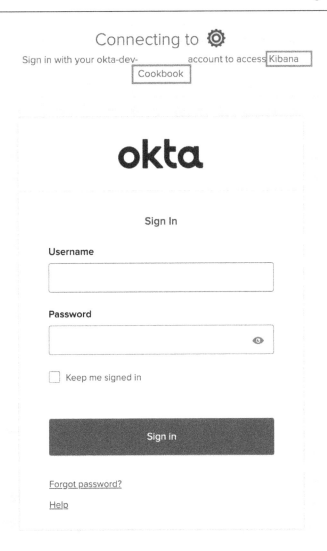

Figure 11.46 – SSO Kibana Okta redirect

3. Enter your username and password, then click the **Sign in** button. You will encounter the screen shown in *Figure 11.47*:

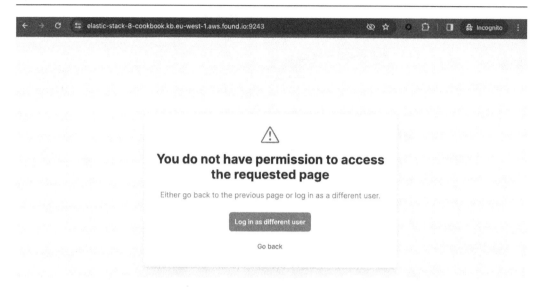

Figure 11.47 – Permission error

While it may appear to be a setback, encountering permissions issues after authenticating with Okta but not having access to Kibana is expected at this stage. By default, users authenticated through OpenID Connect are not assigned any roles. We will address this issue in the next recipe, which covers how to properly grant Kibana access.

4. To verify the authentication, check the logs of your app integration on the Okta dashboard. On the application's detail page, click **View Logs**, as shown in *Figure 11.48*:

Figure 11.48 – Viewing application logs

5. Search the system log for an entry confirming authentication success, as shown in *Figure 11.49*:

Figure 11.49 – Logs for successful authentication

How it works...

In the context of SSO using OpenID Connect with Okta as the OP, the interaction between Elasticsearch and Okta follows a specific flow, involving authentication and authorization processes. Here's an overview of how it works, highlighting the roles of Elasticsearch and Okta:

- Elasticsearch functions as the **relying party** (**RP**). It relies on the OP (Okta, in this case) to authenticate users. When a user attempts to access Elasticsearch, it doesn't handle the login credentials directly. Instead, it redirects the user to the OP (Okta) for authentication.

- Okta acts as the OP. It's responsible for verifying the identity of users and providing this information to Elasticsearch. When a user tries to log in to Elasticsearch, they are redirected to Okta, where they enter their credentials.

The following figure shows the high-level interaction between Elasticsearch (RP), and Okta (OP):

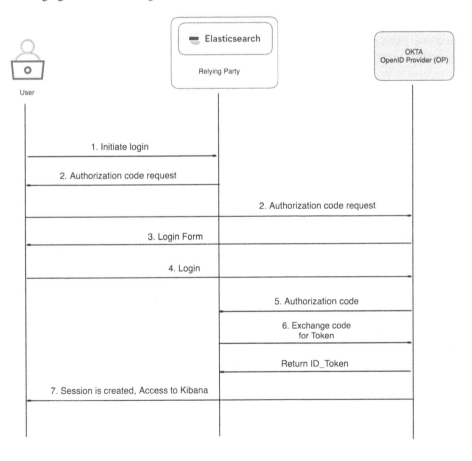

Figure 11.50 – High-level view of SSO with Okta and Elasticsearch

Let us go through each step:

1. **User authentication request**: When a user attempts to access Elasticsearch, it initiates an authentication request by redirecting the user to Okta.

2. **Login and authentication at Okta**: The user logs in using their Okta credentials. Okta authenticates the user and generates an ID token and an access token.

3. **Tokens sent to Elasticsearch**: These tokens are then sent back to Elasticsearch. The ID token contains the identity information of the user, while the access token may be used to access user details or other resources within Okta.

4. **Verification and access granting**: Elasticsearch verifies the token's validity with Okta and, upon successful verification, grants the user access to its services based on the information in the token, such as user identity and group memberships.

In summary, Elasticsearch delegates the responsibility of user authentication to Okta. Upon successful authentication, Okta communicates this back to Elasticsearch, which then provides the appropriate access to the user. This system streamlines the authentication process, improves security by centralizing user credentials in Okta, and enhances user experience by reducing the need for multiple logins.

There's more...

You may have observed that when configuring Elasticsearch, the claims were already incorporated:

```
rp.requested_scopes: ["openid", "groups", "profile", "email"]
claims.principal: email
claims.name: name
claims.mail: email
claims.groups: groups
```

It's important to note, however, that the range of claims supported can vary significantly based on the chosen OP. To streamline this aspect, you can utilize a utility known as **oidc-tester**. This tool enables you to view the ID token and, consequently, the claims it comprises. This visibility facilitates the mapping of these claims to the user properties in your environment.

Attentive observers may have noticed the absence of a logo on our SSO integration on the login page. If you're wondering how to add a visually appealing logo next to your SSO realms, the process is quite simple. All that's required is an update to the Kibana configuration, specifically by incorporating the `icon` setting. This can be demonstrated in the following example code. You have the option to use the default logo of Elastic Stack components or to specify a URL for an SVG image. The snippet can be found at `https://github.com/PacktPublishing/Elastic-Stack-`

8.x-Cookbook/blob/main/Chapter11/snippets.md#kibana-user-settings-for-oidc-provider-with-icon-and-hint:

```
xpack.security.authc.providers:
    oidc:
        oidc1:
            order: 2
            realm: oidc-okta
            description: SSO with Okta via OIDC
            hint: "For business and observability users"
            icon: "logoKibana"
    basic:
        basic1:
            order: 3
```

We have also included the `hint` parameter to guide users in selecting the right login option. By applying this configuration, your login page will be updated as shown in *Figure 11.51*:

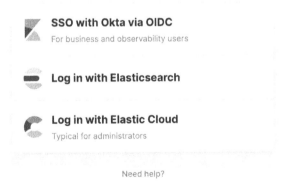

Figure 11.51 – Added the Kibana icon to the OpenID Connect realm

See also

To learn more about `oidc-tester`, read this excellent article: `https://medium.com/application-security/openid-connect-tester-for-developers-63427324faea`.

Mapping users and groups to roles

In the previous recipe, we successfully implemented SSO by delegating the user authentication process to an OpenID Connect provider, specifically Okta. During the test phase, we noticed that, despite successful authentication, access to Kibana was still restricted. This recipe aims to complete the authorization process by mapping roles to grant appropriate access.

Getting ready

To successfully implement this recipe, it's essential to have already completed the preceding *Configuring single sign-on* recipe, as it provides the necessary foundation.

Another required recipe is *Managing and securing access to Kibana spaces*, as we will reuse the `cookbook_business_reader` role and its associated space.

How to do it...

Implementing role mapping is a straightforward process. There are two main approaches: through the Kibana interface or using the Stack's API for those who prefer a programmatic approach. In this recipe, we'll focus on the first method, using Kibana, to demonstrate the process.

Follow these steps:

1. Log back in to **Kibana** with your default Cloud admin user, head to **Stack Management | Role Mappings**, and click on **Create role mapping**.

2. First, fill the **Role mapping** section with the following:

   - **Mapping name**: `okta_oidc_sso_business_reader`

   - **Roles**: Select the **cookbook_business_reader** role

3. In the **Mapping rules** section, click on **Add rules**; configure the rule with the following, then click **Save Role Mapping**:

   - **User field**: **groups**

   - **Type**: **text**

   - **Value**: `Elastic Group`

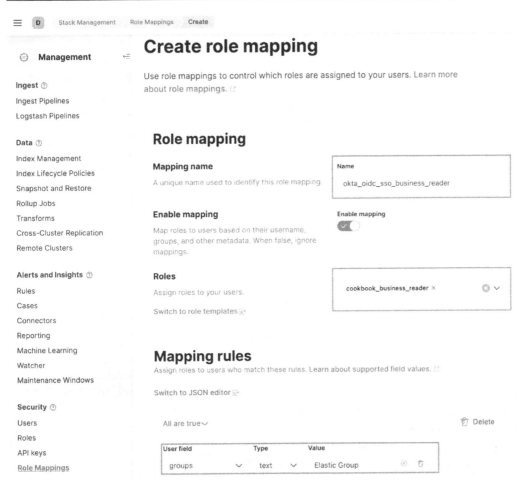

Figure 11.52 – Role mappings setup for SSO

4. Sign out from the current Cloud user, then open an *incognito* or *private* window in your browser and log in to Kibana, opting for SSO using Okta via the OIDC realm. As you authenticate with Okta, you should see the screen shown in *Figure 11.53*:

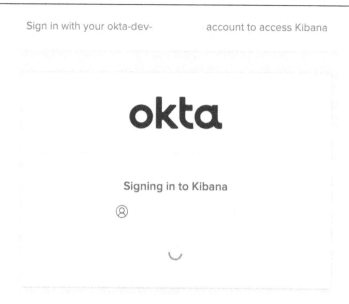

Figure 11.53 – Signing in to Kibana

5. Once the process is complete, you'll gain access to Kibana with the `cookbook_business_reader` role, granting access to the **Traffic analysis** space, as illustrated in *Figure 11.54*:

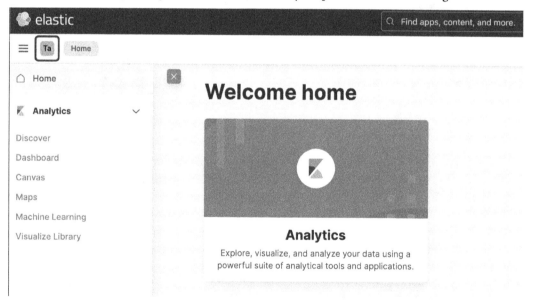

Figure 11.54 – SSO for the Traffic analysis space with the cookbook_business_reader role

As we conclude our exploration of SSO using OpenID Connect within the Elastic Stack, with Okta serving as the OpenID Connect provider, it is important to reflect on how the process was segmented into two fundamental phases: authentication and authorization.

Initially, the focus was on authentication, which involved configuring Okta. This setup included establishing a group and an application, enabling Elastic users to log in via Okta. We also integrated Elasticsearch by setting up OIDC realms based on the data from Okta.

The second phase, which was the core of this recipe, centered on authorization. This entailed mapping an existing role in the Elastic Stack to our group in Okta, thus completing the SSO process. This two-part approach underscores the distinct yet interconnected roles of authentication and authorization in establishing a secure and efficient SSO system.

How it works...

In the context of integrating OpenID Connect with the Elastic Stack and Okta, role mapping plays a crucial role in the authorization process. Role mapping is the mechanism through which user groups defined in Okta are linked to specific roles within the Elastic Stack. When a user successfully authenticates via Okta (the OpenID Connect provider), Elastic Stack then determines what level of access and permissions this user should have based on their group membership in Okta. This is where claims mapping is also very important.

The main purpose of role mapping is to streamline user permission management within the Elastic Stack. By linking Okta groups to specific roles in the Elastic Stack, administrators can control access to various resources and operations in a centralized, efficient manner. As a result, role mapping serves as a bridge between the authentication provided by Okta and the fine-grained access control within the Elastic Stack.

There's more...

When defining role mappings, you can use the role mapping API for a more programmatic approach to how roles should be granted to the user. To learn more about this API, please check the official Elastic documentation page: `https://www.elastic.co/guide/en/elasticsearch/reference/current/security-api-put-role-mapping.html`.

12

Elastic Stack Operation

As we approach the final chapters of our exploration of the Elastic Stack, it's time to ensure that the stack is correctly managed. In this chapter, you will learn the essential recipes for operational best practices such as index lifecycle management, backup strategies, and cluster management automation. In this chapter, we're going to cover the following topics:

- Setting up an index lifecycle policy
- Optimizing time series data streams with downsampling
- Managing the snapshot lifecycle
- Configuring Elastic Stack components with Terraform
- Enabling and configuring cross-cluster search

Technical requirements

In prior chapters, we designed recipes so that you can complete most of them on an Elastic Cloud trial deployment. This offers the easiest way to provision the most up-to-date Elastic Stack deployment, without needing to deploy on your own infrastructure and without cost considerations, as the trial instance is free for 14 days on Elastic Cloud. In this chapter, the following recipes involve multiple Elastic Stack deployments, for which a paid subscription is necessary:

- *Configuring Elastic Stack components with Terraform*
- *Enabling and configuring cross-cluster search*

Nonetheless, the rest of the recipes can be completed using Elastic Cloud trial deployments, just like in previous chapters.

Setting up an index lifecycle policy

Throughout the course of this book, we have seen how to ingest various types of data into the Elastic Stack and utilize it for diverse use cases such as observability, business analytics, and search. In real-world scenarios, a common question that arises is, "How do I manage the lifecycle and retention of my data? More importantly, how can I do so efficiently to optimize and reduce hardware costs while still fully leveraging the data?" This is where **Index Lifecycle Management** (**ILM**) comes into play. ILM is an extremely useful feature that streamlines the orchestration of your data within an Elasticsearch cluster.

In this recipe, we will put this concept into practice by defining an ILM policy for our Logs data stream. You will learn how to configure a policy and along the way, gaining insight into key aspects such as rollover, ILM phases, and much more.

Getting ready

Make sure that you have completed the *Monitoring Kubernetes environments with Elastic Agent* recipe of *Chapter 10*.

ILM policies rely on the concepts of data tiers, which we previously touched upon in *Adding data tiering to your deployment* in *Chapter 1*. For this recipe, your cluster should include the following additional data tiers: **Warm**, **Cold**, and **Frozen**.

> **Important note regarding autoscaling**
>
> If your deployment has autoscaling enabled for data and machine learning, the following steps are not necessary. The autoscaling feature will automatically create the required nodes based on where the data needs to be moved according to the lifecycle policy.

If you don't have autoscaling enabled on your Elastic Cloud deployment, follow these steps to add the required data tiers (note that the specified values are accurate at the time of writing this book but may change over time):

1. Head to **Cloud Console** and click on **Manage** right beside your deployment name. On the deployment page, in the left menu, click **Edit**.

2. Find **Warm data tier** below **Hot data tier** and click on **Add capacity**. In the drop-down list, select **380 GB storage | 2 GB RAM | Up to 2.5 vCPU** for the size per zone. As for the value for **Availability zones**, select **2**.

3. Find **Cold data tier** right below **Warm data tier** and click on **Add capacity**. In the drop-down list, select **380 GB storage | 2 GB RAM | Up to 2.5 vCPU** for the size per zone. As for the value for **Availability zones**, select **1**.

4. Go to **Frozen data tier**, click on **Add capacity**, and leave the default **Size per zone** value of **6.25 TB storage | 4GB RAM | Up to 2.5 vCPU**. Scroll down the **Edit** page and click on the **Save** button. Click **Confirm** on the popup to start applying the configuration changes.

Wait for the configuration settings change to be applied before moving on to the next section.

How to do it...

The objective of this recipe is to configure an ILM policy for our Kubernetes logs. We will change the default managed policy to implement the following lifecycle:

1. Retain data in the hot tier for 7 days before rolling over to a new index.

2. After rollover, move the data to the warm tier and retain it for 30 days.

3. After 30 days in the warm phase, logs are transferred to the cold tier to be retained for an additional 60 days.

4. Following the cold phase, data is transitioned to the frozen tier, where it is kept for 90 days.

5. Delete the data after a total of 180 days.

Let us start configuring the outlined policy:

1. Go to **Kibana | Stack Management | Index Lifecycle Policies** and toggle the **Include managed system policies** option as shown in the following figure:

Figure 12.1 – Include managed system policies

2. In the search box, enter `logs` and select the **logs** policy. Upon navigating to the policy detail page, a warning message will be displayed, cautioning you that editing a managed policy can potentially disrupt Kibana functionality. We will disregard this warning for the moment. The modifications we intend to make are straightforward, minimizing concerns about editing the policy. As you can see, the policy comprises only the hot phase, where the data is kept forever. This is far from ideal and does not reflect a real production use case. Click on **Advanced settings** in the hot phase to collapse the available options.

3. In the **Rollover** section, uncheck the **Use recommended defaults** option to enable other settings. Then, change **Maximum age** from 3 0 days to 7 days. This adjustment instructs Elasticsearch to begin writing to a new index once the current one reaches an age of 7 days, a process known as **rollover**. For the time being, leave the remaining options at their default settings.

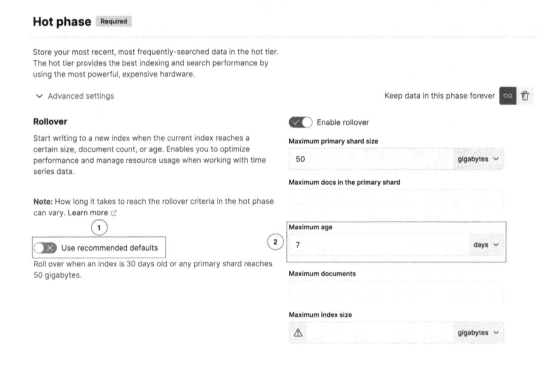

Figure 12.2 – ILM Hot phase settings

Important consideration for rollover

You might be curious about the setting of a 50 GB maximum primary shard size. The two conditions for triggering a rollover—time and shard size—are mutually exclusive, which means the rollover occurs as soon as one condition is met. This means that our data might spend fewer than 7 days in the hot phase if it reaches the 50 GB limit first, for instance. Therefore, considering your data ingestion rate is crucial when designing your ILM policy.

4. Now, we will activate the warm phase by setting 0 days for **Move data into phase when**. You might wonder why. The reason is that we aim to transition the data to the warm phase immediately after the index rolls over in the hot phase, which occurs after 7 days. So, this number corresponds to the age of the data at rollover.

Figure 12.3 – ILM Warm phase settings

5. Scroll down to activate **Cold phase** and enter 30 for the number of days. To leverage the **Searchable Snapshot** feature, enable the **Convert to fully-mounted index** option. This action replaces replicas with a copy of the data in the snapshot repository. The primary benefit is a 50% increase in storage space compared to using replicas on disk:

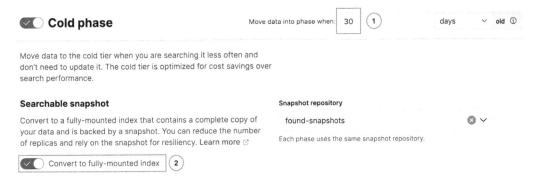

Figure 12.4 – ILM Cold phase settings

6. The next stage is **Frozen phase**, where data is moved after spending 60 days in **Cold phase**. Therefore, set the **Move data into phase when** option to 60 days. The frozen tier is best suited for long-term retention while keeping the ability to query the data through **partially mounted indices** (another feature powered by **Searchable Snapshot**). Since this is the final data tier, proceed to activate the **Delete phase**. Under **Frozen phase**, find and select the **Keep data in this phase forever** option, then click on the trash icon to reveal the **Delete phase** panel. In the **Move data into phase when** field, enter 180 days (about 6 months). This means the data will stay frozen for 90 days (180 – 30 days in warm phase – 60 days in cold phase) Afterward, scroll down and click on **Save policy** to apply the changes:

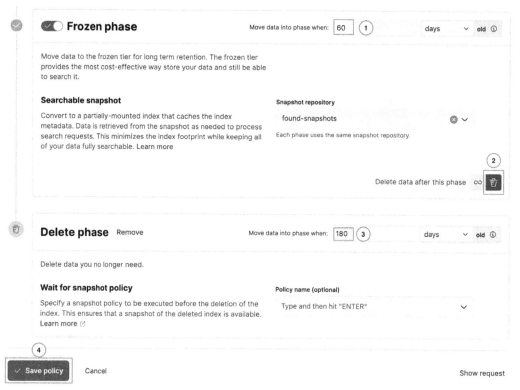

Figure 12.5 – Frozen phase and Delete phase settings

We have just changed the default logs policy to match a more realistic scenario. Instead of keeping the data in the hot phase forever, we have updated the policy to keep the data only for 7 days in the hot phase and move right away into the warm phase, where the data sits for an additional 30 days. Index data is then transferred to the cold tier, where we also activated the full mounted index to reduce the storage footprint while maintaining resiliency. Afterward, data is moved to the frozen phase and kept for 90 days. After that, upon reaching 180 days of retention since rollover, the data gets deleted.

7. To check that your ILM policy is being applied and that the data is being transitioned through the data tiers, go to **Index Management**, locate the **Lifecycle phase** dropdown, and select any of the following data tiers:

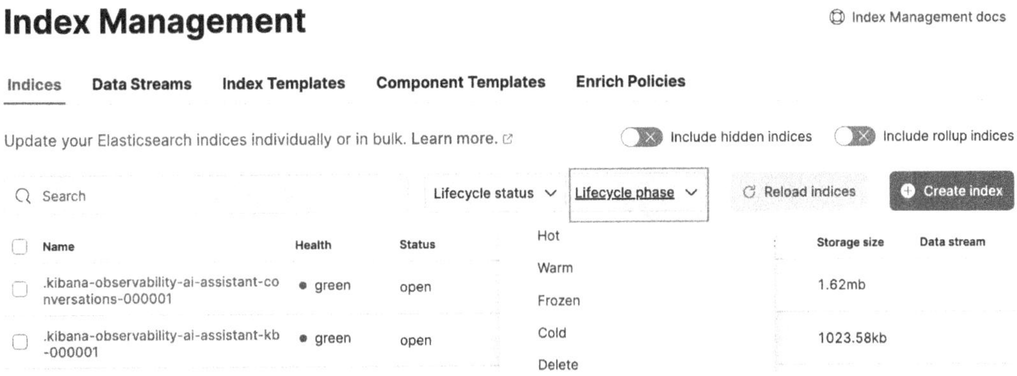

Index Management

Index Management docs

Indices	Data Streams	Index Templates	Component Templates	Enrich Policies

Update your Elasticsearch indices individually or in bulk. Learn more.   ⊗ Include hidden indices   ⊗ Include rollup indices

Q Search   Lifecycle status ∨   Lifecycle phase ∨   ↻ Reload indices   ⊕ Create index

Name	Health	Status		Storage size	Data stream
			Hot		
.kibana-observability-ai-assistant-conversations-000001	● green	open	Warm	1.62mb	
			Frozen		
.kibana-observability-ai-assistant-kb-000001	● green	open	Cold	1023.58kb	
			Delete		

Figure 12.6 – Select a phase to check indices

> **Note**
> Modifying the logs management policy will impact all data streams, which may not be desirable if a more granular approach is preferred. Later in this guide, we will detail alternative methods to achieve more targeted policies.

How it works...

An ILM policy manages the lifecycle of an index from its creation to its eventual deletion. It automates routine tasks based on predefined criteria, ensuring efficient data management and storage optimization across different stages of data's life. These stages, or phases, include hot, warm, cold, and delete, each tailored to the data's access patterns and storage costs. The following table outlines the purpose of each phase and the recommended use cases:

Phase	Purpose	Recommended use
Hot	For active indexing and providing fast access to recent data.	For data that is being actively written to and queried. Ideal for real-time search and analytics.
Warm	To hold data that is accessed less frequently but still needs to be readily available and rarely updated.	For data that is less frequently accessed. Suitable for short-term analysis and historical comparisons.
Cold	To store data that is rarely accessed and does not require fast retrieval times. The index is read-only at this stage.	For infrequent access data. Cost-effective for long-term storage where access speed is not critical. Utilizes searchable snapshots to optimize storage space.
Frozen	To archive data that is seldom accessed, optimizing for storage cost over access speed.	For data that is accessed very rarely. Utilizes searchable snapshots for minimal storage costs.
Delete	To permanently remove data that is no longer needed, freeing up storage space.	For data that has reached the end of its lifecycle and is no longer required for compliance or analysis.

Table 12.1 – ILM phases and use cases

Data tiers support this lifecycle by providing different storage classes (hot, warm, cold, and frozen) that align with the lifecycle phases. Data automatically transitions between tiers as it ages, according to the rules defined in the ILM policy.

Under the hood, ILM uses routing allocation to manage the movement of indices across different data tiers efficiently. Routing allocation is a mechanism that controls where shards of an index are placed within the cluster, based on specific criteria such as node attributes.

As we have discussed previously, Elasticsearch nodes can be configured with specific roles or custom attributes that align with the data tiers (e.g., `data_hot`, `data_warm`, `data_cold`, and `data_frozen`). When an index transitions from one phase to another, ILM updates the index's settings to include allocation rules. These rules specify that the index's shards should only be allocated to nodes with matching attributes (e.g., moving from `data_hot` to `data_cold`).

There's more...

We have briefly touched upon several settings available in the hot phase and warm phase. Those settings are very important when it comes to tuning lifecycle policies according to your needs, so let's look at some of them:

- **Force merge**: This is very important, especially when dealing with time series. Essentially, it reduces the number of segments to 1, which in turn reduces resource usage. As a best practice, always activate this setting in the hot phase to take advantage of the higher computing resources available on hot nodes.

- **Shrink**: Reduce the number of shards. If you've set multiple primary shards to speed indexing, for example, you can use the **Shrink** parameter to reduce primary shards after rollover.

- **Searchable snapshots**: This one might seem surprising as this feature is usually associated with the cold and frozen tiers, but you can indeed leverage a fully mounted index in the hot phase and replace replicas. While this sounds great in terms of storage, it will hurt your query latency and resiliency as you might lose data if an outage happens while the index is being actively written into.

- **Downsample**: This is for time series data streams and rolling up documents to reduce storage footprint. We will discuss this in detail in the *Optimizing time series data streams with downsampling* recipe.

- **Read only**: This is self-explanatory—it deactivates the writing of new data to the index.

- **Index priority**: This one is especially interesting for recovery scenarios. It allows you to define the indices you want to give priority to during a node restart.

- **Replicas**: This one is available only in the warm phase and it allows you to define the number of replicas for your indices. As data moves to the warm tier, it is generally a best practice to reduce the number of replicas, assuming you've configured your indices with more than one, simply because the data is generally less queried.

Warning for when modifying managed ILM policies

While editing a managed ILM policy, you might have seen a warning advising against directly modifying managed policies. For simplicity, we have intentionally chosen to work with managed ILM policies in this and next recipes. To apply a custom ILM policy to an integration's data stream, please consult the following documentation: `https://www.elastic.co/guide/en/fleet/current/data-streams-ilm-tutorial.html`.

See also

- Troubleshooting ILM can sometimes be a challenging task, but we've got you covered with an excellent guide: `https://www.elastic.co/guide/en/elasticsearch/reference/current/index-lifecycle-error-handling.html`

- If you're looking for a deeper dive into the intricacy of ILM and data tiers, check out this blog article: `https://www.elastic.co/blog/elasticsearch-data-lifecycle-management-with-data-tiers`

Optimizing time series data streams with downsampling

In the previous recipe, we learned how ILM automates the process of moving older data to less expensive storage tiers, optimizing costs without affecting data availability. In this recipe, we will explore how to use downsampling to reduce the granularity of data by aggregating it over larger time intervals, leading to further cost savings in both storage and operational expenses.

Getting ready

Make sure you have completed the recipes in *Chapter 10*.

In this recipe, we will explore how to downsample metrics data.

How to do it...

In this recipe, you will learn how to configure downsampling with ILM policies, verify that downsampling is correctly triggered, and examine the effects of these operations on your data at the document level and in terms of overall data visualization:

1. Go to **Kibana | Stack Management | Index Lifecycle Policies**, and type `metric` in the search bar. Then toggle on **Include managed system policies**:

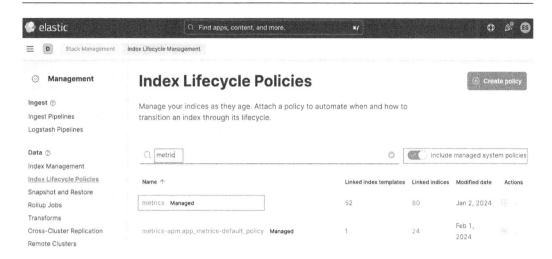

Figure 12.7 – Finding the metrics index lifecycle policy

2. Once on the **Index Lifecycle Policies** configuration page, scroll down to the **Warm phase** section and toggle it to activate the warm phase. For the **Move data into phase when** condition, set the threshold to 3 days old. Note that depending on when you completed the recipes in *Chapter 10*, this value can be adjusted to less than 3 days old; it will not alter the flow of this recipe. After setting the condition for the warm phase, enable downsampling by toggling **Enable downsampling** on in the **Downsample** section and set **Downsampling interval** to 5 minutes, as shown in *Figure 12.8*:

Figure 12.8 – Configuring downsampling in the warm phase

3. After configuring the ILM policy for the metrics, if your metrics data has sufficient retention and meets the condition set in the previous step, the downsampling process will start. To verify the downsampling indices, go to **Stack Management | Index Management**, toggle **Include hidden indices**, and in the search bar, type `downsample kubernetes`. As illustrated in *Figure 12.9*, you should see indices prefixed with **downsample-5m** automatically created for the **metrics-kubernetes.pod-default** data stream. This indicates that downsampling is occurring as expected:

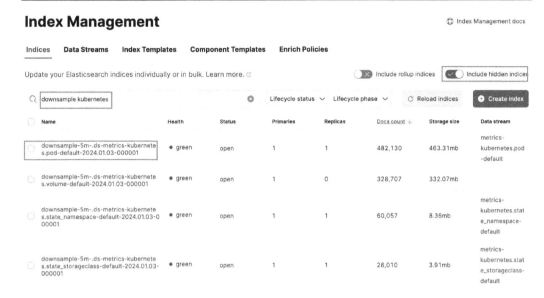

Figure 12.9 – Verifying the downsample indices

Now, let's examine what is happening at the document level and the effect of downsampling.

4. First, let's check the documents that have not yet been downsampled, which remain in the hot phase. Go to **Analytics | Discover** and select the **metrics**- data view in the top-left corner. Then, in the search bar, type `kubernetes.pod.cpu.usage.node.pct: *` as shown in *Figure 12.10*; this query will filter out the **metrics-kubernetes.pod-default** data stream.

5. To finish, select a time range that falls within the hot phase of your metrics data. For instance, if you set the transition to the warm phase to 3 days ago, you might choose **Last 24 hours** to inspect documents still in the hot phase. If your condition differs from 3 days ago, adjust the time range accordingly:

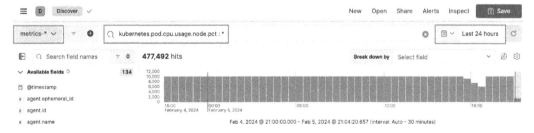

Figure 12.10 – Inspecting the unsampled metrics documents in the hot phase

6. Inspect the details of any document by clicking on the toggle symbol next to the document row, which will reveal a flyout with detailed information. Scroll down to the `kubernetes.pod.cpu.usage.node.pct` metric to see the original unsampled value, as shown in *Figure 12.11*:

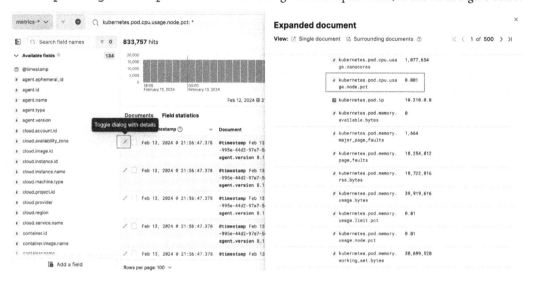

Figure 12.11 – Inspecting the unsampled document details

7. It is worthwhile to compare this with the downsampled documents. Select a time range that falls within the warm phase—for example, from 5 days ago to 4 days ago if you set the transition to 3 days ago. This allows you to inspect documents that have been moved to the warm phase and have been downsampled (adjust the time range if your conditions are different). After selecting the new time range, you may notice a reduction in document counts over the same 24-hour interval due to downsampling, as shown in *Figure 12.12*:

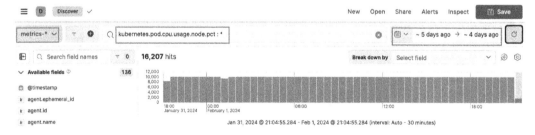

Figure 12.12 – Inspecting the downsampled documents

8. Check the details of one of the documents, as you did previously, and look for the `kubernetes.pod.cpu.usage.limit.pct` metric. The downsampled document should now contain different aggregations, including **min**, **max**, **sum**, and **values_count**, over the 5-minute downsampling interval we defined earlier in *step 2*. We can also see that some other metrics, such as `kubernetes.pod.cpu.usage.nanocores`, have also been aggregated:

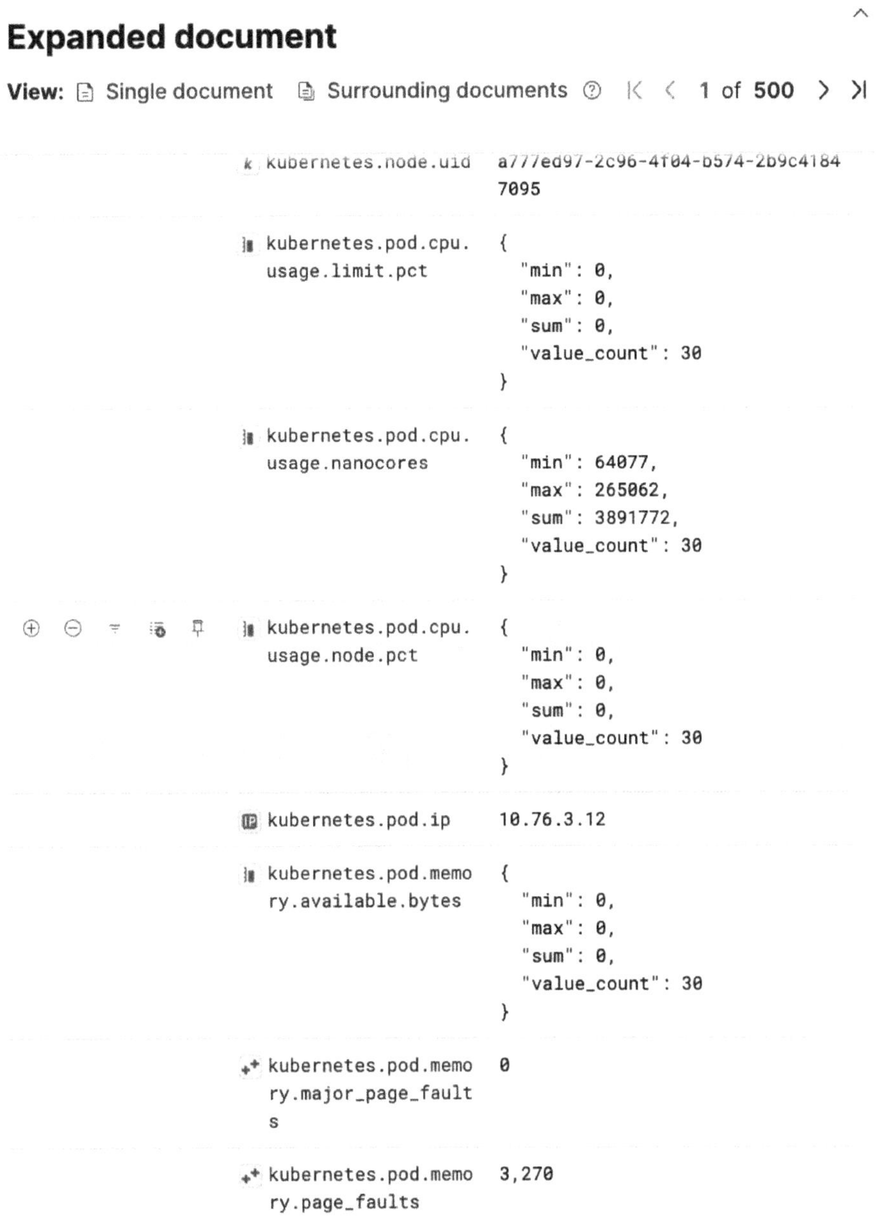

Expanded document

View: ▤ Single document   ▤ Surrounding documents   ⑦ |< ‹ 1 of **500** › >|

```
k kubernetes.node.uid    a777ed97-2c96-4f04-b574-2b9c4184
                         7095

je kubernetes.pod.cpu.   {
   usage.limit.pct          "min": 0,
                            "max": 0,
                            "sum": 0,
                            "value_count": 30
                         }

je kubernetes.pod.cpu.   {
   usage.nanocores          "min": 64077,
                            "max": 265062,
                            "sum": 3891772,
                            "value_count": 30
                         }

⊕  ⊖  ≂  ⓘ  ⎙   je kubernetes.pod.cpu.   {
               usage.node.pct          "min": 0,
                                       "max": 0,
                                       "sum": 0,
                                       "value_count": 30
                                    }

⑫ kubernetes.pod.ip     10.76.3.12

je kubernetes.pod.memo   {
   ry.available.bytes       "min": 0,
                            "max": 0,
                            "sum": 0,
                            "value_count": 30
                         }

₊⁺ kubernetes.pod.memo   0
   ry.major_page_fault
   s

₊⁺ kubernetes.pod.memo   3,270
   ry.page_faults
```

Figure 12.13 – Inspecting the downsampled document details

9. Finally, let's compare the impact of downsampling on data exploration. Go to **Analytics |
    Dashboard** and open the **[Metrics Kubernetes] Pods** dashboard. First, set the time range to
    Last 24 hours to match the period within the hot phase (adjust as necessary based on *step 4*).
    Then scroll down to examine the Pod CPU and memory visualizations, as shown in *Figure 12.14*:

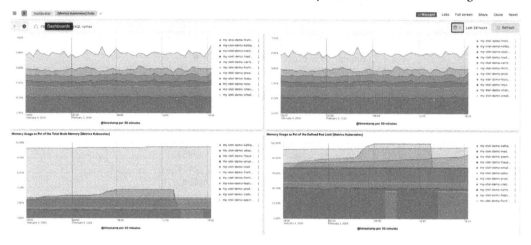

Figure 12.14 – An overview visualization of the unsampled documents

10. Change the time range to a 24-hour period that falls within the warm phase and examine the
    Pod CPU and memory visualizations again. Despite the downsampling, the visualizations should
    retain their accuracy with a 5-minute downsampling interval, as indicated in *Figure 12.15*:

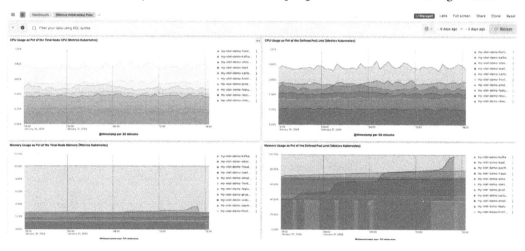

Figure 12.15 – An overview of the visualization of downsampled documents

How it works...

In the *Downsampling time series data* recipe in *Chapter 5*, we explored the fundamentals of downsampling within the context of time series data. Please refer back to that chapter for a foundational understanding of how downsampling works.

This time, we applied the concept of downsampling to metrics collected by Elastic Agent. As a practical example, we used data from the Kubernetes cluster's metrics from the OpenTelemetry demo that we introduced in *Chapter 10*.

Now, let's see how to find whether a specific Elastic Agent integration supports time series data streams and how to customize built-in ILM policies.

Elastic Agent TSDS support for downsampling

For downsampling to be effective, the Elastic Agent integration must support **time series data streams** (**TSDS**). To verify whether your installed integration has this support, navigate to **Kibana** and go to **Management | Integrations**. Then, click on the **Installed integrations** tab and use the search bar to find the integration you want to check, as illustrated in *Figure 12.16*. In this example, we will look at the **Kubernetes** integration to align with our recipe:

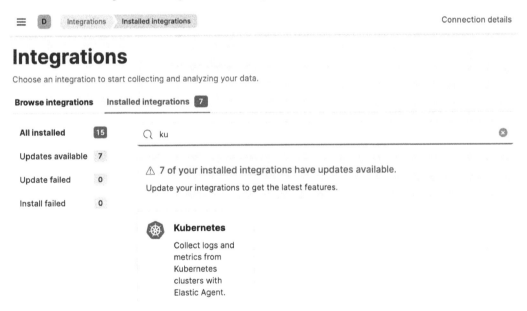

Figure 12.16 – Locating the installed integration

1. After clicking on the integration of interest, click **View Changelog** on the right side of the integration page to verify the support status, as shown in *Figure 12.17*:

Screenshots |< < **1** of **1** > >|

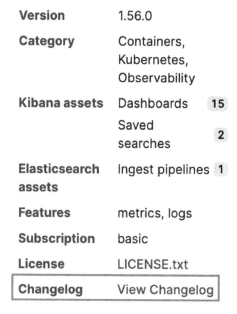

Details

Version	1.56.0
Category	Containers, Kubernetes, Observability
Kibana assets	Dashboards `15`
	Saved searches `2`
Elasticsearch assets	Ingest pipelines `1`
Features	metrics, logs
Subscription	basic
License	LICENSE.txt
Changelog	View Changelog

Figure 12.17 – Details of the integration

2. In the changelog popup, search for `tsds` (you can use *Ctrl/Cmd + F* in your browser). The support for TSDS is explicitly recorded in the changelog:

Changelog

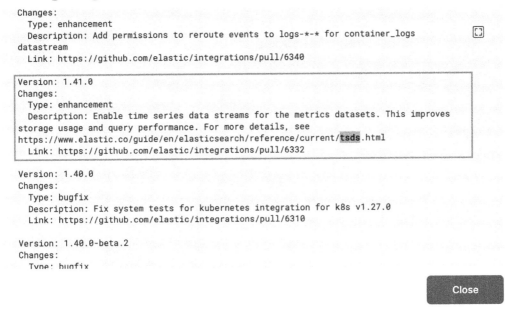

```
Changes:
  Type: enhancement
  Description: Add permissions to reroute events to logs-*-* for container_logs
datastream
  Link: https://github.com/elastic/integrations/pull/6340
```

```
Version: 1.41.0
Changes:
  Type: enhancement
  Description: Enable time series data streams for the metrics datasets. This improves
storage usage and query performance. For more details, see
https://www.elastic.co/guide/en/elasticsearch/reference/current/tsds.html
  Link: https://github.com/elastic/integrations/pull/6332
```

```
Version: 1.40.0
Changes:
  Type: bugfix
  Description: Fix system tests for kubernetes integration for k8s v1.27.0
  Link: https://github.com/elastic/integrations/pull/6310
```

```
Version: 1.40.0-beta.2
Changes:
  Type: bugfix
```

Close

Figure 12.18 – Elastic Agent integration changelog

Customizing the built-in ILM policies

Both the previous recipe and this one begin with identifying and modifying existing ILM policies for logs and metrics. The Elastic Stack provides a set of default policies for indices and data streams created using Elastic Agent, Beats, or Logstash output plugins, intended to simplify the configuration of ILM policies for your data streams and indices.

You can easily locate the list of utilized built-in ILM policies by navigating to **Stack Management | Index Lifecycle Policies**. There, activate the toggle for **Include Managed System Policies** and click **Linked indices** to sort the linked indices by the number using each ILM policy, from highest to lowest. As indicated in *Figure 12.19*, the Elastic Stack's built-in ILM policies are tagged as **Managed**:

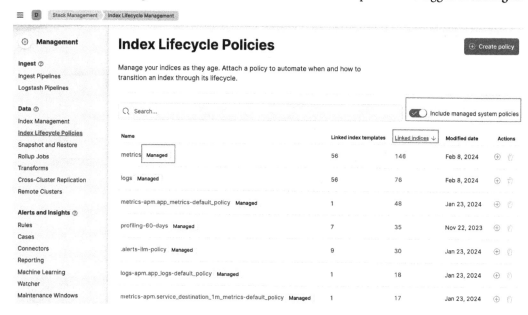

Figure 12.19 – Finding the most used built-in ILM policies

There's more...

Downsampling is linked to data streams that are managed by ILM policies. The Elastic Stack enables you to configure downsampling operations across the hot, warm, and cold phases, allowing the application of different granularity strategies for each phase. A practical example follows:

- Hot phase retention, 7 days (no downsampling)

- Warm phase retention, 1 month (downsampling at 1-hour intervals)

- Cold phase retention, 6 months (downsampling at 1-day intervals)

See also

- The following blog post explains in detail how downsampling can help to increase storage and performance efficiency: `https://www.elastic.co/blog/70-percent-storage-savings-for-metrics-with-elastic-observability`

Managing the snapshot lifecycle

Snapshots are a crucial aspect of operating your stack in a production environment. It allows you to have frequent backups of your whole cluster and be able to use backups to restore your system and data in the case of a serious failure. Elasticsearch comes with a built-in mechanism to streamline the creation and retention of snapshots: **Snapshot Lifecycle Management** (**SLM**). It is a very important feature that removes the burden of worrying about taking regular snapshots.

In this recipe, we go through the process of setting up an SLM policy and applying it to our deployment.

Getting ready

Make sure to have completed the *Monitoring Kubernetes environments with Elastic Agent* recipe in *Chapter 10*.

For this recipe, we'll be creating a repository for the snapshot. Follow the *Setting up a snapshot repository to register a new repository on AWS* recipe in *Chapter 1*. If you happen to have set up the repository on any other cloud provider, that's perfectly fine. To simplify the process, please name your repository `cookbook-slm-repository`.

How to do it...

The goal of this recipe is to define an SLM policy in a new repository. The policy will take daily snapshots of the `logs-*` indices and retain them for 30 days.

> **Note**
>
> As mentioned in the *Getting ready* section, we assume that you have set up and registered a repository for snapshot storage.

1. To create a policy, head to **Kibana | Stack Management | Snapshot | Restore | Policies**. Then, click on **Create policy**:

Figure 12.20 – Snapshot policies

2. Enter the following information on the policy creation page:

    -   **Name**: `cookbook-snapshots`

    -   **Snapshot name**: `daily-cookbook-snapshots`

    -   **Repository**: Select the newly created repository (**cookbook-slm-repository**)

- **Schedule**: Here you define the desired schedule of the policy. In our case, we will use a cron expression by clicking on **Create cron expression**. Then, set the cron expression to 0   50 0   *   *   ?.

Create policy

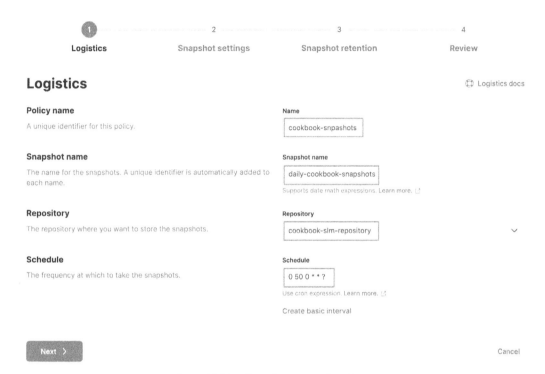

Figure 12.21 – Snapshot general settings

3. Click **Next** to proceed to the snapshot settings. Uncheck the **All data streams and indices** option, and in the **Index patterns** field, enter `logs-*`. Then, click **Next**:

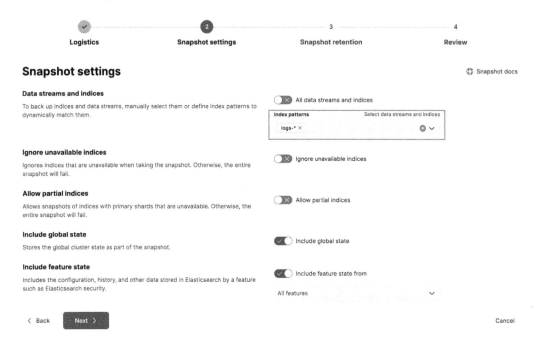

Figure 12.22 – Snapshot data stream and indices settings

4. On the **Snapshot retention (optional)** page, set **Delete After** to 30 days and click **Next**:

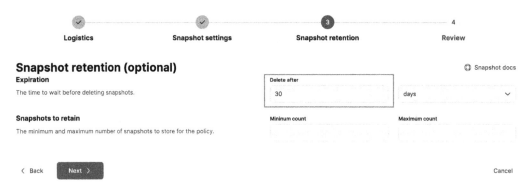

Figure 12.23 – Snapshot retention

5.   Review the settings, and if everything looks satisfactory, proceed by clicking on **Create policy**:

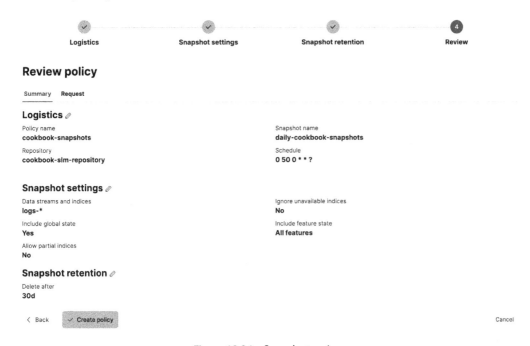

Figure 12.24 – Snapshot review

6.   To initiate the snapshot for the `logs-*` indices in the new repository, click on the play icon to run the policy:

Figure 12.25 – Initiating manually a snapshot

7. After the policy has executed, click on the **cookbook-snapshots** policy. In the details panel under the **Summary** tab, you will find useful information, such as the number of snapshots taken, any failures, and more:

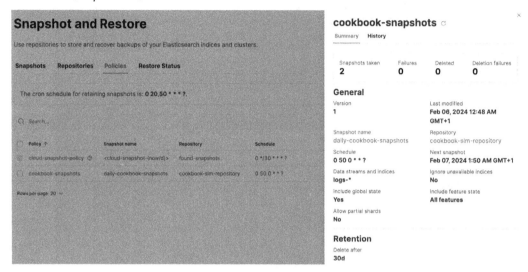

Figure 12.26 – Snapshot policy summary

While the **Snapshot and Restore** feature provides valuable information on the status of your snapshot policies, we can leverage another Elastic Stack feature to have a more responsive approach: alerting. Let us set up a quick rule that will alert us whenever our policy fails to execute successfully.

8. Head to **Dev Tools** and execute the following API call to create a data view for the indices where SLM policies history is stored (the snippet can be found at the following address: https://github.com/PacktPublishing/Elastic-Stack-8.x-Cookbook/blob/main/Chapter12/snippets.md#dev-tools-command-to-create-a-data-view-for-slm-history-indices):

```
POST kbn:/api/data_views/data_view
{
  "data_view": {
    "title": ".slm-history-*",
    "name": "SLM data view",
    "timeFieldName": "@timestamp"
  }
}
```

You should obtain an HTTP 200 response with information on the data view:

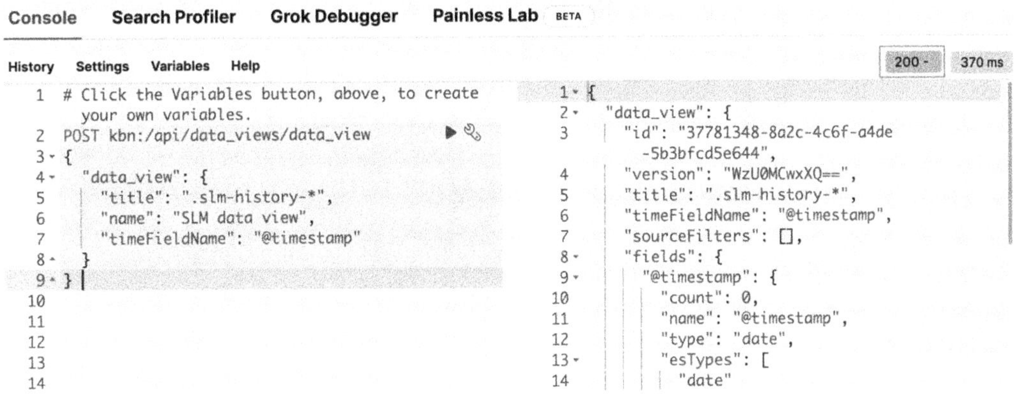

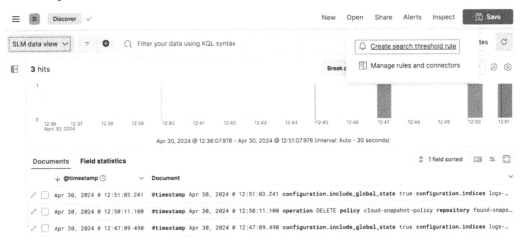

Figure 12.27 – SLM data view creation

9. Open **Discover** in a new browser tab and select the SLM data view you just created. Enter `policy: cookbook-snapshots` into the search bar. With the data view's creation validated, we'll move on to setting up the alert. Recall that our goal is to establish a rule to notify us if our policy fails to run. Locate the **Alerts** button in the upper right, click it, and from the drop-down menu, select **Create search threshold rule** to access the rule configuration flyout:

Figure 12.28 – Create a rule from Discover

In the rule configuration flyout, enter the following information:

- **Name**: Detect failure on cookbook snapshots policy.
- **Tags**: SLM.

- In the **Define your query** section, configure the query `policy: cookbook-snapshots` and `success: false`, as shown in *Figure 12.29*:

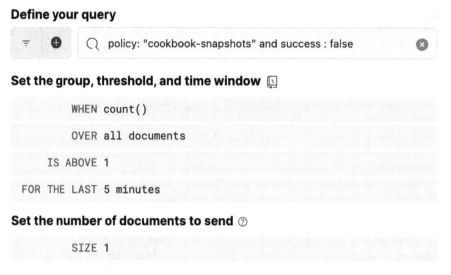

Figure 12.29 – Define the query for an alert on snapshot policy failure

- Set **Check every** to 5 minutes for testing purposes.

- Add an **Email connector** action type and select **Elastic-Cloud-SMTP** as the email connector. In the **To** field, input the email address where you would like to receive notifications. For **Subject**, type in `Cookbook snapshot policy execution failure alert`. In the **Message** field, insert the following snippet (you can find the snippet here: `https://github.com/PacktPublishing/Elastic-Stack-8.x-Cookbook/blob/main/Chapter12/snippets.md#message-for-alert-on-snapshot-policy-failure`):

```
Elasticsearch rule '{{rule.name}}' is active:
- Message: {{context.message}}
- Value: {{context.value}}
- Conditions Met: {{context.conditions}} over {{rule.params.
timeWindowSize}}{{rule.params.timeWindowUnit}}
- Timestamp: {{context.date}}
- Link: {{context.link}}
```

Your configuration should look as shown in *Figure 12.30*:

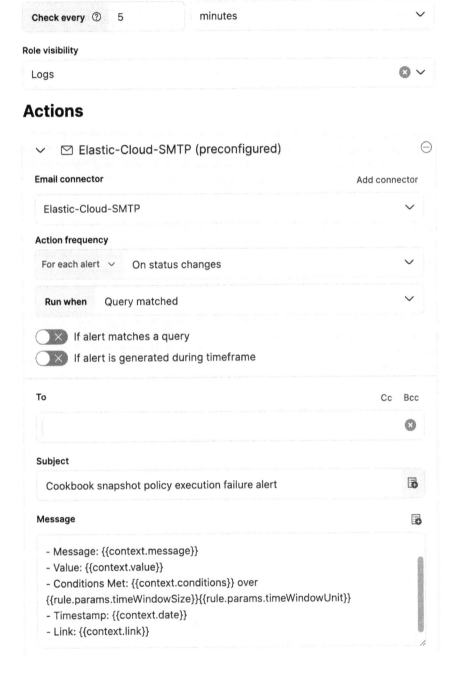

Figure 12.30 – Schedule and Action configuration

10. Finally, click **Save** to save the rule. To test the effectiveness of our rule, consider temporarily modifying the repository's configuration, for example, by changing the repository's base path to an invalid value. Then, attempt to execute the snapshot policy. By doing this, the next time the policy tries to run, you should receive an alert notification at the email address you configured with the following message:

Elasticsearch query rule 'Detect failure on cookbook snapshots policy' is active:

- Message: Document count is 2 in the last 1h in SLM data view data view. Alert when greater than 1.
- Value: 2
- Conditions Met: Number of matching documents is greater than 1 over 1h
- Timestamp: 2024-02-06T21:58:46.353Z
- Link: https://elastic-stack-8-cookbook.kb.eu-west-1.aws.found.io:9243/app/r?l=DISCOVER_APP_LOCATOR&v=8.12.0&lz=
N4IgJghgLhBqCWBTA7gZQA6IMYgFynjDxAE4AjCAZgCYBWLAMwFoA2LAdhKYBYWBGMkwAcYdkKZkADEOrtEJWrTJYW
IADQgo8KABtExAHQBnHQFsmAC3hGoAewBOATyYAqdZvinEAMSQ6wAHIQXsQAAIpeNsHo7ka2AK72WD7wOICI9kZ4AN
oAuhoMfmDeDqbQWfgAvhr28QB2ESmI%2FgCyEDFVGhA6OrblAbYAknVgiAAeeAzdRogadcH6uCCoADItAASQMOsAbkj
I7t29yAAShKN1k9OI1SCFaRkVeRoAjvEZjnigbx%2FE6LY6eBYRy4dYAHRAWFstgA1mRoTCmEZ5ugjBZbFAjBD1hARusjPEsMkjE
Z1qCpjoZut3DpcQBzeIQOmLEAw95OEC3RoAJXpi1ADHstIMxGokmo3CY4qILAAKmLcLQhLheAZKLRKAAtdx2UXiyXS
yRy6h8RXK1XqrWcjTWACCensUG5iAJaQqUFqNyAA%3D%3D%3D

This message was sent by Elastic. View rule in Kibana.

Figure 12.31 – Email notification for the snapshot rule

11. By clicking on the link, you will be redirected to the **Discover** page, where you can expand the document that triggered the alert. There, examine the `error_details` field for a better understanding of the issue:

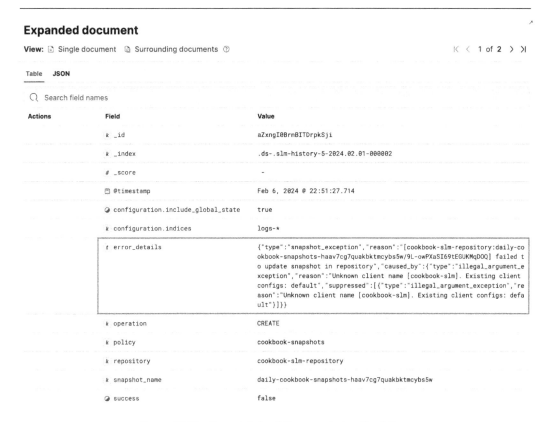

Expanded document

View: 📄 Single document   📄 Surrounding documents ⑦ |< < 1 of 2 > >|

Table   JSON

🔍 Search field names

Actions	Field	Value
	k _id	aZxngI0BrnBITDrpkSji
	k _index	.ds-.slm-history-5-2024.02.01-000002
	# _score	-
	🕓 @timestamp	Feb 6, 2024 @ 22:51:27.714
	◉ configuration.include_global_state	true
	k configuration.indices	logs-*
	t error_details	{"type":"snapshot_exception","reason":"[cookbook-slm-repository:daily-cookbook-snapshots-haav7cg7quakbktmcybs5w/9L-owPXaSI69tEGUKMqDOQ] failed to update snapshot in repository","caused_by":{"type":"illegal_argument_exception","reason":"Unknown client name [cookbook-slm]. Existing client configs: default","suppressed":[{"type":"illegal_argument_exception","reason":"Unknown client name [cookbook-slm]. Existing client configs: default"}]}}
	k operation	CREATE
	k policy	cookbook-snapshots
	k repository	cookbook-slm-repository
	k snapshot_name	daily-cookbook-snapshots-haav7cg7quakbktmcybs5w
	◉ success	false

Figure 12.32 – Error details of the snapshot policy failure

That's all, folks! In this recipe, we've successfully set up a snapshot policy to automate regular data backups in our Elastic Stack deployment. Additionally, we created a rule and configured alerts to notify us in the case of a failure.

How it works...

SLM in the Elastic Stack automates the process of taking snapshots of your Elasticsearch indices at regular intervals, ensuring that data and system information (e.g., cluster state) are backed up and can be restored in the case of data loss or corruption.

SLM allows administrators to define policies that specify when and how often snapshots are taken, which indices are included, and where these snapshots are stored (e.g., in a local repository or cloud-based storage). These policies can be tailored to meet various retention requirements, enabling the efficient management of snapshot storage space by automatically deleting old snapshots according to the defined rules. The following figure illustrates the modus operandi of an SLM policy:

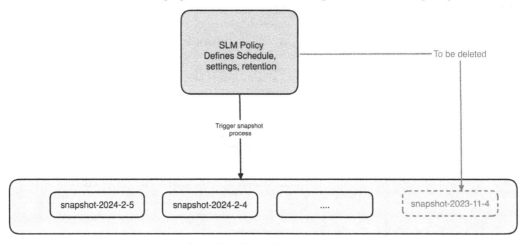

Figure 12.33 – SLM policy overview

Through Kibana, as we did in the recipe, or by using an API, you can easily create, manage, and monitor snapshot policies, making it straightforward to implement a robust data backup and recovery strategy within the Elastic Stack.

Different pieces are needed to automate the snapshot lifecycle; *Figure 12.34* recapitulates the necessary steps:

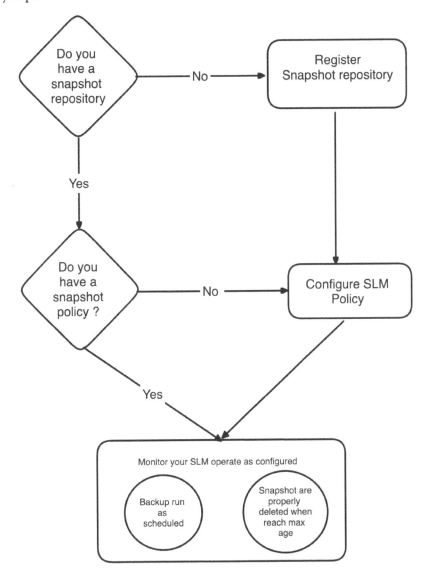

Figure 12.34 – Process for setting up and maintaining an SLM policy

There's more...

To ensure that your SLM policy is running as expected in the Elastic Stack, you can utilize several tools and features:

- **Elasticsearch APIs**: The SLM status API allows you to check the operational status of your snapshot lifecycle policies, ensuring they are active and functioning as intended. Additionally, the Get Snapshot API can be used to retrieve information about the snapshots that have been taken, including their success or failure status.

- **Kibana management UI**: As we saw during the course of this recipe, Kibana provides a user-friendly interface for managing snapshot lifecycle policies. Through the **Stack Management** section, you can create, view, edit, and delete policies, as well as see their execution history. This visual interface makes it easier to monitor and adjust your snapshot policies.

- **Alerting and monitoring**: Implement alerting and monitoring to get notifications about the success or failure of snapshot operations. The Elastic Stack's alerting framework can be configured to send alerts via email, Slack, or other channels if a snapshot fails or if there are issues with the snapshot repository.

Here are some best practices to avoid running into issues with your snapshot policies:

- Regularly review your snapshot and repository statistics through the Elasticsearch APIs or Kibana to ensure that snapshots are completing successfully and storage usage is within expected limits.

- Test your snapshot and restore processes periodically to confirm that you can reliably restore data from snapshots when necessary.

- Adjust your snapshot frequency and retention settings based on your data volume, change rate, and storage capacity to optimize performance and cost.

Configuring Elastic Stack components with Terraform

In this recipe, we will explore the usage of Terraform with Elastic Stack components, particularly on Elastic Cloud. Terraform, an open source **infrastructure as code (IaC)** software tool created by HashiCorp, enables users to define and provision a cloud or even on-premises infrastructure using a high-level declarative configuration language. By leveraging Terraform, we can automate the deployment and management of Elastic Stack components.

We will guide you through setting up Terraform to work effectively with Elastic Cloud by using the Elastic Cloud Terraform Provider, and you'll learn how to define IaC, allowing for the automated provisioning of Elastic Stack resources and how to manage their lifecycle effectively.

Getting ready

Be sure to meet the following requirements:

- You will need Terraform installed on your machine; if you don't, we'll advise you to follow the steps outlined in the Terraform official documentation (`https://developer.hashicorp.com/terraform/install`) in order to have everything up and running. The version we used for our recipe is 1.8.0.

- As we're going to provision deployment on Elastic Cloud, we will need an API key that'll be passed to Terraform. In order to get such a key, follow the steps described in the official Elastic documentation here: `https://www.elastic.co/guide/en/cloud/current/ec-api-authentication.html`. We will refer to this key as `api_key` later in the recipe.

- The snippets for this recipe can be found at this address: `https://github.com/PacktPublishing/Elastic-Stack-8.x-Cookbook/blob/main/Chapter12/snippets.md#configuring-elastic-stack-components-with-terraform`

How to do it...

The objective is to execute a Terraform plan that will create an Elastic deployment on the cloud.

1. Download the following files in the following folder in the GitHub repository: `https://github.com/PacktPublishing/Elastic-Stack-8.x-Cookbook/blob/main/Chapter12/terraform/base`:

   - `main.tf`: The main file with the desired configuration

   - `terraform.tfvars`: The file to store environment variables

2. Open the file in your preferred editor, add your `api_key`, and then save the file.

3. To install the required dependencies, execute the following command in the folder you previously downloaded:

   ```
   $ terraform init
   ```

4. Execute `terraform plan` to validate our configuration can be created, with the following command:

   ```
   $ terraform plan
   ```

You should obtain the following output (the figure does not display the full output):

```
Terraform will perform the following actions:

  # ec_deployment.terraform will be created
  + resource "ec_deployment" "terraform" {
      + alias                 = (known after apply)
      + apm_secret_token      = (sensitive value)
      + deployment_template_id = "gcp-io-optimized"
      + elasticsearch         = {
          + autoscale     = true
          + cloud_id      = (known after apply)
          + frozen        = {
              + autoscaling                = {
                  + max_size          = (known after apply)
                  + max_size_resource = (known after apply)
                  + min_size          = (known after apply)
                  + min_size_resource = (known after apply)
                  + policy_override_json = (known after apply)
                }
              + instance_configuration_id = (known after apply)
              + node_roles                = (known after apply)
              + node_type_data            = (known after apply)
              + node_type_ingest          = (known after apply)
              + node_type_master          = (known after apply)
              + node_type_ml              = (known after apply)
              + size                      = (known after apply)
              + size_resource             = "memory"
              + zone_count                = (known after apply)
            }
          + hot           = {
              + autoscaling                = {
                  + max_size          = "8g"
                  + max_size_resource = "memory"
                  + min_size          = (known after apply)
                  + min_size_resource = (known after apply)
                  + policy_override_json = (known after apply)
                }
              + instance_configuration_id = (known after apply)
              + node_roles                = (known after apply)
              + node_type_data            = (known after apply)
              + node_type_ingest          = (known after apply)
              + node_type_master          = (known after apply)
              + node_type_ml              = (known after apply)
              + size                      = "4g"
              + size_resource             = "memory"
              + zone_count                = (known after apply)
            }
```

Figure 12.35 – Terraform plan output example

5. Having confirmed that our plan can be successfully applied, we can now proceed to deploy the plan on Elastic Cloud using the following command:

    ```
    $ terraform apply
    ```

    After executing the command, Terraform will prompt you to confirm the actions it is about to take. Type yes to proceed. Terraform will then begin provisioning the resources according to your configuration, which may take approximately 5 minutes. The output in the console will resemble what is shown in *Figure 12.36*:

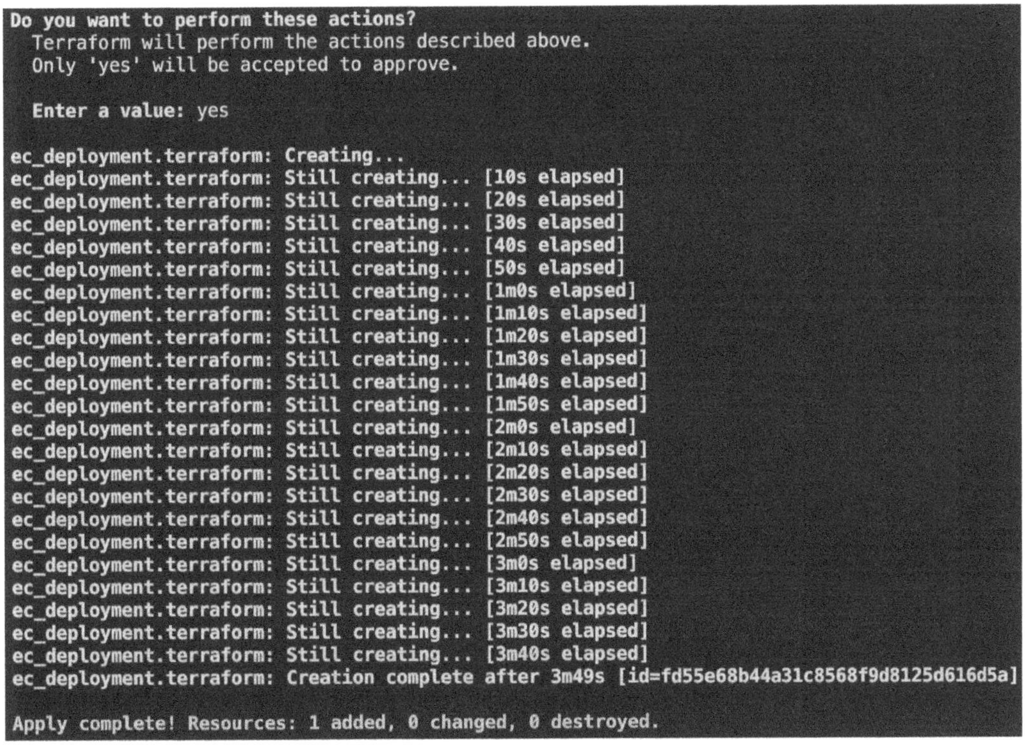

```
Do you want to perform these actions?
  Terraform will perform the actions described above.
  Only 'yes' will be accepted to approve.

  Enter a value: yes

ec_deployment.terraform: Creating...
ec_deployment.terraform: Still creating... [10s elapsed]
ec_deployment.terraform: Still creating... [20s elapsed]
ec_deployment.terraform: Still creating... [30s elapsed]
ec_deployment.terraform: Still creating... [40s elapsed]
ec_deployment.terraform: Still creating... [50s elapsed]
ec_deployment.terraform: Still creating... [1m0s elapsed]
ec_deployment.terraform: Still creating... [1m10s elapsed]
ec_deployment.terraform: Still creating... [1m20s elapsed]
ec_deployment.terraform: Still creating... [1m30s elapsed]
ec_deployment.terraform: Still creating... [1m40s elapsed]
ec_deployment.terraform: Still creating... [1m50s elapsed]
ec_deployment.terraform: Still creating... [2m0s elapsed]
ec_deployment.terraform: Still creating... [2m10s elapsed]
ec_deployment.terraform: Still creating... [2m20s elapsed]
ec_deployment.terraform: Still creating... [2m30s elapsed]
ec_deployment.terraform: Still creating... [2m40s elapsed]
ec_deployment.terraform: Still creating... [2m50s elapsed]
ec_deployment.terraform: Still creating... [3m0s elapsed]
ec_deployment.terraform: Still creating... [3m10s elapsed]
ec_deployment.terraform: Still creating... [3m20s elapsed]
ec_deployment.terraform: Still creating... [3m30s elapsed]
ec_deployment.terraform: Still creating... [3m40s elapsed]
ec_deployment.terraform: Creation complete after 3m49s [id=fd55e68b44a31c8568f9d8125d616d5a]

Apply complete! Resources: 1 added, 0 changed, 0 destroyed.
```

Figure 12.36 – Terraform apply output

6. Once the plan is complete, the outputs will be displayed in your console as illustrated here:

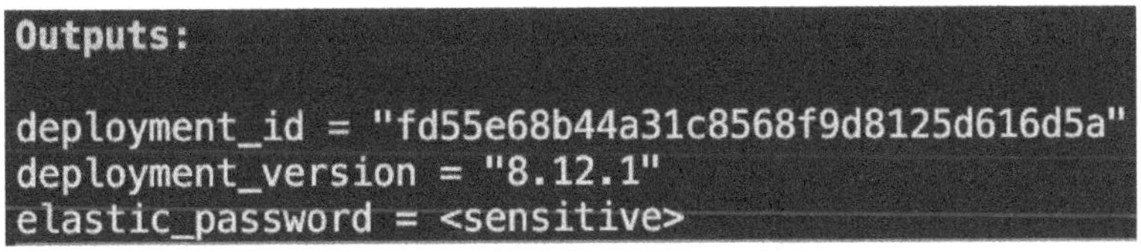

```
Outputs:

deployment_id = "fd55e68b44a31c8568f9d8125d616d5a"
deployment_version = "8.12.1"
elastic_password = <sensitive>
```

Figure 12.37 – Terraform apply final outputs

You will be able to find the Elatics Stack password in the `.tfstate` file under the `outputs` directory.

7. To confirm the successful implementation of the plan, go to the Elastic Cloud console (`https://cloud.elastic.co`) and search for a deployment named `terraform-deployment`:

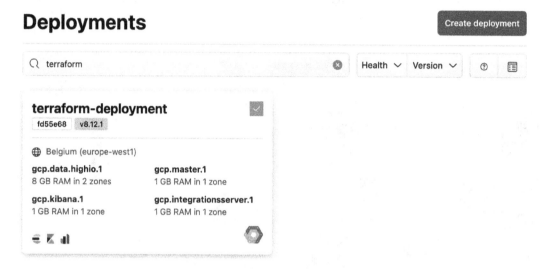

Figure 12.38 – Terraform deployment created in Elastic Cloud

8. Now that we have confirmed that our deployment is up and running, we will use `terraform apply` to delete the deployment with the following command:

```
$ terraform apply --destroy
```

You will be prompted to validate the deletion of the resources, enter `yes` to proceed.

By executing this command, we are telling Terraform to delete the plan we have just applied. Wait a couple of minutes as it unfolds, and you should see the following output in your terminal once the deletion of the resources is complete:

```
Do you really want to destroy all resources?
  Terraform will destroy all your managed infrastructure, as shown above.
  There is no undo. Only 'yes' will be accepted to confirm.

  Enter a value: yes

ec_deployment.custom-ccs-id: Destroying... [id=30467f99cdd4e158cd726531ed29b974]
ec_deployment.custom-ccs-id: Still destroying... [id=30467f99cdd4e158cd726531ed29b974, 10s elapsed]
ec_deployment.custom-ccs-id: Still destroying... [id=30467f99cdd4e158cd726531ed29b974, 20s elapsed]
ec_deployment.custom-ccs-id: Still destroying... [id=30467f99cdd4e158cd726531ed29b974, 30s elapsed]
ec_deployment.custom-ccs-id: Still destroying... [id=30467f99cdd4e158cd726531ed29b974, 40s elapsed]
ec_deployment.custom-ccs-id: Still destroying... [id=30467f99cdd4e158cd726531ed29b974, 50s elapsed]
ec_deployment.custom-ccs-id: Still destroying... [id=30467f99cdd4e158cd726531ed29b974, 1m0s elapsed]
ec_deployment.custom-ccs-id: Still destroying... [id=30467f99cdd4e158cd726531ed29b974, 1m10s elapsed]
ec_deployment.custom-ccs-id: Still destroying... [id=30467f99cdd4e158cd726531ed29b974, 1m20s elapsed]
ec_deployment.custom-ccs-id: Still destroying... [id=30467f99cdd4e158cd726531ed29b974, 1m30s elapsed]
ec_deployment.custom-ccs-id: Destruction complete after 1m38s

Apply complete! Resources: 0 added, 0 changed, 1 destroyed.
```

Figure 12.39 – Terraform destroy plan output in terminal

How it works...

The Terraform provider for Elastic Cloud facilitates the automated provisioning, updating, and management of Elastic Cloud resources, including Elasticsearch clusters, Kibana instances, and integrations. In the configuration used for this recipe, the Terraform provider is mentioned in the following section:

```
// Here we specify which providers Terraform should use, in this case the Elastic Cloud Provider, ec
  required_providers {

/*
  Unqualified names will pull from the Github backed Terraform Registry;
  https://registry.terraform.io/providers/elastic/ec
  https://github.com/elastic/terraform-provider-ec
*/
    ec = {
      source  = "elastic/ec"
      version = "0.9.0" // Pinning to a specific version ensures compatibility and avoids nasty surprises
    }
  }
}
```

Figure 12.40 – Elastic Cloud provider reference

This provider acts as a bridge between Terraform's declarative configuration language and Elastic Cloud's API, allowing developers and operations teams to define their Elastic Stack infrastructure using code. By specifying the desired state of your resources in Terraform configurations, you can apply changes systematically, ensuring that their infrastructure matches the defined state. Additionally, maintaining Terraform code in source version control provides visibility into any changes that occur.

The part that describes the desired Elastic deployment configuration is shown in *Figure 12.41*:

```
// Resources require a type (ec_deployment) and an arbitrary name (workshop).
resource "ec_deployment" "terraform" {

  name                   = "terraform-deployment"
  version                = data.ec_stack.latest.version
  region                 = "gcp-europe-west1"
  deployment_template_id = "gcp-io-optimized"

  elasticsearch = {

    autoscale = "true"
    hot = {
        size = "4g"
        autoscaling = {
          max_size = "8g",
          max_size_resource = "memory"
        }
    }
    frozen = {
      autoscaling = {}
    }
    ml = {
      autoscaling = {}
    }
  }

  kibana = {}

  integrations_server = {}
}
```

Figure 12.41 – Elastic resources description in Terraform configuration

Under the hood, the Terraform Elastic Cloud provider leverages the Elastic Cloud API to create, manage, and delete resources. When a Terraform plan is executed, it communicates with the Elastic Cloud API, translating the high-level Terraform configurations into API calls that provision the specified resources. This integration supports a range of configurations, from basic cluster setups to more complex deployments involving multiple clusters, cross-cluster search, and advanced security settings.

There's more...

In this recipe, we've used the Elastic Cloud provider, but there is another provider called the Elastic Stack provider that is designed to configure and manage all Elastic Stack resources (alerting rules, ingest pipelines, index templates, and more) as code. For more information, go to the Terraform registry: `https://registry.terraform.io/providers/elastic/elasticstack/latest`.

See also

- To get the complete list of features provided by the Elastic Cloud provider, refer to the official Terraform documentation: `https://registry.terraform.io/providers/elastic/ec/latest/docs`

- To learn how you can use Terraform to deploy an Elastic Stack cluster on GKE, check out this article: `https://www.elastic.co/blog/installing-eck-with-terraform-on-gcp`

- For a complete example of leveraging terraform in a multi-cloud environment see this repository: `https://github.com/felix-lessoer/elastic-terraform-examples/tree/main?tab=readme-ov-file`

Enabling and configuring cross-cluster search

Cross-cluster search (CCS) is an important feature provided by the Elastic Stack, enabling users to query multiple clusters simultaneously and obtain an aggregated view from multiple data sources. It can be employed in various scenarios such as data isolation/aggregation, resource efficiency management, combined data analysis, hybrid environments, and more. In this recipe, we will learn how to establish trust relationships and explore how CCS can be beneficial through concrete examples of data isolation and combined data analysis.

Getting ready

Ensure you have a running deployment on Elastic Cloud, which you have previously used to complete the recipes for monitoring the OpenTelemetry demo in *Chapter 10*. At this point, we need to confirm that the existing cluster is collecting application traces from the OpenTelemetry demo.

In Kibana, under your existing cluster, navigate to **Observability | APM** and verify that traces are being collected, as shown in *Figure 12.42*. This cluster will act as our main cluster for implementing CCS on our observability data:

	Health	Name ↑	Latency (avg.)	Throughput	Failed transaction rate
Cases					
Logs	Unknown	GO accountingser...	0.0 ms	9.1 tpm	N/A
Explorer BETA	Healthy	adservice	0.7 ms	13.7 tpm	N/A
Stream					
Anomalies	Healthy	.NET cartservice	2.4 ms	86.3 tpm	0%
Categories	Healthy	GO checkoutservice	53 ms	9.1 tpm	N/A
Infrastructure	Healthy	C++ currencyservice	0.1 ms	25.8 tpm	0%
Inventory	Critical	emailservice	3.9 ms	9.1 tpm	0%
Metrics Explorer					
Hosts BETA	Healthy	featureflagser...	1.4 ms	37.6 tpm	0%
	Unknown	frauddetection...	0.2 ms	9.1 tpm	N/A
APM	Healthy	frontend	9.6 ms	317.3 tpm	0%
Services	Healthy	frontend-proxy	11 ms	167.7 tpm	0.0%
Traces					
Dependencies	Unknown	kafka	0 ms	N/A	N/A

Figure 12.42 – Checking the application traces on the main cluster

Before proceeding, ensure you've finished the previous recipe, *Configuring Elastic Stack components with Terraform*, to familiarize yourself with the Terraform provider for Elastic Cloud. In this recipe, we will use Terraform once more to create an additional cluster, which will serve as the remote cluster for the CCS scenario.

The snippets for this recipe can be found at this address: `https://github.com/PacktPublishing/Elastic-Stack-8.x-Cookbook/blob/main/Chapter12/snippets.md#enabling-and-configuring-cross-cluster-search`

How to do it...

In this recipe, you'll learn how to expand your Elastic Stack by creating a new cluster using Terraform and establishing a trust relationship with the existing main cluster. We'll then use CCS to merge observability data from multiple sources, simplify monitoring configurations, and enhance data analysis capabilities.

1. We'll begin by creating a new cluster named `new-team-deployment`, using Terraform as we did in the previous recipe. If you have not yet done so, clone the book's GitHub repository (`https://github.com/PacktPublishing/Elastic-Stack-8.x-Cookbook`), and then navigate to the `Chapter12/terraform/ccs` directory (the main Terraform

module `main.tf` for this recipe is in this folder). Update the `main.tfvars` file and set the API key as we did in the previous recipe. Execute the following commands in the terminal to validate the Terraform module and create our Elastic Cloud deployment. The deployment creation will begin, and after a while, you should see the resource creation message, as shown in *Figure 12.43*:

```
$ terraform init
$ terraform plan
$ terraform apply
```

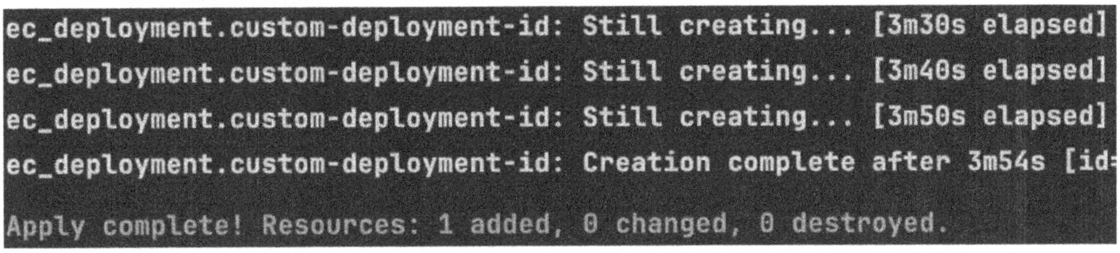

```
ec_deployment.custom-deployment-id: Still creating... [3m30s elapsed]
ec_deployment.custom-deployment-id: Still creating... [3m40s elapsed]
ec_deployment.custom-deployment-id: Still creating... [3m50s elapsed]
ec_deployment.custom-deployment-id: Creation complete after 3m54s [id=

Apply complete! Resources: 1 added, 0 changed, 0 destroyed.
```

Figure 12.43 – Creating a new CCS deployment with Terraform

2. We can now check the new deployment in the Elastic Cloud console at `https://cloud.elastic.co`, as shown in *Figure 12.4*. At this stage, you should see both clusters up and running:

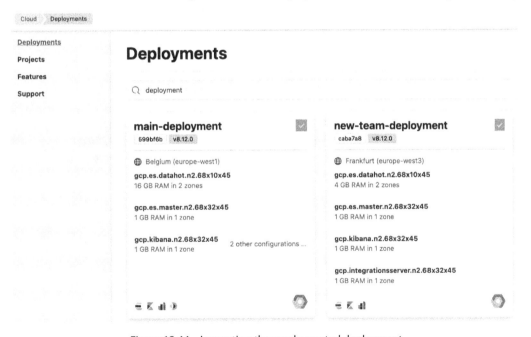

Figure 12.44 – Inspecting the newly created deployment

3. Before establishing a trust relationship between these two deployments, we need to note the **Organization ID** value of the Elastic Cloud account that we will use in the next steps. Click on the user icon in the top-right corner of the Elastic Cloud console, then select **Organization** from the drop-down menu and jot down the **Organization ID** value, as shown in *Figure 12.45*:

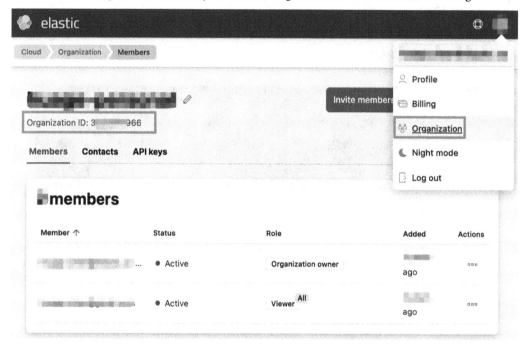

Figure 12.45 – Checking the cloud organization ID

4. The next step is to establish a trust relationship between these two deployments. Return to the home page of the Elastic Cloud console and click on **new-team-deployment**, then navigate to **Security** in the left-hand menu. Scroll down to the **Remote Connections** section and click on **Add trusted environment**, as shown in *Figure 12.46*:

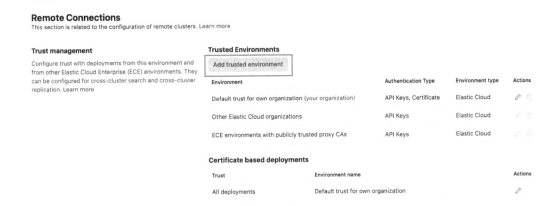

Figure 12.46 – Adding a trusted environment

5. On the first screen, select **Elastic Cloud** and click **Next**. On the second screen, select **Certificates** and click **Next**. You will reach the ultimate step to add the trusted environment, as illustrated in *Figure 12.47*. Here, enter the organization ID that you noted in *step 3*:

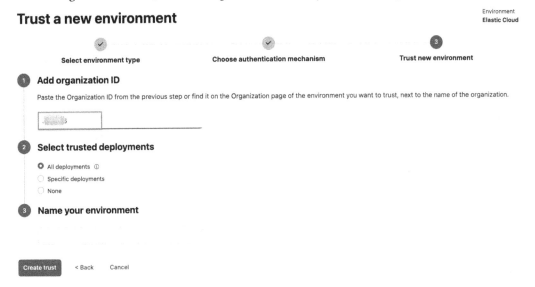

Figure 12.47 – Filtering out the environments using the organization ID

6. Choose **Specific deployments** and use the search bar to locate the main deployment, if necessary. Then select the main deployment as shown in *Figure 12.48*. Click on the **Create trust** button. A confirmation page indicating that the environment is now trusted should appear:

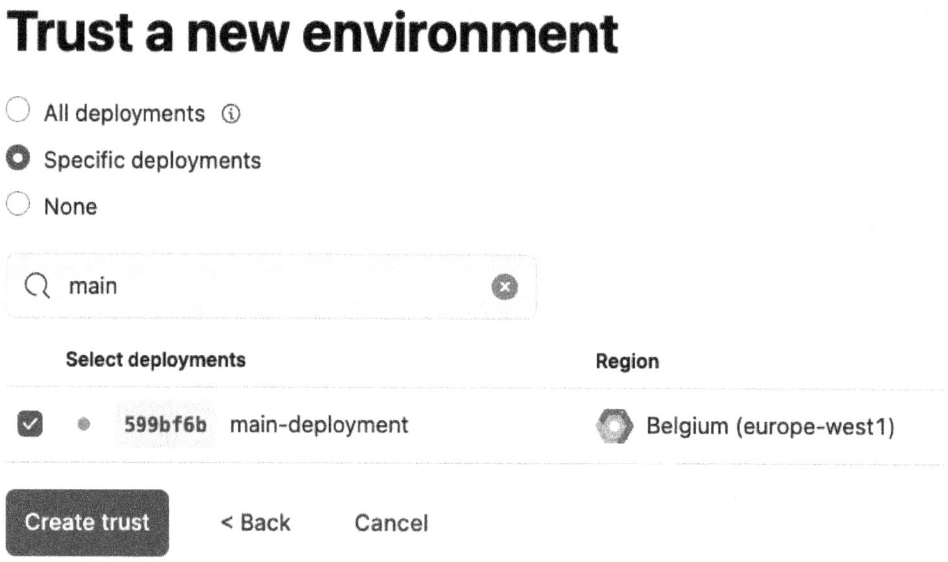

Trust a new environment

○ All deployments ⓘ

● Specific deployments

○ None

Q main   ✕

Select deployments   **Region**

☑   ●   **599bf6b**   main-deployment   ⬡ Belgium (europe-west1)

Create trust   < Back   Cancel

Figure 12.48 – Filtering out the environment

7. Scroll down to the end of the page, find the **Remote cluster parameters** section, and click on **Proxy address** to copy the URL address. Record it temporarily or in your notes, as we will use the connection parameter for the main cluster in the next step:

Remote cluster parameters   📋 Proxy address   The endpoint of the remote connection used by Kibana.

Connect clusters from a trusted environment to clusters of
this deployment using these parameters. Learn more   📋 Server name   The host name string to use for remote connections.

Figure 12.49 – Remote cluster parameters

8. Open Kibana from the main cluster, navigate to **Stack Management | Remote Clusters**, and click on **Add a remote cluster**. On the setup page, name the cluster `new-team-deployment` and fill the **Elasticsearch endpoint URL** field with the **Proxy address** value you recorded earlier. Click on **Next** to complete the configuration:

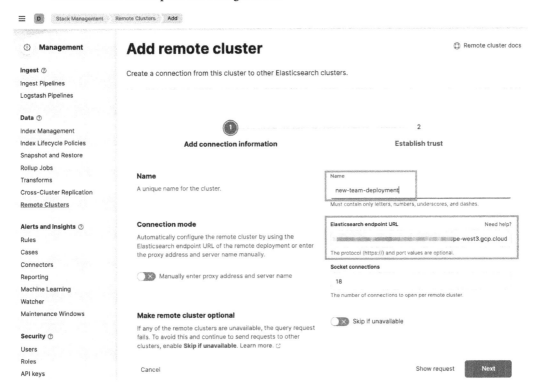

Figure 12.50 – Configuring the remote cluster

9. Validate the trust by selecting **Add remote cluster**. Confirm the setup by checking the **Yes, I have set up trust** box and proceed with validation:

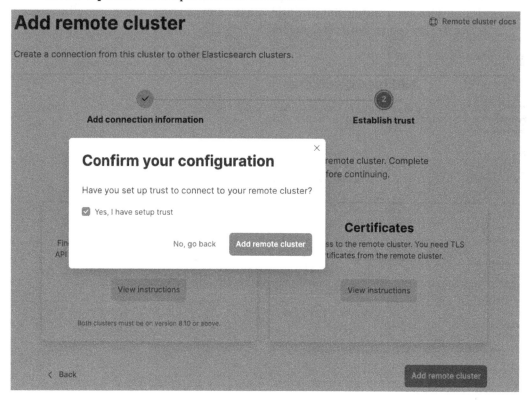

Figure 12.51 – Validating the remote cluster creation

You should see the remote cluster listed with a **Connected** status as displayed in *Figure 12.52*:

Remote Clusters

🌐 Remote Clusters docs

🔍 Search...

⊕ Add a remote cluster

☐ Name ↑	Status	Mode	Addresses	Authentication type	Connections	Actions
☐ new-team-deployment	Connected	proxy	668903f65acd45a	Certificate ⓘ	18	🖉 🗑

Figure 12.52 – Remote cluster status

At this stage, we have configured both clusters and established a trust relationship between them. Let's now onboard some observability data to the remote cluster.

10. We will revisit the Elastiflix demo application example from *Chapter 10*. Follow the *Setting up real user monitoring* recipe from *Chapter 10*, but this time, direct the **Application Traces** and **Real User Monitoring** (**RUM**) data to the remote cluster. Once the data is successfully onboarded, navigate to **Observability | APM** in the Kibana instance of the remote cluster and verify that the data has been correctly collected, as illustrated in *Figure 12.53*:

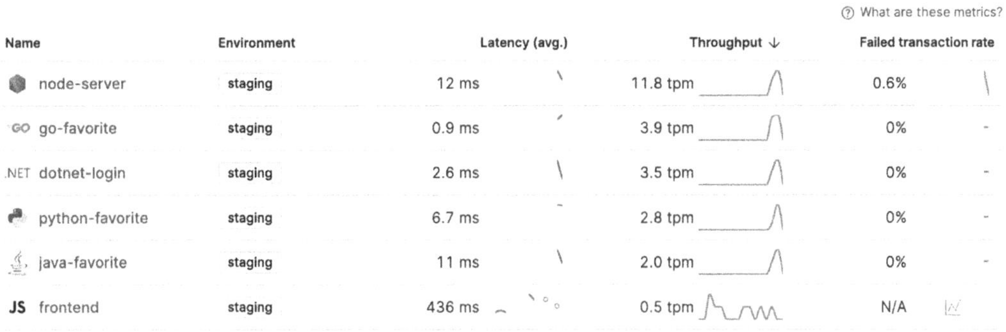

⑦ What are these metrics?

Name	Environment	Latency (avg.)	Throughput ↓	Failed transaction rate
node-server	staging	12 ms	11.8 tpm	0.6%
go-favorite	staging	0.9 ms	3.9 tpm	0%
.NET dotnet-login	staging	2.6 ms	3.5 tpm	0%
python-favorite	staging	6.7 ms	2.8 tpm	0%
java-favorite	staging	11 ms	2.0 tpm	0%
JS frontend	staging	436 ms	0.5 tpm	N/A

Figure 12.53 – APM services on the remote cluster

11. We can now configure the APM indices on the main cluster to simultaneously inspect the OpenTelemetry Demo application traces from the main cluster and Elastiflix application traces from the remote cluster. In the Kibana instance of the main cluster, navigate to **Observability | APM** and click on **Settings** in the top menu, then select the **Indices** tab. Fill in the form with the following values, copying the default values and adding the remote cluster prefix to each as shown in *Figure 12.54*, then click on the **Apply changes** button (you can copy the setting from the snippets at this address: `https://github.com/PacktPublishing/Elastic-Stack-8.x-Cookbook/blob/main/Chapter12/snippets.md#apm-indices-settings`):

- **Error Indices**: `logs-apm*,apm-*,new-team-deployment:logs-apm*,new-team-deployment:apm-*`

- **Onboarding Indices**: `apm-*,new-team-deployment:apm-*`

- **Span Indices**: `traces-apm*,apm-*,new-team-deployment:traces-apm*,new-team-deployment:apm-*`

- **Transaction Indices**: `traces-apm*,apm-*,new-team-deployment:apm-*,new-team-deployment:traces-apm*`

- **Metrics Indices**: `metrics-apm*,apm-*,new-team-deployment:metrics-apm*,new-team-deployment:apm-*`

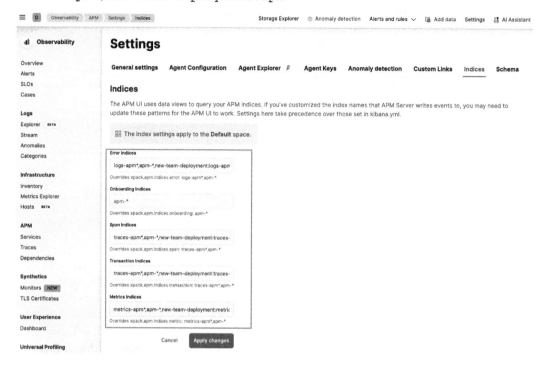

Figure 12.54 – Configuring the APM indices for main and remote clusters

> **Note**
>
> Instead of listing the remote deployment explicitly to the prefix, as in `new-team-deployment:apm-*`, it's also possible to use a wildcard to include all the trusted remote deployment, as in `*:apm-*`.

12. Let's return to **Observability | APM | Services**. Here, we can see that in addition to the existing OpenTelemetry Demo service, we now have direct access to the Elastiflix services via CCS as shown in *Figure 12.55*:

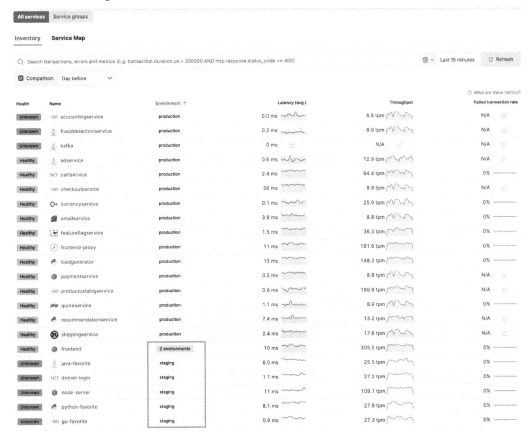

Figure 12.55 – Inspecting remote APM services via CCS

13. Once the remote APM services have been correctly configured via CCS, we can interact with them as if they were local services. For example, we can create anomaly detection jobs for service latencies. Click on **Anomaly detection** in the menu bar, then click on the **Create jobs** button. From the available service environments, select **staging** and create the machine learning anomaly detection job for the remote services associated with the **staging** environment, as shown in *Figure 12.56*:

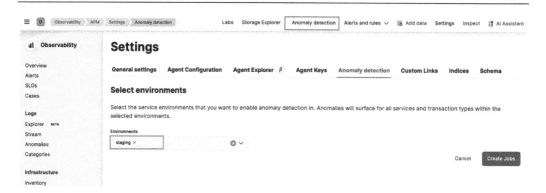

Figure 12.56 – Configuring the anomaly detection jobs for the remote services

How it works...

In this recipe, we've learned to use CCS within an observability context. Prior to the recipe, our **main-deployment** cluster was already operational, gathering application traces from the OpenTelemetry demo using OpenTelemetry Agents and the OpenTelemetry Collector. Our scenario unfolded in four key steps:

1. Creating a new remote cluster using Terraform.

2. Establishing a trust relationship between the main cluster and the remote cluster.

3. Onboarding application traces from Elastiflix with the APM Agent to the remote cluster and verifying trace collection in the remote cluster's Kibana.

4. Configuring the remote services in the APM settings of the main cluster and accessing an aggregated view of both OpenTelemetry Demo and Elastiflix traces by leveraging CCS:

   The aforementioned steps are shown in Figure 12.57:

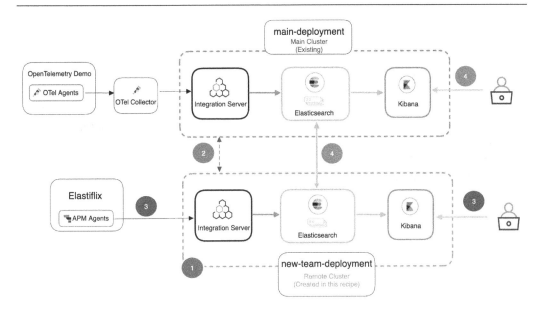

Figure 12.57 – Scenario for this recipe

There's more...

In our recipe, we illustrated how CCS can help in an observability use case to achieve the following goals:

- **Aggregate data and enhance analysis**: Consolidate search results from multiple clusters
- **Manage multi-tenancy**: Enable searches across various tenant data while preserving privacy

CCS can also address broader operational demands:

- **Increasing resource efficiency**: Enable advanced searches across large datasets using smaller clusters.

- **Streamlining post-merger and acquisition integration**: Combine new clusters without needing immediate infrastructure changes.

- **Optimizing hybrid cloud environments**: Ease searches across both on-premises and cloud environments. Starting from version 8.3, CCS now supports searching data across Elastic Cloud, self-managed clusters, ECK, and ECE clusters.

In our recipe, we have shown the flexible configuration of CCS for APM traces. The Elastic Stack provides both CCS APIs and native integrations with solutions such as Logs, Metrics, and Security. This allows us to expand the CCS for other use cases, such as security/SIEM, data analytics, and observability (logs, metrics, etc.). You can find the references in the *See also* section of this recipe.

There are also some useful features that have been released for version 8 of the Elastic Stack, including API key authentication and Terraform support for CCS/CCR.

API key authentication

In our recipe, we used TLS certificate authentication to establish a trust relationship between the main and remote clusters. With the release of Elastic Stack version 8.10, a new authentication mechanism based on API keys was introduced. The complete documentation for this feature is available here: `https://www.elastic.co/guide/en/elasticsearch/reference/current/remote-clusters-api-key.html#remote-clusters-api-key`.

CCS with Terraform

In our recipe, we used Terraform to create our remote deployment. We then established trust relationships between the remote cluster and the main cluster via the Cloud Console and Kibana. The reason for this is that the main cluster had been created previously and was already ingesting observability data prior to this recipe.

If we had been in a situation where both the remote and main clusters were being created from scratch, we could have used Terraform to set up both clusters and establish their trust relationships directly. This would eliminate the need to configure them manually through the Cloud Console and Kibana. An example of this approach can be found at the following address: `https://registry.terraform.io/providers/elastic/ec/latest/docs/resources/deployment#with-cross-cluster-search-settings`

Cross-cluster replication (CCR)

In addition to using CCS to enhance APM tracing in observability scenarios, another extremely useful Elastic Stack feature that could further enhance your data strategy is **cross-cluster replication (CCR)**. While CCS provides a unified query interface across multiple clusters, CCR can be used to duplicate data from one cluster to another. This replication ensures that your clusters are resilient, with real-time data redundancy, which is critical for disaster recovery and high-availability setups.

CCR is especially useful in these scenarios:

- **Disaster recovery**: Replicating indices to a remote cluster can provide a failover in case the primary cluster experiences issues.
- **Enhanced performance**: By replicating data closer to end users, CCR can improve search performance for geographically distributed deployments.
- **Data locality**: For businesses managing international data centers, CCR can ensure local clusters serve local data, adhering to location-based data laws.

See also

- CCS API documentation can be found here: `https://www.elastic.co/guide/en/elasticsearch/reference/current/modules-cross-cluster-search.html`

- Cross-cluster replication documentation can be found here: `https://www.elastic.co/guide/en/elasticsearch/reference/current/xpack-ccr.html`

- Elasticsearch 8.10 provides a lot of new features on CCS; make sure to check the release blog at this address: `https://www.elastic.co/blog/whats-new-elasticsearch-platform-8-10-0`

- You can find how to configure the Security app to use CCS in the following blog post: `https://www.elastic.co/blog/elastic-on-elastic-configuring-the-security-app-to-use-cross-cluster-search`

13
Elastic Stack Monitoring

As the concluding chapter of our comprehensive journey through the Elastic Stack, let's turn our attention to Elastic Stack monitoring. A strong monitoring strategy ensures the health and performance of the stack and offers the insights needed for quick troubleshooting and smart decision-making. In this chapter, you will learn essential techniques for stack monitoring and troubleshooting.

In this chapter, we are going to cover the following main topics:

- Setting up Stack Monitoring
- Building custom visualizations for monitoring data
- Monitoring cluster health via an API
- Enabling audit logging

Technical requirements

This chapter has the following recipes involving multiple elastic deployments, for which **a paid subscription is necessary**:

- Setting up Stack Monitoring
- Building custom visualizations for monitoring data

Setting up Stack Monitoring

In this recipe, we will explore the details of setting up Stack Monitoring within the Elastic Stack. Monitoring is a crucial component that provides insights into the health, performance, and availability of your Elastic Stack deployment, including Elasticsearch, Kibana, Integrations Server, Elastic Agent, Beats, and Logstash.

We'll start by guiding you through the initial setup steps, including configuring your Elastic Stack deployments for monitoring. This involves enabling monitoring features and specifying the collection of metrics and logs that will give you visibility into your stack's operation. Next, we'll delve into the use of Kibana for visualizing monitoring data. Kibana offers a dedicated monitoring UI where you can view metrics and logs, analyze the health of your nodes and indices, and track performance issues across your deployment.

Additionally, we'll cover advanced topics, such as setting up alerting rules for automated notifications about potential issues within your stack. This proactive approach ensures that you can respond swiftly to any anomalies before they impact your operations.

Getting ready

Make sure to have the following requirements met:

- Terraform installed on your machine; if that is not the case, we'll advise you to follow the steps outlined in the Terraform official documentation in order to have everything up and running.

- An Elastic Cloud deployment up and running. To quickly spin up a deployment, please refer to the *Configuring Elastic Stack components with Terraform* recipe in *Chapter 12*.

- As we're going to provision a deployment on Elastic Cloud, we will need an API key that'll be passed to Terraform. In order to get such a key, follow the steps described in the official Elastic documentation here: `https://www.elastic.co/guide/en/cloud/current/ec-api-authentication.html`. We will refer to this key as `api_key` later in the recipe.

The snippets for this recipe can be found here: `https://github.com/PacktPublishing/Elastic-Stack-8.x-Cookbook/blob/main/Chapter13/snippets.md#setting-up-stack-monitoring`

> **Important note**
> While the setup for this guide follows the recommended best practice of having a dedicated monitoring deployment, you can also opt for self-monitoring by using your main deployment to collect monitoring data.

How to do it...

The objective of this recipe is to set up a monitoring cluster beside our main deployment. We will use Terraform to deploy the additional monitoring cluster and set up the routing of logs and metrics from our main cluster to the monitoring cluster.

Figure 13.1 shows a simplified architecture of the setup:

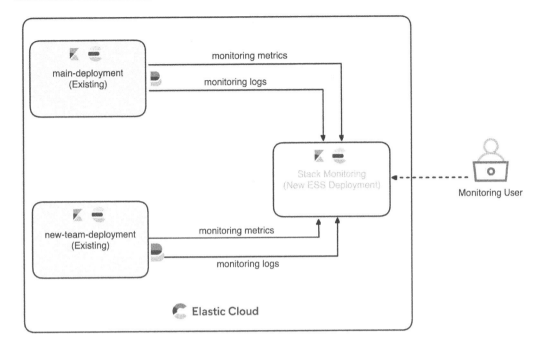

Figure 13.1 – Monitoring setup overview

Note on monitoring in Elastic Cloud

For a deployment in Elastic Cloud to send logs and metrics to a dedicated monitoring cluster, both the main deployment and the monitoring cluster must be situated in the same cloud region. As of the current writing, cross-region monitoring is not supported. This requirement ensures that data transfer remains efficient and reduces latency, but it also means planning your deployment strategy to accommodate these limitations if monitoring across regions is a necessity for your operations.

1. Download the following Terraform configuration files from the book's official repository: `https://github.com/PacktPublishing/Elastic-Stack-8.x-Cookbook/blob/main/Chapter13/terraform/monitoring`.

2. Next, open the `.tfvar` file in your preferred editor, add your `api_key` key, and then save the file.

3. As we did in the *Configuring Elastic Stack components with Terraform* recipe in *Chapter 12*, run the following commands to get a sense of what will be deployed in terms of components:

    ```
    $ terraform init
    $ terraform plan
    ```

4. Next, execute the configuration with the following command:

```
$ terraform apply
```

Upon successful completion of the command, head to the Elastic Cloud console and look for the deployment named `terraform-monitoring`.

5. Having established the deployment that will act as a repository for monitoring data, we must now configure our main deployments to forward logs and monitoring metrics. Navigate to the cloud console and select the deployment you wish to monitor. Then, find the **Logs and metrics** option in the left menu and click on it. On the **Logs and metrics** page, click **Enable**. From the drop-down menu, choose the `terraform-monitoring` deployment and click **Save** to apply and save the configuration changes:

Logs and metrics

Ship to a deployment

Ship logs and monitoring metrics to a deployment where they are stored separately. Then, you can enable additional log types, search the logs, configure retention periods, and use Kibana to view monitoring visualizations. Learn more

Ship data to

Data being shipped
☑ Logs
☑ Metrics

| Save | Discard changes |

Figure 13.2 – Logs and metrics: Ship to a deployment setup

Once the configuration has been applied, the **Logs and metrics** page will look as pictured in *Figure 13.3*. Repeat the same operation for the `new-team` deployment as well:

Logs and metrics

Ship to a deployment

Ship logs and monitoring metrics to a deployment where they are stored separately. Then, you can enable additional log types, search the logs, configure retention periods, and use Kibana to view monitoring visualizations. Learn more

Figure 13.3 – Ship to a deployment activated

Now that we have successfully configured the shipping of logs and metrics to our `terraform-monitoring` deployment, let's explore how to utilize this data within the stack monitoring application.

6. Head to Kibana for the `terraform-monitoring` deployment and go to **Stack Monitoring** under the **Management** section:

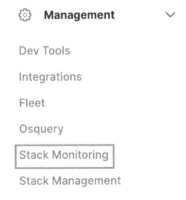

Figure 13.4 – Stack Monitoring app in the Kibana menu

7. Upon arriving at the **Stack Monitoring** page, you may encounter a popup about the creation of out-of-the-box rules. We will address this later, so for now, please click on **Remind me later**. You will then be presented with a list of clusters being monitored:

Figure 13.5 – Stack Monitoring cluster listing

8. Click on the `main-deployment` cluster in the list to go to the monitoring overview page:

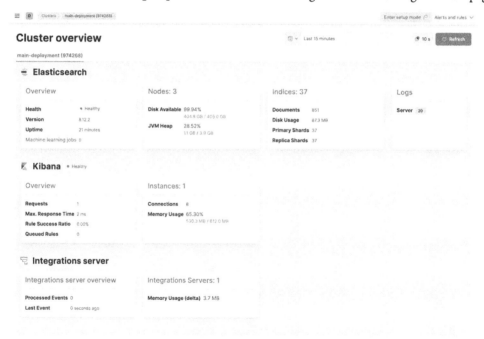

Figure 13.6 – Stack Monitoring cluster overview

9. We will start with the **Elasticsearch overview| Overview** page. From here, you can zoom in on key aspects of the health and status of your clusters:

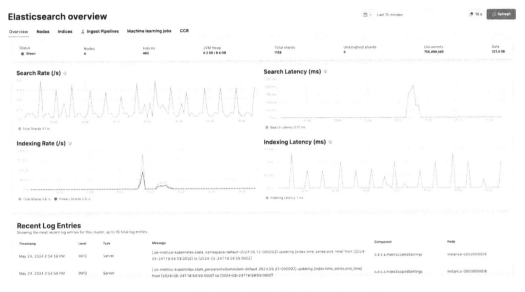

Figure 13.7 – Stack Monitoring Elasticsearch overview

The Elasticsearch overview page is divided into two sections:

- The upper half of *Figure 13.7* focuses on key search and indexing metrics. Leverage those dashboards to quickly spot performance issues in your deployment.

- The lower half of *Figure 13.7* focuses on recent log entries. Here, you can catch any error or suspicious activity from the nodes.

You will spend most of your time in the **Nodes** and **Indices** tabs. The former is incredibly useful to have a good understanding of how resources are used and allocated on each node of your deployment and quickly pinpoint issues such as node hot spotting, over usage of CPU, and disk saturation, as illustrated in *Figure 13.8*:

Figure 13.8 – Node monitoring overview

Nodes preceded by a star designate the master node. You can click on each node to get more detailed information and metrics for specific nodes, as shown in *Figure 13.9*:

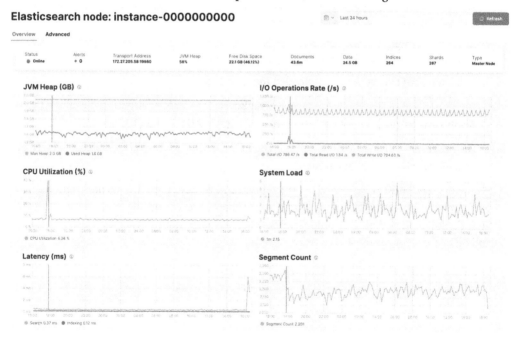

Figure 13.9 – Instance monitoring details

Those critical metrics can help you answer the following questions:

- Is my cluster or a particular node under heavy resource usage?

- What is the bottleneck in terms of CPU, I/O, or JVM?

- How does the latency experienced by my users correlate with the resource's usage?

The **Indices** tab is more focused on the data store side of things. You will find useful indicators such as the document count, index and search rates, and any unassigned shards that might make your cluster unhealthy:

Elasticsearch indices

📅 ∨   Last 15 minutes   🔎 10 s   ↻ Refresh

Overview   Nodes | Indices | ⚠ Ingest Pipelines   Machine learning jobs   CCR

Status	Alerts	Nodes	Indices	JVM Heap	Total shards	Unassigned shards	Documents	Data
● Green	● 0	5	291	6.0 GB / 10.4 GB	590	0	47,929,855	50.0 GB

☑ Filter for system indices

🔍 Filter Indices...

Name ↑	Alerts	Status	Document Count	Data	Index Rate	Search Rate	Unassigned Sha...
.apm-agent-configuration	● Clear	● Green	0	499.0 B	0 /s	0.03 /s	0
.apm-custom-link	● Clear	● Green	0	499.0 B	0 /s	0 /s	0
.apm-source-map	● Clear	● Green	0	500.0 B	0 /s	0.07 /s	0
.async-search	● Clear	● Green	0	524.0 B	0 /s	0 /s	0
.ds-.kibana-event-log-ds-2024.01.26-000011	● Clear	● Green	386k	125.6 MB	0 /s	0 /s	0

Figure 13.10 – Stack Monitoring Indices tab overview

10. By clicking on the name of an index in the table, you can zoom in to view metrics specific to that index. You will have the same metrics but for that specific index. You can use the information available here to troubleshoot a specific index experiencing slow indexing or search issues.

11. The **Ingest Pipelines** tab is a quite recent addition to the Stack Monitoring application, and it focuses on giving some great insights into the throughput and performance of overall ingest processing. It is especially valuable if you have a lot of ingest pipelines running in your cluster:

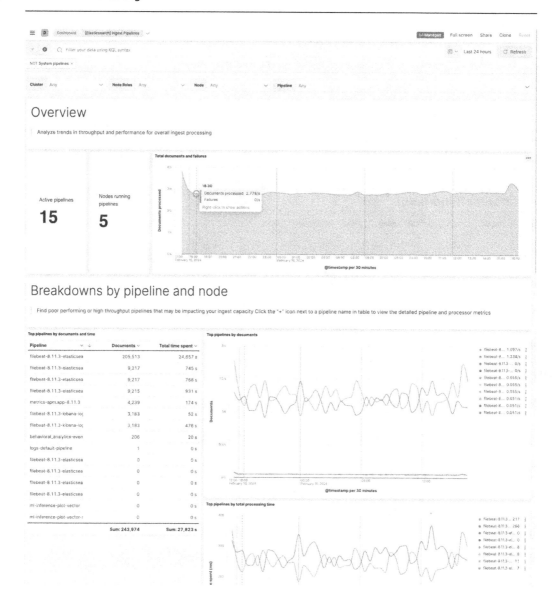

Figure 13.11 – Ingest Pipelines monitoring overview

> **Note on Ingest Pipelines monitoring**
>
> The first time you try to access the **Ingest Pipelines** tab, you will be prompted to install the Elasticsearch integration. Click on the **Install** button in the popup window to proceed with the installation.

The last two tabs are dedicated to specific features of the stack:

- **Machine learning jobs** gives some high-level information on **machine learning (ML)** nodes and key metrics on jobs, such as their state, processed records, the node on which they're running, and model size

- **Cross-Cluster Replication (CCR)** monitors and manages the health, performance, and status of cross-cluster replication operations

Now, while those dashboards and visual information on the clusters are unbelievably valuable, the great benefit of having Stack Monitoring enabled is the ability to receive alerts if your deployment is showing signs of degradation. Stack Monitoring comes with a set of prebuilt rules, designed by the experts of the Elastic Stack based on the recommended best practices.

12. To activate the rules, on the top right of the overview page, locate the **Alerts and rules** dropdown, and once opened, choose the **Create default rules** option:

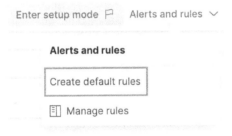

Figure 13.12 – Activating default rules for Stack Monitoring

13. If you are presented with a popup regarding the migration of watches, click on **Create** and wait for a few seconds for the rules to be activated. Once that is the case, you will notice alerts if you have any.

14. It is also worth mentioning that you can edit the provided rules if you wish, but keep in mind that those rules are based on best practices and recommendations from the stack experts. To do so, locate the **Enter setup mode** button on the top right of the monitoring page. Click on it and select the rules you wish to adjust. Editing the rules is also the best approach to add actions and get notifications by leveraging connectors such as email, Slack, or more:

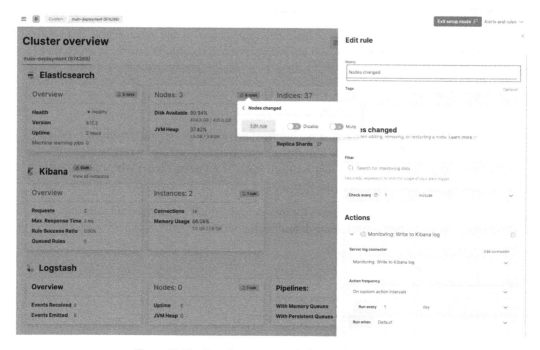

Figure 13.13 – Entering setup mode for stack monitoring

15. While we have been focusing on Elasticsearch, notice that Kibana and Integration Servers monitoring data is also available. Go back to the cluster overview page and click on **Overview** under the **Kibana** section. As pictured in the following figure, this dashboard shows client (user) activity metrics (requests and responses) as well as some important data on queue usage for all your Kibana instances:

Kibana overview

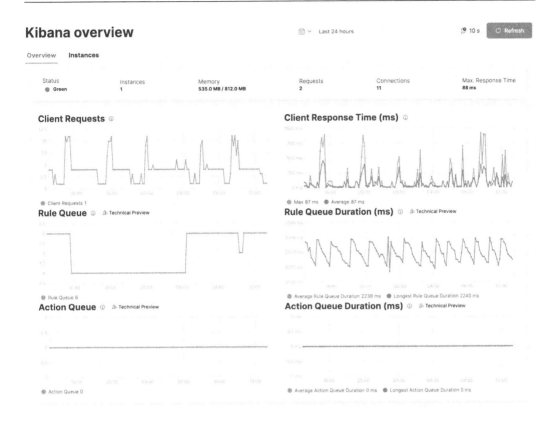

Figure 13.14 – Stack Monitoring: Kibana overview

The Kibana **Instances** view gives some metrics about client (user) activities on Kibana, plus details on HTTP connections and memory size broken down by Kibana instances.

16. The **Integration Servers** section also focuses on resource usage and key metrics related to component-specific ingest activities. You will find information such as event rates, and HTTP requests and responses emitted by the server.

The advantage of monitoring the Elastic Stack using Elastic's own tools is that monitoring data is stored as regular indices, which can be utilized for personalized analysis.

As demonstrated in the *Building custom visualizations for monitoring data* recipe, this approach enables you to delve into various aspects of your clusters' operations and usage. For instance, you can analyze the volume of data ingested daily and the quantity of data queried or identify the most common time ranges used in queries. All these insights can be gained by simply leveraging the monitoring data available to you.

How it works...

To understand how Stack Monitoring operates, we've broken down its inner workings into the following sections.

Data collection

At the core of Stack Monitoring is the collection of metrics and logs from various components of the Elastic Stack. This is facilitated by Metricbeat and Filebeat, which are configured to collect detailed operational data. Metricbeat gathers metrics from each component, such as CPU usage, memory usage, and node health, whereas Filebeat collects logs that provide insights into operational events. These Beats are designed to work seamlessly with Elastic Cloud, ensuring data is efficiently captured and transmitted for analysis. You can also use Elastic Agent since 8.5 to collect stack monitoring events: `https://www.elastic.co/guide/en/elasticsearch/reference/current/configuring-elastic-agent.html`.

Shipping and storage

Once collected, the data is shipped to the dedicated monitoring cluster within Elastic Cloud. This separation of the monitoring data from the production data ensures that monitoring activities do not impact the performance of the production environment. The data is stored in indices, just as with any other data within Elasticsearch, making it accessible for analysis and visualization. When running on-premise, it is also considered a best practice to isolate the monitoring cluster.

Analysis and visualization

The analysis and visualization of collected data are primarily conducted through Kibana, which offers a specialized Stack Monitoring UI. This UI presents users with a comprehensive dashboard that visualizes the health, performance, and logs of Elastic Stack components. Users can drill down into specific metrics, view historical trends, and identify patterns or anomalies that may indicate issues or opportunities for optimization.

Alerting

Elasticsearch's alerting framework is an integral part of Stack Monitoring, allowing users to define alerts based on specific conditions within the monitoring data. These alerts can notify users of potential issues, such as a sudden drop in performance or a node going offline, enabling rapid response to ensure the stability and reliability of the Elastic Stack.

There's more...

Security in Stack Monitoring is tightly integrated with Elastic Cloud's overall security model. Access to monitoring data and functionalities is controlled through **role-based access control** (**RBAC**), ensuring that only authorized users can view or manipulate monitoring configurations and data.

For optimal monitoring, it is recommended to do the following:

- Regularly review the health and performance dashboards to stay ahead of potential issues

- Configure alerts to proactively manage the Elastic Stack environment

- Utilize detailed metrics and logs to perform **root cause analysis (RCA)** of any operational issues

Logstash can also be monitored through the Stack Monitoring application. To do so, you can leverage the Logstash integrations for Elastic Agents. By deploying an elastic agent on the infrastructure where Logstash is running, you will be able to collect monitoring data and ship it to your monitoring cluster. Check out the official Elastic documentation for more information on this feature: `https://www.elastic.co/guide/en/logstash/current/monitoring-with-elastic-agent.html`.

In this recipe, we have manually set up the shipping of monitoring data through the cloud console; if you are using Terraform, you can quickly set up a monitoring deployment in your configuration with the **Observability** settings. Have a look at the documentation if you are interested in knowing more: `https://registry.terraform.io/providers/elastic/ec/latest/docs/resources/deployment#with-observability`.

See also

- For a complete list of Stack monitoring features from the official Elastic documentation, check this address: `https://www.elastic.co/guide/en/kibana/current/xpack-monitoring.html`

- If you are looking for guidelines to configure Stack Monitoring on a self-managed Elasticsearch setup, check the following documentation: `https://www.elastic.co/guide/en/elasticsearch/reference/current/monitor-elasticsearch-cluster.html`

Building custom visualizations for monitoring data

In the previous recipe, we learned how to set up a dedicated monitoring cluster and how to use the Stack Monitoring UI to explore and understand monitoring data. In this recipe, you will learn how to create custom visualizations for this monitoring data to gain additional key performance indicators (KPIs). This will help you better understand the business and operational values of your deployment and provide more insights for troubleshooting.

Getting ready

Make sure to have completed the previous recipe, *Setting up Stack Monitoring*.

How to do it...

In this recipe, we will create a data view for the monitoring indices and build a custom visualization to track daily storage changes per index. At the end of the recipe, we will present an example of a custom monitoring dashboard that integrates the newly created visualization.

1.   Let's begin by creating a data view for the monitoring indices. Open Kibana on the monitoring cluster. Navigate to **Stack Management** | **Kibana** | **Data Views** and click on the **Create data view** button. This will open a flyout to create a data view. First, toggle on **Allow hidden and system indices**, as shown in *Figure 13.15*. Then, set both **Name** and **Index pattern** to `.monitoring-es-*` and choose **@timestamp** as the **Timestamp field** value. Then, click on the **Save data view to Kibana** button:

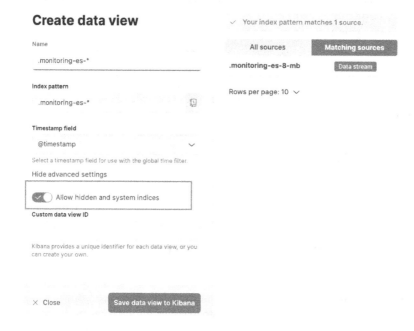

Figure 13.15 – Creating a data view for monitoring data

We can now create a visualization to better understand daily changes in storage for the indices.

2.   In Kibana, navigate to **Analytics** | **Visualize Library** and click the **Create visualization** button. Then, select **Lens**. In the Kibana Lens editor, ensure that the `.monitoring-es-*` data view is selected.

3. Next, choose **Table** as the type of visualization, as shown in *Figure 13.16*. In the next steps, we will define details for the table visualization, including rows, columns, and metrics, using the options on the right side of the editor:

Figure 13.16 – Creating Kibana Lens visualization for indices storage

4. Click on **Add or drag-and-drop a field** under **Rows**, and a flyout will appear. Select **Top values** from the **Functions** section and `index_stats.index` under **Fields**. Then, set the **Number of values** field to 200 to determine the number of values displayed, as shown in *Figure 13.17*. Close the flyout once done:

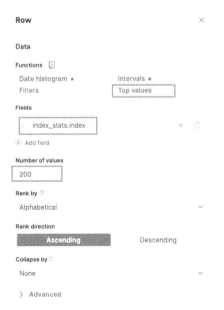

Figure 13.17 – Setting the table row data

5. Next, click on **Add or drag-and-drop a field** under **Split metrics by**, and a flyout will appear. Select **Date histogram** from the **Functions** section and **@timestamp** under **Fields**. Then, set the **Minimum interval** field to **Day** to ensure data is aggregated daily, as shown in *Figure 13.18*. Close the flyout once done:

Figure 13.18 – Setting the table metrics' split data

6. Let's complete the visualization setup by clicking on **Add or drag-and-drop a field** under **Metrics,** and a flyout will appear. Switch to the **Formula** tab and enter the following snippet; the following formula combines the last recorded value of the index size per day with the number of shards to calculate the total storage for each index across the cluster:

```
last_value(index_stats.total.store.size_in_
bytes)*max(elasticsearch.index.shards.total)
```

7. Next, customize the display by setting the name to `Last index storage`, selecting **Bytes (1024)** as the value format, and specifying 2 for the number of decimal places, as shown in *Figure 13.19*. Close the flyout once you have finished:

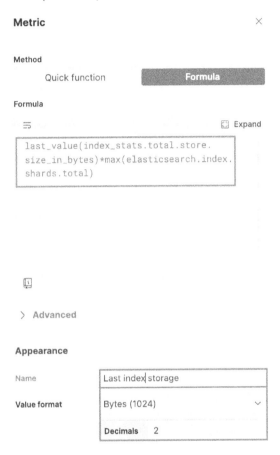

Figure 13.19 – Setting the table metrics

8. Observe the result of the visualization. Adjust the time range to **Last 3 days**, then click on the column for the last day and select **Sort descending**, as shown in *Figure 13.20*. This will display daily changes in storage per index, ranked by storage size. You can now save the visualization:

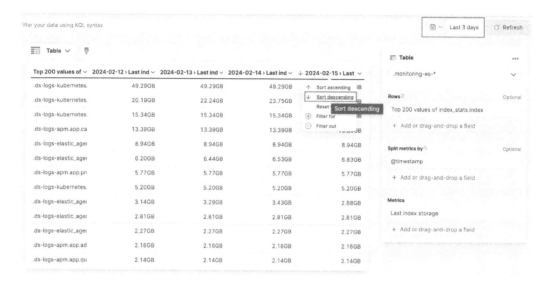

Figure 13.20 – Kibana Lens visualization for daily changes in storage per index

At this stage, you have successfully created a visualization based on monitoring data. To build various valuable dashboards, we can create additional visualizations using the methods we have just learned.

For the sake of simplicity, we have built an example for you, which you can download from the following address: `https://github.com/PacktPublishing/Elastic-Stack-8.x-Cookbook/blob/main/Chapter13/monitoring-dashboard/export_cluster_indices_ingest_query.ndjson`.

Download the file and import the dashboard into Kibana (refer to *Chapter 11's Managing and securing access to spaces* recipe to learn about importing saved objects into Kibana). After importing, you should be able to see the dashboard with monitoring data, as shown in *Figure 13.21*:

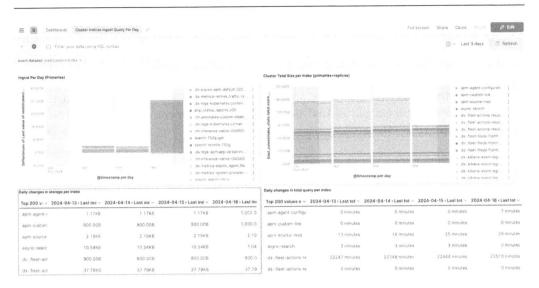

Figure 13.21 – Custom monitoring dashboard

The visualization at the bottom left corresponds to the one we created earlier in this recipe. Feel free to edit and inspect the other visualizations to understand their configurations and consider creating additional visualizations that may help in your specific use case.

How it works...

In this recipe, we have explored how to access monitoring data to build custom visualizations. We learned most of the Kibana visualization techniques in *Chapter 6*. Building visualizations on system indices is no different from working with regular business data. The most challenging part is identifying the correct fields that contain useful metrics to begin with. You can find a list of fields and their meanings at this address: `https://www.elastic.co/guide/en/beats/metricbeat/current/exported-fields-elasticsearch.html`.

There's more...

Starting with version 8.10, in addition to the dedicated Stack Monitoring UI, Elastic now also includes built-in dashboards related to data ingestion metrics through ingest pipelines. We learned how to activate these metrics in the previous recipe, *Setting up Stack Monitoring*. Once these data ingestion metrics are enabled, you can navigate to **Analytics | Dashboards** to discover some very useful built-in dashboards, as depicted in *Figure 13.22*:

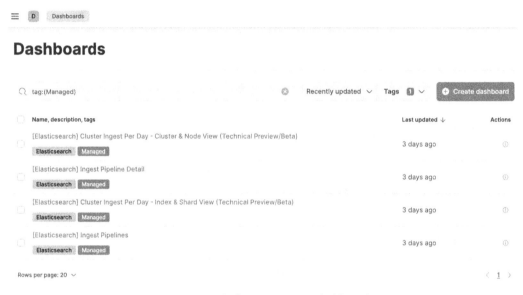

Figure 13.22 – Built-in monitoring dashboards

Figure 13.23 provides an example of one of these dashboards. Examine the dashboard and its underlying visualizations and try to inspect how these dashboards are constructed. This could offer you some ideas and inspiration on how to leverage monitoring data effectively and how you can create your own dashboards:

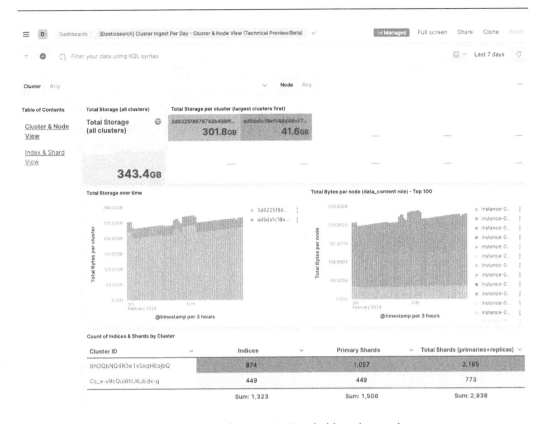

Figure 13.23 – Built-in monitoring dashboard example

Monitoring cluster health via an API

In previous recipes, we learned how to use a dedicated monitoring cluster to obtain telemetry and KPIs for deployments, how to set up alerting rules, and how to build custom visualizations and dashboards to gain insights into the performance of these deployments. In this recipe, we will explore how the newly introduced Health API (available since version 8.7) can assist us in health checking, diagnostics, and troubleshooting of Elastic deployments.

Getting ready

Make sure to have a running Elastic deployment.

The sample commands of this recipe can be found at this address: `https://github.com/PacktPublishing/Elastic-Stack-8.x-Cookbook/blob/main/Chapter13/snippets.md#monitoring-cluster-health-via-api`

How to do it...

In this recipe, we will demonstrate how to use various Health API commands in different situations. We will execute the command examples in the Kibana Dev Tools of our main cluster (please ensure you are not using the dedicated monitoring cluster as we inspect directly from the main cluster).

1. In Kibana, go to **Management | Dev Tools** and execute the following command:

    ```
    GET _health_report
    ```

    If your cluster is in a green status, you will see a result like the one shown in *Figure 13.24*. In addition to the status, you can observe other important indicators such as master_is_stable, repository integrity, and so on:

```
 1▾ {
 2      "status": "yellow",
 3      "cluster_name": "3d9225f8678742b498ff349826e8410d",
 4▾     "indicators": {
 5▾         "master_is_stable": {
 6              "status": "green",
 7              "symptom": "The cluster has a stable master node",
 8▾             "details": {
 9▾                 "current_master": {
10                      "node_id": "TiKKgIqRToyHOMnugTOUPQ",
11                      "name": "instance-0000000022"
12▾                 },
13▾                 "recent_masters": [
14▾                     {
15                          "node_id": "TiKKgIqRToyHOMnugTOUPQ",
16                          "name": "instance-0000000022"
17▾                     }
18▾                 ]
19▾             }
20▾         },
21▾         "repository_integrity": {
22              "status": "green",
23              "symptom": "No corrupted snapshot repositories.",
24▾             "details": {
25                  "total_repositories": 1
26▾             }
27▾         },
28▾         "disk": {
29              "status": "green",
30              "symptom": "The cluster has enough available disk space.",
31▾             "details": {
32                  "indices_with_readonly_block": 0,
33                  "nodes_with_enough_disk_space": 10,
34                  "nodes_with_unknown_disk_status": 0,
35                  "nodes_over_high_watermark": 0,
36                  "nodes_over_flood_stage_watermark": 0
37▾             }
38▾         },
39▾         "shards_capacity": {
40              "status": "green",
41              "symptom": "The cluster has enough room to add new shards.",
42▾             "details": {
43▾                 "data": {
44                      "max_shards_in_cluster": 15000
```

Figure 13.24 – Health API returns a green status

Figure 13.25 shows an example when the cluster status is `yellow`. In addition to that, we notice some interesting information related to symptoms and impacts, such as shard unavailability:

```
{
  "cluster_name": "3d9225f8678742b498ff349826e8410d",
  "indicators": {
    "shards_availability": {
      "status": "yellow",
      "symptom": "This cluster has 853 unavailable replica shards.",
      "details": {
        "started_primaries": 3140,
        "unassigned_replicas": 853,
        "initializing_replicas": 0,
        "creating_primaries": 0,
        "restarting_replicas": 0,
        "unassigned_primaries": 0,
        "started_replicas": 2313,
        "creating_replicas": 0,
        "initializing_primaries": 0,
        "restarting_primaries": 0
      },
      "impacts": [
        {
          "id": "elasticsearch:health:shards_availability:impact
            :replica_unassigned",
          "severity": 2,
          "description": "Searches might be slower than usual. Fewer
            redundant copies of the data exist on 853 indices [.ds-metrics
            -elastic_agent.elastic_agent-default-2023.05.22-000001, .ds
            -metrics-elastic_agent.filebeat_input-default-2023.10.23-000005,
            .ds-metrics-elastic_agent.filebeat_input-default-2023.11.29
            -000014, .ds-metrics-elastic_agent.filebeat_input-default-2023.12
            .02-000017, .ds-metrics-elastic_agent.filebeat_input-default-2023
            .12.05-000020, .ds-metrics-elastic_agent.filebeat_input-default
            -2023.12.08-000023, .ds-metrics-elastic_agent.filebeat_input
            -default-2023.12.17-000031, .ds-metrics-elastic_agent
            .filebeat_input-default-2023.12.18-000032, .ds-metrics
            -elastic_agent.filebeat_input-default-2023.12.22-000036, .ds
            -metrics-elastic_agent.filebeat_input-default-2024.01.01-000046,
            ...].",
          "impact_areas": [
            "search"
          ]
        }
      ],
```

Figure 13.25 – Health API returns a yellow status with an unassigned shard

2. To verify the shard availability in Dev Tools, we can use the following command:

```
GET _health_report/shards_availability
```

In our case, we receive a detailed explanation as to why the cluster status is `yellow`, its impacts, and additional diagnostic suggestions. For instance, running `GET _cluster/allocation/explain` for the affected index will help us understand the reasons behind the unassignment of the replicas:

```
GET _cluster/allocation/explain
{
    "index": "<index-name>",
    "shard": 0,
    "primary": false
}
```

```
{
"cluster_name": "3d9225f8678742b498ff349826e8410d",
"indicators": {
  "shards_availability": {
    "status": "yellow",
    "symptom": "This cluster has 2 unavailable replica shards.",
    "details": {
      "creating_replicas": 0,
      "started_replicas": 3952,
      "unassigned_primaries": 0,
      "restarting_replicas": 0,
      "creating_primaries": 0,
      "initializing_replicas": 0,
      "unassigned_replicas": 2,
      "started_primaries": 3928,
      "restarting_primaries": 0,
      "initializing_primaries": 0
    },
    "impacts": [
      {
        "id": "elasticsearch:health:shards_availability:impact:replica_unassigned",
        "severity": 2,
        "description": "Searches might be slower than usual. Fewer redundant copies of the data exist on 2
            indices [.ds-metrics-kubernetes.state_pod-default-2023.12.19-000028, downsample-1h-.ds-metrics-system
            .network-default-2023.12.23-000172].",
        "impact_areas": [
          "search"
        ]
      }
    ],
    "diagnosis": [
      {
        "id": "elasticsearch:health:shards_availability:diagnosis:explain_allocations",
        "cause": "Elasticsearch isn't allowed to allocate some shards from these indices to any of the nodes in
            the cluster.",
        "action": "Diagnose the issue by calling the allocation explain API for an index [GET _cluster
            /allocation/explain]. Choose a node to which you expect a shard to be allocated, find this node in
            the node-by-node explanation, and address the reasons which prevent Elasticsearch from allocating the
            shard.",
        "help_url": "https://ela.st/diagnose-shards",
        "affected_resources": {
          "indices": [
            ".ds-metrics-kubernetes.state_pod-default-2023.12.19-000028",
            "downsample-1h-.ds-metrics-system.network-default-2023.12.23-000172"
          ]
        }
      }
    ]
  }
}
}
```

Figure 13.26 – Health report on shard availability

3. If you are using an Elastic Cloud deployment, you can navigate to the cloud console and go to **Deployments** | **Your deployment** | **Monitoring** | **Health**. The console will show you a green status along with a checklist, as shown in *Figure 13.27*:

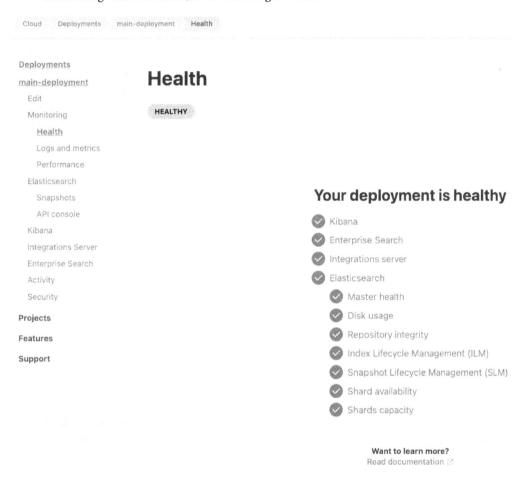

Figure 13.27 – Cloud console health check

4. The following example shows you the same health reporting in the cloud console, this time in a `yellow` status. It shows the reason for the health issues and, more importantly, gives us information on the affected capabilities and guidelines for troubleshooting, as shown in *Figure 13.28*:

Figure 13.28 – Cloud console health check troubleshooting

How it works...

As demonstrated in this recipe, there are two ways to leverage the new Health API capabilities that Elasticsearch offers:

- Health report API (full documentation is available at this address: `https://www.elastic.co/guide/en/elasticsearch/reference/current/health-api.html`)
- Cloud console

The Health API offers an intuitive method for inspecting, diagnosing, and troubleshooting your elastic deployments.

There's more...

While the new Health API is very useful for understanding deployment issues, their impacts, and subsequent steps to be taken, there may be situations where you need deeper insights into the health of indices and data streams. For these cases, you can still use the traditional Cluster Health API (`https://www.elastic.co/guide/en/elasticsearch/reference/current/cluster-health.html`). In practice, it is advisable to start with the Health API and then use the Cluster Health API for additional information if necessary.

See also

- This blog post provides some more examples of the Health API and additional explanations of the concrete meaning behind different statuses: `https://www.elastic.co/blog/cluster-health-diagnosis-elasticsearch-health-api`

- Full documentation on cluster APIs is available at this address: `https://www.elastic.co/guide/en/elasticsearch/reference/current/cluster.html`

Enabling audit logging

Another crucial aspect of monitoring, which is closely linked to regulatory compliance, is ensuring every action performed by a user on the platform can be traced. This is the purpose of audit logs. In this recipe, we will see how you can activate them and use them to increase your visibility.

Getting ready

Make sure to have an up-and-running Elastic Cloud deployment and complete this chapter's *Setting up Stack Monitoring* recipe.

The snippets for this recipe can be found at this address: `https://github.com/PacktPublishing/Elastic-Stack-8.x-Cookbook/blob/main/Chapter13/snippets.md#enabling-audit-logging`

How to do it...

Audit logging activation, as with many security-related configurations in the Elastic Stack, is twofold: first, we activate audit logging on Elasticsearch events, and afterward, we do this in Kibana. You can activate only the Kibana or Elasticsearch side based on your need, but generally speaking, it's good practice to have both components generate audit logs for greater visibility as you can correlate events from Kibana down to Elasticsearch. Let's start by activating audit logs in Elasticsearch.

1. Navigate to the cloud console and select **Manage** next to the name of your main deployment. From the left-hand menu, find and click the **Edit** button. On the ensuing page, locate and click on **Manage user settings and extensions**, adjacent to Elasticsearch. This action will open the user settings panel. Here, you need to input specific options to enable audit logs (the snippet can be found at this address: `https://github.com/PacktPublishing/Elastic-Stack-8.x-Cookbook/blob/main/Chapter13/snippets.md#elasticsearch-user-settings`):

    ```
    xpack.security.audit.enabled: true
    xpack.security.audit.logfile.events.emit_request_body: false
    ```

    Click on the **Back** button at the bottom to save the settings.

2. Navigate downward to the **Kibana** section and select **Edit user settings** to access the configuration panel. Then, append the provided configuration snippet at the end of the settings file (the snippet is available at `https://github.com/PacktPublishing/Elastic-Stack-8.x-Cookbook/blob/main/Chapter13/snippets.md#kibana-user-settings`):

```
xpack.security.audit.enabled: true
```

3. To save and retain the changes you have made to the user settings, click **Back**.

4. Then, scroll down to find and select the **Save** button.

5. Afterward, a popup will appear for you to review the changes. Click **Confirm** to proceed. Once done, wait for both Elasticsearch and Kibana to restart, as this step is necessary for the changes to take effect:

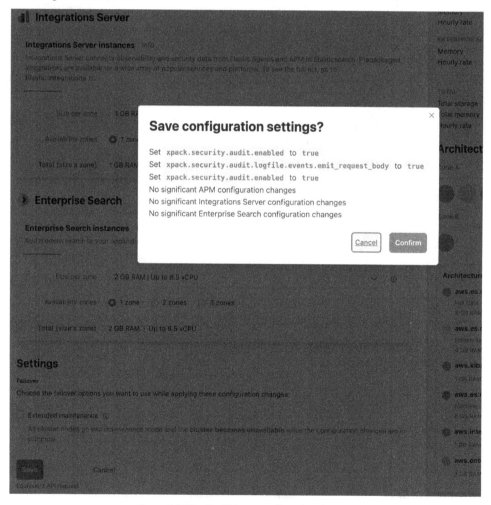

Figure 13.29 – Audit logs confirm the change

6. After the changes are implemented, sign in to Kibana for your monitoring deployment using your default cloud admin credentials. Navigate to **Analytics | Discover** to examine the audit logs. Here, click on the **Data View** drop-down menu and select **Create Data View**. A panel for creating a new data view will appear, where you should enter the following information:

    *   **Name**: `elastic-clouds-logs-8`

    *   **Index pattern**: `elastic-cloud-logs-8`

7. Click on **Save data view to Kibana** and go back to **Discover**. To ease our exploration, we are going to explicitly filter on audit logs. Click on the ± sign right beside the data view name to add a filter with the following parameters:

    *   **Select a field**: **event.dataset**

    *   **Select operator**: **is one of**

    *   **Values**: **kibana.audit** and **elasticsearch.audit**

8. Click on **Add filter**:

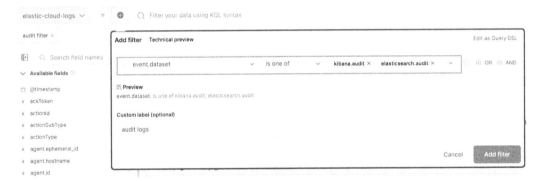

Figure 13.30 – Elastic Cloud logs filter in Discover

We are ready to explore our audit logs. In the search bar, type your username to find all relevant events. This is straightforward but gives you every action related to your users' audit.

9. Now, let us explore the audit logs with the **Elasticsearch Query Language (ES|QL)**. Imagine the following scenario: As a platform admin, I want to know how many times users tried to log in, in which space, and what was the outcome. This a common use case for audit logs, and with the power of ES|QL, we can have the answer with just a single query (the snippet can be found at this address: `https://github.com/PacktPublishing/Elastic-Stack-8.x-Cookbook/blob/main/Chapter13/snippets.md#esql-snippet-for-user-login-attempts`):

```
from elastic-cloud-logs-8
    | where event.type == "access"
    | stats attempts = count(event.type) by user.name, kibana.
space_id, event.outcome
    | limit 20
```

As a result, you should get a chart and an associated table, as shown in the following figure:

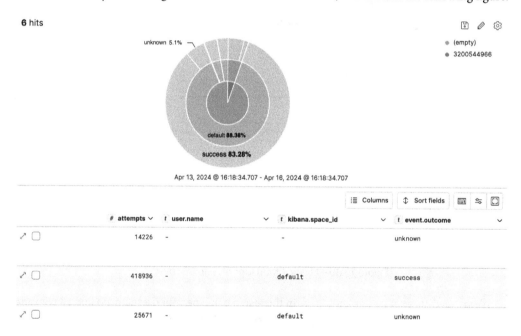

Figure 13.31 – ES|QL access attempts' audit logs

10. In the same vein, imagine we want to know who was recently granted access to the system and on which resources. You can use the following query (the snippet can be found at this address: `https://github.com/PacktPublishing/Elastic-Stack-8.x-Cookbook/blob/main/Chapter13/snippets.md#esql-snippet-for-user-login-granted-attempts`):

```
from elastic-cloud-logs-8
    | where event.action == "access_granted"
    | stats attempts = count(event.action) by user.
name,elasticsearch.audit.user.roles
    | sort attempts desc
    | limit 50
```

The following figure shows the results you should expect after running the aforementioned query:

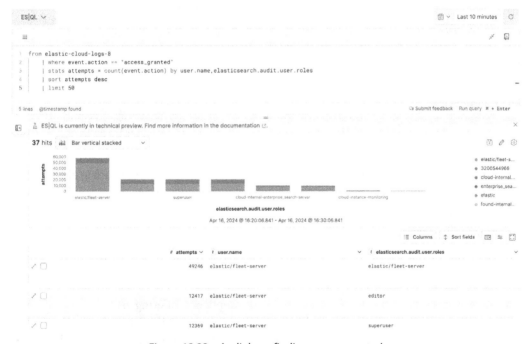

Figure 13.32 – Audit logs: finding access granted

You can go from there to get more insights into user behavior on the platform. Of course, as audit data is stored in regular indices, you can also leverage it to build powerful dashboards to track activity in the platform. This is particularly useful for any regulated environment and for most production-grade clusters.

How it works...

Audit logs are essential for tracking and monitoring security-related activities. These logs, which need to be explicitly enabled by administrators, record events such as authentication failures, access attempts, and administrative changes, including user and role management.

Here is a brief overview of some of the most significant audit events from both:

Elasticsearch audit events:

- **Authentication success and failure**: Records successful and failed attempts to authenticate, providing insights into potential security breaches or access issues

- **Access granted and denied**: Logs instances where access to a resource is either allowed or denied, crucial for monitoring unauthorized access attempts

- **Index creation and deletion**: Tracks the creation and deletion of indices, important for understanding changes in data structures and potential unauthorized modifications

- **API key creation and deletion**: Monitors the management of API keys, which is vital for ensuring that API access remains secure

Kibana audit events:

- **Login and logout events**: Captures user logins and logouts, essential for tracking user access and identifying abnormal access patterns

- **Saved object read and write**: Logs when users read or write to saved objects such as dashboards and visualizations, important for monitoring changes and usage

- **Space creation and deletion**: Keeps track of the creation and deletion of Kibana spaces, helping in overseeing the organization and access control within Kibana

- **User and role management**: Records changes to user accounts and roles, pivotal for maintaining secure and appropriate access levels across the system

These audit events play a crucial role in security and compliance, offering visibility into system operations, user activities, and changes within both Elasticsearch and Kibana environments. They are invaluable for system administrators and security teams for monitoring, troubleshooting, and ensuring the integrity of the Elastic Stack deployment.

There's more...

When activating audit logs, it is important to be mindful of their potential verbosity. To manage the volume of events generated, there are two options available for controlling the output:

- `xpack.security.audit.logfile.events.include`: To specify explicitly which events you want to include. Use that to reduce or extend the default values.

- `xpack.security.audit.logfile.events.exclude`: Explicitly exclude specific events from the logs.

- Specific events, particularly `xpack.security.audit.logfile.events.emit_request_body`, can generate a significant amount of data. This setting controls whether to include the body of REST requests in the audit logs for specific events. While logging request bodies can provide valuable insights into user behavior, such as the nature of requests being made or the most frequently accessed indices, it also has the capacity to rapidly fill up your logging cluster.

Therefore, it is advisable to use this feature judiciously to avoid overwhelming your logging infrastructure. One tip if it is imperative for you to keep audit events longer: leverage an **Index Lifecycle Management (ILM)** policy to optimize retention and resource usage. ILM is covered in the *Setting up an index lifecycle policy for your data* recipe in *Chapter 12*.

See also

- To set up audit logs on a self-managed cluster, check the official documentation: `https://www.elastic.co/guide/en/elasticsearch/reference/current/enable-audit-logging.html`

Index

packtpub.com

Subscribe to our online digital library for full access to over 7,000 books and videos, as well as industry leading tools to help you plan your personal development and advance your career. For more information, please visit our website.

Why subscribe?

- Spend less time learning and more time coding with practical eBooks and Videos from over 4,000 industry professionals

- Improve your learning with Skill Plans built especially for you

- Get a free eBook or video every month

- Fully searchable for easy access to vital information

- Copy and paste, print, and bookmark content

Did you know that Packt offers eBook versions of every book published, with PDF and ePub files available? You can upgrade to the eBook version at packtpub.com and as a print book customer, you are entitled to a discount on the eBook copy. Get in touch with us at customercare@packtpub.com for more details.

At www.packtpub.com, you can also read a collection of free technical articles, sign up for a range of free newsletters, and receive exclusive discounts and offers on Packt books and eBooks.

Other Books You May Enjoy

If you enjoyed this book, you may be interested in these other books by Packt:

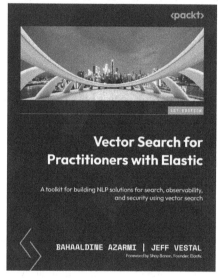

Vector Search for Practitioners with Elastic

Bahaaldine Azarmi, Jeff Vestal

ISBN: 978-1-80512-102-2

- Optimize performance by harnessing the capabilities of vector search
- Explore image vector search and its applications
- Detect and mask personally identifiable information
- Implement log prediction for next-generation observability
- Use vector-based bot detection for cybersecurity
- Visualize the vector space and explore Search.Next with Elastic
- Implement a RAG-enhanced application using Streamlit

Getting Started with DuckDB

Simon Aubury, Ned Letcher

ISBN: 978-1-80324-100-5

- Understand the properties and applications of a columnar in-process database
- Use SQL to load, transform, and query a range of data formats
- Discover DuckDB's rich extensions and learn how to apply them
- Use nested data types to model semi-structured data and extract and model JSON data
- Integrate DuckDB into your Python and R analytical workflows
- Effectively leverage DuckDB's convenient SQL enhancements
- Explore the wider ecosystem and pathways for building DuckDB-powered data applications

Packt is searching for authors like you

If you're interested in becoming an author for Packt, please visit `authors.packtpub.com` and apply today. We have worked with thousands of developers and tech professionals, just like you, to help them share their insight with the global tech community. You can make a general application, apply for a specific hot topic that we are recruiting an author for, or submit your own idea.

Share Your Thoughts

Now you've finished *Elastic Stack 8.x Cookbook*, we'd love to hear your thoughts! Scan the QR code below to go straight to the Amazon review page for this book and share your feedback or leave a review on the site that you purchased it from.

`https://packt.link/r/1-837-63429-7`

Your review is important to us and the tech community and will help us make sure we're delivering excellent quality content.

Download a free PDF copy of this book

Thanks for purchasing this book!

Do you like to read on the go but are unable to carry your print books everywhere?

Is your eBook purchase not compatible with the device of your choice?

Don't worry, now with every Packt book you get a DRM-free PDF version of that book at no cost.

Read anywhere, any place, on any device. Search, copy, and paste code from your favorite technical books directly into your application.

The perks don't stop there, you can get exclusive access to discounts, newsletters, and great free content in your inbox daily

Follow these simple steps to get the benefits:

1. Scan the QR code or visit the link below

https://packt.link/free-ebook/978-1-83763-429-3

2. Submit your proof of purchase
3. That's it! We'll send your free PDF and other benefits to your email directly